UNDER NEC® 2002

Volume II:
Commercial and Industrial

POCKET GUIDE TO ELECTRICAL INSTALLATIONS UNDER NEC® 2002

Volume II:
Commercial and Industrial

CHARLES R. MILLER
Lighthouse Educational Services

National Fire Protection Association
Quincy, Massachusetts

Product Manager: Charles Durang
Editorial-Production Services: Omegatype Typography, Inc.
Interior Design: Omegatype Typography, Inc.
Composition: Omegatype Typography, Inc.
Cover Design: Natasha Bumbeck
Manufacturing Manager: Ellen Glisker
Printer: Rose Printing

Copyright © 2001
National Fire Protection Association, Inc.
One Batterymarch Park
Quincy, Massachusetts 02269

All rights reserved. No part of the material protected by this copyright notice may be reproduced or utilized in any form without acknowledgment of the copyright owner nor may it be used in any form for resale without written permission from the copyright owner.

NFPA No.: PGNECIND-02
ISBN: 0-87765-458-1
Library of Congress Control Number: 2001093273

Printed in the United States of America

05 04 03 02   5 4 3 2

IMPORTANT NOTICES AND DISCLAIMERS OF LIABILITY

This *Pocket Guide* is a compilation of extracts from the 2002 edition of the *National Electrical Code®* (*NEC®*). These extracts have been selected and arranged by the Author of the *Pocket Guide*, who has also supplied introductory material generally located at the beginning of each chapter.

The *NEC*, like all NFPA codes and standards, is a document that is developed through a consensus standards development process approved by the American National Standards Institute. This process brings together volunteers representing varied viewpoints and interests to achieve consensus on fire, electrical, and other safety issues. While the NFPA administers the process and establishes rules to promote fairness in the development of consensus, it does not independently test, evaluate, or verify the accuracy of any information or the soundness of any judgments contained in its codes and standards.

The NFPA and the author disclaim liability for any personal injury, property or other damages of any nature whatsoever, whether special, indirect, consequential, or compensatory, directly or indirectly resulting from the publication, use of, or reliance on the *NEC* or this *Pocket Guide*. The NFPA and the Author also make no guaranty or warranty as to the accuracy or completeness of any information published herein.

In issuing and making the *NEC* and this *Pocket Guide* available, the NFPA and the Author are not undertaking to render professional or other services for or on behalf of any person or entity. Nor is the NFPA or the Author undertaking to perform any duty owed by any person or entity to someone else. Anyone using these materials should rely on his or her own independent judgment or, as appropriate, seek the advice of a competent professional in determining the exercise of reasonable care in any given circumstances.

For more important information and notices concerning the use of NFPA codes and standards, please consult the introductory materials accompanying all published versions of these documents.

NOTICE CONCERNING CODE INTERPRETATIONS

The introductory materials and the selection and arrangement of the *NEC* extracts contained in this *Pocket Guide* reflect the personal opinions and judgments of the Author and do not necessarily represent the official position of the NFPA (which can only be obtained through Formal Interpretations processed in accordance with NPFA rules).

®Registered trademarks of the National Fire Protection Association, Inc.

CONTENTS

CHAPTER 1	INTRODUCTION TO THE NATIONAL ELECTRICAL CODE	1
CHAPTER 2	GENERAL INSTALLATION REQUIREMENTS	5
CHAPTER 3	TEMPORARY WIRING	17
CHAPTER 4	BRANCH-CIRCUIT, FEEDER, AND SERVICE CALCULATIONS	23
CHAPTER 5	SERVICES	31
CHAPTER 6	FEEDERS	47
CHAPTER 7	SWITCHBOARDS, PANELBOARDS, AND DISCONNECTS	53
CHAPTER 8	OVERCURRENT PROTECTION	63
CHAPTER 9	GROUNDED (NEUTRAL) CONDUCTORS	81
CHAPTER 10	GROUNDING	85
CHAPTER 11	BOXES AND ENCLOSURES	127
CHAPTER 12	CABLES	145
CHAPTER 13	RACEWAYS	153

vii

CONTENTS

CHAPTER 1	INTRODUCTION TO THE NATIONAL ELECTRICAL CODE	1
CHAPTER 2	GENERAL INSTALLATION REQUIREMENTS	5
CHAPTER 3	TEMPORARY WIRING	17
CHAPTER 4	BRANCH-CIRCUIT, FEEDER, AND SERVICE CALCULATIONS	23
CHAPTER 5	SERVICES	31
CHAPTER 6	FEEDERS	47
CHAPTER 7	SWITCHBOARDS, PANELBOARDS, AND DISCONNECTS	53
CHAPTER 8	OVERCURRENT PROTECTION	63
CHAPTER 9	GROUNDED (NEUTRAL) CONDUCTORS	81
CHAPTER 10	GROUNDING	85
CHAPTER 11	BOXES AND ENCLOSURES	127
CHAPTER 12	CABLES	145
CHAPTER 13	RACEWAYS	153

CHAPTER 14	WIRING METHODS AND CONDUCTORS	187
CHAPTER 15	BRANCH CIRCUITS	215
CHAPTER 16	RECEPTACLE PROVISIONS	223
CHAPTER 17	SWITCHES	231
CHAPTER 18	LIGHTING REQUIREMENTS	235
CHAPTER 19	MOTORS, MOTOR CIRCUITS, AND CONTROLLERS	251
CHAPTER 20	AIR-CONDITIONING AND REFRIGERATING EQUIPMENT	291
CHAPTER 21	GENERATORS	305
CHAPTER 22	TRANSFORMERS	309
CHAPTER 23	SWIMMING POOLS, FOUNTAINS, AND SIMILAR INSTALLATIONS	321
CHAPTER 24	EMERGENCY SYSTEMS	347
CHAPTER 25	CONDUIT AND TUBING FILL TABLES	359

INTRODUCTION

The *Pocket Guide to Electrical Installations Under NEC® 2002, Volume II: Commercial and Industrial,* was compiled to provide a portable, handy reference source for field practitioners: installers, contractors, and inspectors. With its companion volume, *Pocket Guide to Electrical Installations Under NEC® 2002, Volume I: Residential,* it gathers together all the *National Electrical Code®* rules you need for most standard installations, in the order you need them.

The code requirements are not organized in this *Pocket Guide* in the same sequence found in the *NEC®*. Rather, they are presented in an order similar to that of an actual electrical system installation process. A few introductory chapters are followed by the requirements that pertain to temporary wiring. Next, the *Guide* covers certain calculations that must be performed before the actual electrical installation (e.g., branch circuits, feeders, and services). The book then continues with provisions that are essential for safeguarding persons and property from the hazards arising from the use of electricity. Because this book is literally a pocket guide, space does not permit an in-depth reference to and explanation of *NEC* regulations. Other instructional material, such as NFPA's *National Electrical Code Handbook,* can provide a better insight into and understanding of the *Code.*

State or local agencies may—and often do—adopt regulations above and beyond the *National Electrical Code.* Local authorities can adopt the *NEC* exactly as written, or they can amend the *Code* or supplement it with more stringent regulations. For example, additional restrictions may limit the number of lights or receptacles on a circuit. Obtain a copy of any additional requirements for your city, county, and/or state.

These *Pocket Guides* are not intended to be a substitute for the *NEC* itself, which should always be consulted to ensure full *Code* compliance. The choice of material included in this *Pocket Guide* was based on the problems and situations most frequently encountered in the field, where it is not practical to carry bulky reference material.

Remember, however, that important provisions of the *Code,* such as the more specialized rules for special occupancies, special equipment, special conditions, or communications circuits, while not included in this *Pocket Guide,* are important to know and understand and should be consulted where appropriate.

It is hoped that this *Pocket Guide* will prove to be a valuable addition to your toolbox, glove compartment, briefcase, or, as the name implies, your pocket.

POCKET GUIDE TO ELECTRICAL INSTALLATIONS UNDER NEC® 2002

Volume II:
Commercial and Industrial

CHAPTER 1

INTRODUCTION TO THE NATIONAL ELECTRICAL CODE

INTRODUCTION

This chapter contains provisions from Article 90—Introduction. That article covers basic provisions essential for understanding and applying the *National Electrical Code*®. Besides providing the purpose of the *Code*, Article 90 specifies the document scope, describes how the *Code* is arranged, provides information on enforcement, and contains other provisions that are essential to the proper use and application of the *Code*. The first sentence states that "the purpose of this *Code* is the practical safeguarding of persons and property from hazards arising from the use of electricity." Safety, for both property and people, is the main objective for the provisions found in the *Code*. Compliance with these provisions and proper maintenance will result in an installation essentially free from electrical hazards but not necessarily efficient, convenient, or adequate for good service or future expansion of electrical use.

Although state and local jurisdictions generally adopt the *Code* exactly as written, they may also amend the *Code* or incorporate local requirements. Therefore, obtain a copy of any amendments or additional rules and regulations for your area.

ARTICLE 90
Introduction

90.1 Purpose.

(A) Practical Safeguarding. The purpose of this *Code* is the practical safeguarding of persons and property from hazards arising from the use of electricity.

(B) Adequacy. This *Code* contains provisions that are considered necessary for safety. Compliance therewith and proper maintenance will result in an installation that is essentially free from hazard but not necessarily efficient, convenient, or adequate for good service or future expansion of electrical use.

90.2 Scope.

(A) Covered. This *Code* covers the installation of electric conductors, electric equipment, signaling and communications conductors and equipment, and fiber optic cables and raceways for the following:

(1) Public and private premises, including buildings, structures, mobile homes, recreational vehicles, and floating buildings
(2) Yards, lots, parking lots, carnivals, and industrial substations
(3) Installations of conductors and equipment that connect to the supply of electricity
(4) Installations used by the electric utility, such as office buildings, warehouses, garages, machine shops, and recreational buildings, that are not an integral part of a generating plant, substation, or control center

90.4 Enforcement.

This *Code* is intended to be suitable for mandatory application by governmental bodies that exercise legal jurisdiction over electrical installations, including signaling and communications systems, and for use by insurance inspectors. The authority having jurisdiction for enforcement of the *Code* has the responsibility for making interpretations of the rules, for deciding on the approval of equipment and materials, and for granting the special permission contemplated in a number of the rules.

Introduction to the National Electrical Code

By special permission, the authority having jurisdiction may waive specific requirements in this *Code* or permit alternative methods where it is assured that equivalent objectives can be achieved by establishing and maintaining effective safety.

This *Code* may require new products, constructions, or materials that may not yet be available at the time the *Code* is adopted. In such event, the authority having jurisdiction may permit the use of the products, constructions, or materials that comply with the most recent previous edition of this *Code* adopted by the jurisdiction.

90.5 Mandatory Rules, Permissive Rules, and Explanatory Material.

(A) Mandatory Rules. Mandatory rules of this *Code* are those that identify actions that are specifically required or prohibited and are characterized by the use of the terms *shall* or *shall not*.

(B) Permissive Rules. Permissive rules of this *Code* are those that identify actions that are allowed but not required, are normally used to describe options or alternative methods, and are characterized by the use of the terms *shall be permitted* or *shall not be required*.

(C) Explanatory Material. Explanatory material, such as references to other standards, references to related sections of this *Code*, or information related to a *Code* rule, is included in this *Code* in the form of fine print notes (FPNs). Fine print notes are informational only and are not enforceable as requirements of this *Code*.

90.7 Examination of Equipment for Safety.
For specific items of equipment and materials referred to in this *Code*, examinations for safety made under standard conditions provide a basis for approval where the record is made generally available through promulgation by organizations properly equipped and qualified for experimental testing, inspections of the run of goods at factories, and service-value determination through field inspections. This avoids the necessity for repetition of examinations by different examiners, frequently with inadequate facilities for such work,

and the confusion that would result from conflicting reports on the suitability of devices and materials examined for a given purpose.

It is the intent of this *Code* that factory-installed internal wiring or the construction of equipment need not be inspected at the time of installation of the equipment, except to detect alterations or damage, if the equipment has been listed by a qualified electrical testing laboratory that is recognized as having the facilities described in the preceding paragraph and that requires suitability for installation in accordance with this *Code*.

90.8 Wiring Planning.

(A) Future Expansion and Convenience. Plans and specifications that provide ample space in raceways, spare raceways, and additional spaces allow for future increases in the use of electricity. Distribution centers located in readily accessible locations provide convenience and safety of operation.

(B) Number of Circuits in Enclosures. It is elsewhere provided in this *Code* that the number of wires and circuits confined in a single enclosure be varyingly restricted. Limiting the number of circuits in a single enclosure minimizes the effects from a short circuit or ground fault in one circuit.

CHAPTER 2

GENERAL INSTALLATION REQUIREMENTS

INTRODUCTION

This chapter contains provisions from Article 110—Requirements for Electrical Installations. Article 110 covers provisions that are essential to the installation of conductors and equipment and provides fundamental safety requirements that apply to all types of electrical installations. Although Article 110 is relatively short, do not underestimate its importance. It contains general requirements for the approval, examination, identification, installation, use, access to, and spaces about electrical equipment (and conductors). Since these are general installation requirements, they are applicable to virtually every electrical installation.

Compliance with the requirements covering the working space around electric equipment is essential. Sufficient access and working space must be provided and maintained around all electrical equipment to permit ready and safe operation and maintenance of such equipment. Section 110.26 provides specific dimensions for the height, width, and depth of the working space. Parts of that section pertain to all equipment that is likely to require examination, adjustment, servicing, or maintenance while energized; other parts pertain only to switchboards, panelboards, distribution boards, and motor control centers.

Because of the limited space within this *Pocket Guide*, the provisions pertaining to conductors and equipment used on circuits exceeding 600 volts, nominal, are not included. Reference the *2002 National Electrical Code* for requirements that specifically address installations over 600 volts.

ARTICLE 110
Requirements for Electrical Installations

I. GENERAL

110.1 Scope. This article covers general requirements for the examination and approval, installation and use, access to and spaces about electrical conductors and equipment, and tunnel installations.

110.2 Approval. The conductors and equipment required or permitted by this *Code* shall be acceptable only if approved.

110.3 Examination, Identification, Installation, and Use of Equipment.

(A) Examination. In judging equipment, considerations such as the following shall be evaluated:

(1) Suitability for installation and use in conformity with the provisions of this *Code*
(2) Mechanical strength and durability, including, for parts designed to enclose and protect other equipment, the adequacy of the protection thus provided
(3) Wire-bending and connection space
(4) Electrical insulation
(5) Heating effects under normal conditions of use and also under abnormal conditions likely to arise in service
(6) Arcing effects
(7) Classification by type, size, voltage, current capacity, and specific use
(8) Other factors that contribute to the practical safeguarding of persons using or likely to come in contact with the equipment

(B) Installation and Use. Listed or labeled equipment shall be installed and used in accordance with any instructions included in the listing or labeling.

110.8 Wiring Methods. Only wiring methods recognized as suitable are included in this *Code*. The recognized methods of wiring shall be permitted to be installed in any type of building or occupancy, except as otherwise provided in this *Code*.

General Installation Requirements

110.9 Interrupting Rating. Equipment intended to interrupt current at fault levels shall have an interrupting rating sufficient for the nominal circuit voltage and the current that is available at the line terminals of the equipment.

Equipment intended to interrupt current at other than fault levels shall have an interrupting rating at nominal circuit voltage sufficient for the current that must be interrupted.

110.10 Circuit Impedance and Other Characteristics. The overcurrent protective devices, the total impedance, the component short-circuit current ratings, and other characteristics of the circuit to be protected shall be selected and coordinated to permit the circuit-protective devices used to clear a fault to do so without extensive damage to the electrical components of the circuit. This fault shall be assumed to be either between two or more of the circuit conductors or between any circuit conductor and the grounding conductor or enclosing metal raceway. Listed products applied in accordance with their listing shall be considered to meet the requirements of this section.

110.11 Deteriorating Agents. Unless identified for use in the operating environment, no conductors or equipment shall be located in damp or wet locations; where exposed to gases, fumes, vapors, liquids, or other agents that have a deteriorating effect on the conductors or equipment; or where exposed to excessive temperatures.

Equipment identified only as "dry locations," "Type 1," or "indoor use only" shall be protected against permanent damage from the weather during building construction.

110.12 Mechanical Execution of Work. Electrical equipment shall be installed in a neat and workmanlike manner.

(A) Unused Openings. Unused cable or raceway openings in boxes, raceways, auxiliary gutters, cabinets, cutout boxes, meter socket enclosures, equipment cases, or housings shall be effectively closed to afford protection substantially equivalent to the wall of the equipment. Where metallic plugs or plates are used with nonmetallic enclosures, they shall be recessed at least 6 mm (1/4 in.) from the outer surface of the enclosure.

(B) Subsurface Enclosures. Conductors shall be racked to provide ready and safe access in underground and subsurface enclosures into which persons enter for installation and maintenance.

(C) Integrity of Electrical Equipment and Connections. Internal parts of electrical equipment, including busbars, wiring terminals, insulators, and other surfaces, shall not be damaged or contaminated by foreign materials such as paint, plaster, cleaners, abrasives, or corrosive residues. There shall be no damaged parts that may adversely affect safe operation or mechanical strength of the equipment such as parts that are broken; bent; cut; or deteriorated by corrosion, chemical action, or overheating.

110.13 Mounting and Cooling of Equipment.

(A) Mounting. Electrical equipment shall be firmly secured to the surface on which it is mounted. Wooden plugs driven into holes in masonry, concrete, plaster, or similar materials shall not be used.

(B) Cooling. Electrical equipment that depends on the natural circulation of air and convection principles for cooling of exposed surfaces shall be installed so that room airflow over such surfaces is not prevented by walls or by adjacent installed equipment. For equipment designed for floor mounting, clearance between top surfaces and adjacent surfaces shall be provided to dissipate rising warm air.

Electrical equipment provided with ventilating openings shall be installed so that walls or other obstructions do not prevent the free circulation of air through the equipment.

110.14 Electrical Connections. Because of different characteristics of dissimilar metals, devices such as pressure terminal or pressure splicing connectors and soldering lugs shall be identified for the material of the conductor and shall be properly installed and used. Conductors of dissimilar metals shall not be intermixed in a terminal or splicing connector where physical contact occurs between dissimilar conductors (such as copper and aluminum, copper and copper-clad aluminum, or aluminum and copper-clad aluminum), unless the device is identified for the purpose and conditions of use. Materials such as solder, fluxes, inhibitors, and compounds, where employed, shall be suitable for

General Installation Requirements

the use and shall be of a type that will not adversely affect the conductors, installation, or equipment.

(A) Terminals. Connection of conductors to terminal parts shall ensure a thoroughly good connection without damaging the conductors and shall be made by means of pressure connectors (including set-screw type), solder lugs, or splices to flexible leads. Connection by means of wire-binding screws or studs and nuts that have upturned lugs or the equivalent shall be permitted for 10 AWG or smaller conductors.

Terminals for more than one conductor and terminals used to connect aluminum shall be so identified.

(B) Splices. Conductors shall be spliced or joined with splicing devices identified for the use or by brazing, welding, or soldering with a fusible metal or alloy. Soldered splices shall first be spliced or joined so as to be mechanically and electrically secure without solder and then be soldered. All splices and joints and the free ends of conductors shall be covered with an insulation equivalent to that of the conductors or with an insulating device identified for the purpose.

Wire connectors or splicing means installed on conductors for direct burial shall be listed for such use.

(C) Temperature Limitations. The temperature rating associated with the ampacity of a conductor shall be selected and coordinated so as not to exceed the lowest temperature rating of any connected termination, conductor, or device. Conductors with temperature ratings higher than specified for terminations shall be permitted to be used for ampacity adjustment, correction, or both.

(1) Equipment Provisions. The determination of termination provisions of equipment shall be based on 110.14(C)(1)(a) or 110.14(C)(1)(b). Unless the equipment is listed and marked otherwise, conductor ampacities used in determining equipment termination provisions shall be based on Table 310.16 as appropriately modified by 310.15(B)(1) through (6).

(a) Termination provisions of equipment for circuits rated 100 amperes or less, or marked for 14 AWG through 1 AWG conductors, shall be used only for one of the following:

(1) Conductors rated 60°C (140°F)
(2) Conductors with higher temperature ratings, provided the ampacity of such conductors is determined based

on the 60°C (140°F) ampacity of the conductor size used

(3) Conductors with higher temperature ratings if the equipment is listed and identified for use with such conductors

(4) For motors marked with design letters B, C, D, or E, conductors having an insulation rating of 75°C (167°F) or higher shall be permitted to be used provided the ampacity of such conductors does not exceed the 75°C (167°F) ampacity.

(b) Termination provisions of equipment for circuits rated over 100 amperes, or marked for conductors larger than 1 AWG, shall be used only for one of the following:

(1) Conductors rated 75°C (167°F)
(2) Conductors with higher temperature ratings, provided the ampacity of such conductors does not exceed the 75°C (167°F) ampacity of the conductor size used, or up to their ampacity if the equipment is listed and identified for use with such conductors

110.15 High-Leg Marking. On a 4-wire, delta-connected system where the midpoint of one phase winding is grounded to supply lighting and similar loads, the conductor or busbar having the higher phase voltage to ground shall be durably and permanently marked by an outer finish that is orange in color or by other effective means. Such identification shall be placed at each point on the system where a connection is made if the grounded conductor is also present.

110.16 Flash Protection. Switchboards, panelboards, industrial control panels, and motor control centers that are in other than dwelling occupancies and are likely to require examination, adjustment, servicing, or maintenance while energized shall be field marked to warn qualified persons of potential electric arc flash hazards. The marking shall be located so as to be clearly visible to qualified persons before examination, adjustment, servicing, or maintenance of the equipment.

FPN No. 1: NFPA 70E-2000, *Electrical Safety Requirements for Employee Workplaces*, provides as-

General Installation Requirements

sistance in determining severity of potential exposure, planning safe work practices, and selecting personal protective equipment.

FPN No. 2: ANSI Z535.4-1998, *Product Safety Signs and Labels*, provides guidelines for the design of safety signs and labels for application to products.

110.22 Identification of Disconnecting Means. Each disconnecting means shall be legibly marked to indicate its purpose unless located and arranged so the purpose is evident. The marking shall be of sufficient durability to withstand the environment involved.

Where circuit breakers or fuses are applied in compliance with the series combination ratings marked on the equipment by the manufacturer, the equipment enclosure(s) shall be legibly marked in the field to indicate the equipment has been applied with a series combination rating. The marking shall be readily visible and state the following:

> CAUTION—SERIES COMBINATION SYSTEM RATED ____ AMPERES. IDENTIFIED REPLACEMENT COMPONENTS REQUIRED.

110.23 Current Transformers. Unused current transformers associated with potentially energized circuits shall be short-circuited.

II. 600 VOLTS, NOMINAL, OR LESS

110.26 Spaces About Electrical Equipment. Sufficient access and working space shall be provided and maintained about all electric equipment to permit ready and safe operation and maintenance of such equipment. Enclosures housing electrical apparatus that are controlled by lock and key shall be considered accessible to qualified persons.

(A) Working Space. Working space for equipment operating at 600 volts, nominal, or less to ground and likely to require examination, adjustment, servicing, or maintenance while energized shall comply with the dimensions of 110.26(A)(1), (2), and (3) or as required or permitted elsewhere in this *Code*.

(1) Depth of Working Space. The depth of the working space in the direction of live parts shall not be less than that specified in Table 110.26(A)(1) unless the requirements of 110.26(A)(1)(a), (b), or (c) are met. Distances shall be measured from the exposed live parts or from the enclosure or opening if the live parts are enclosed.

(a) Dead-Front Assemblies. Working space shall not be required in the back or sides of assemblies, such as dead-front switchboards or motor control centers, where all connections and all renewable or adjustable parts, such as fuses or switches, are accessible from locations other than the back or sides. Where rear access is required to work on nonelectrical parts on the back of enclosed equipment, a minimum horizontal working space of 762 mm (30 in.) shall be provided.

(b) Low Voltage. By special permission, smaller working spaces shall be permitted where all uninsulated parts operate at not greater than 30 volts rms, 42 volts peak, or 60 volts dc.

(c) Existing Buildings. In existing buildings where electrical equipment is being replaced, Condition 2 working

Table 110.26(A)(1) Working Spaces

Nominal Voltage to Ground	Minimum Clear Distance		
	Condition 1	Condition 2	Condition 3
0–150	900 mm (3 ft)	900 mm (3 ft)	900 mm (3 ft)
151–600	900 mm (3 ft)	1 m (3-1/2 ft)	1.2 m (4 ft)

Note: Where the conditions are as follows:

Condition 1—Exposed live parts on one side and no live or grounded parts on the other side of the working space, or exposed live parts on both sides effectively guarded by suitable wood or other insulating materials. Insulated wire or insulated busbars operating at not over 300 volts to ground shall not be considered live parts.

Condition 2—Exposed live parts on one side and grounded parts on the other side. Concrete, brick, or tile walls shall be considered as grounded.

Condition 3—Exposed live parts on both sides of the work space (not guarded as provided in Condition 1) with the operator between.

General Installation Requirements

clearance shall be permitted between dead-front switchboards, panelboards, or motor control centers located across the aisle from each other where conditions of maintenance and supervision ensure that written procedures have been adopted to prohibit equipment on both sides of the aisle from being open at the same time and qualified persons who are authorized will service the installation.

(2) Width of Working Space. The width of the working space in front of the electric equipment shall be the width of the equipment or 750 mm (30 in.), whichever is greater. In all cases, the work space shall permit at least a 90 degree opening of equipment doors or hinged panels.

(3) Height of Working Space. The work space shall be clear and extend from the grade, floor, or platform to the height required by 110.26(E). Within the height requirements of this section, other equipment that is associated with the electrical installation and is located above or below the electrical equipment shall be permitted to extend not more than 150 mm (6 in.) beyond the front of the electrical equipment.

(B) Clear Spaces. Working space required by this section shall not be used for storage. When normally enclosed live parts are exposed for inspection or servicing, the working space, if in a passageway or general open space, shall be suitably guarded.

(C) Entrance to Working Space.

(1) Minimum Required. At least one entrance of sufficient area shall be provided to give access to working space about electrical equipment.

(2) Large Equipment. For equipment rated 1200 amperes or more and over 1.8 m (6 ft) wide that contains overcurrent devices, switching devices, or control devices, there shall be one entrance to the required working space not less than 610 mm (24 in.) wide and 2.0 m (6-1/2 ft) high at each end of the working space. Where the entrance has a personnel door(s), the door(s) shall open in the direction of egress and be equipped with panic bars, pressure plates, or other devices that are normally latched but open under simple pressure.

A single entrance to the required working space shall be permitted where either of the conditions in 110.26(C)(2)(a) or (b) is met.

(a) Unobstructed Exit. Where the location permits a continuous and unobstructed way of exit travel, a single entrance to the working space shall be permitted.

(b) Extra Working Space. Where the depth of the working space is twice that required by 110.26(A)(1), a single entrance shall be permitted. It shall be located so that the distance from the equipment to the nearest edge of the entrance is not less than the minimum clear distance specified in Table 110.26(A)(1) for equipment operating at that voltage and in that condition.

(D) Illumination. Illumination shall be provided for all working spaces about service equipment, switchboards, panelboards, or motor control centers installed indoors. Additional lighting outlets shall not be required where the work space is illuminated by an adjacent light source or as permitted by 210.70(A)(1), Exception No. 1, for switched receptacles. In electrical equipment rooms, the illumination shall not be controlled by automatic means only.

(E) Headroom. The minimum headroom of working spaces about service equipment, switchboards, panelboards, or motor control centers shall be 2.0 m (6-1/2 ft). Where the electrical equipment exceeds 2.0 m (6-1/2 ft) in height, the minimum headroom shall not be less than the height of the equipment.

(F) Dedicated Equipment Space. All switchboards, panelboards, distribution boards, and motor control centers shall be located in dedicated spaces and protected from damage.

Exception: Control equipment that by its very nature or because of other rules of the Code must be adjacent to or within sight of its operating machinery shall be permitted in those locations.

(1) Indoor. Indoor installations shall comply with 110.26(F)(1)(a) through (d).

(a) Dedicated Electrical Space. The space equal to the width and depth of the equipment and extending from the floor to a height of 1.8 m (6 ft) above the equipment or to the structural ceiling, whichever is lower, shall be dedi-

General Installation Requirements

cated to the electrical installation. No piping, ducts, leak protection apparatus, or other equipment foreign to the electrical installation shall be located in this zone.

Exception: Suspended ceilings with removable panels shall be permitted within the 1.8-m (6-ft) zone.

(b) **Foreign Systems.** The area above the dedicated space required by 110.26(F)(1)(a) shall be permitted to contain foreign systems, provided protection is installed to avoid damage to the electrical equipment from condensation, leaks, or breaks in such foreign systems.

(c) **Sprinkler Protection.** Sprinkler protection shall be permitted for the dedicated space where the piping complies with this section.

(d) **Suspended Ceilings.** A dropped, suspended, or similar ceiling that does not add strength to the building structure shall not be considered a structural ceiling.

(2) Outdoor. Outdoor electrical equipment shall be installed in suitable enclosures and shall be protected from accidental contact by unauthorized personnel, or by vehicular traffic, or by accidental spillage or leakage from piping systems. The working clearance space shall include the zone described in 110.26(A). No architectural appurtenance or other equipment shall be located in this zone.

110.27 Guarding of Live Parts.

(A) Live Parts Guarded Against Accidental Contact. Except as elsewhere required or permitted by this *Code*, live parts of electrical equipment operating at 50 volts or more shall be guarded against accidental contact by approved enclosures or by any of the following means:

(1) By location in a room, vault, or similar enclosure that is accessible only to qualified persons.
(2) By suitable permanent, substantial partitions or screens arranged so that only qualified persons have access to the space within reach of the live parts. Any openings in such partitions or screens shall be sized and located so that persons are not likely to come into accidental contact with the live parts or to bring conducting objects into contact with them.

(3) By location on a suitable balcony, gallery, or platform elevated and arranged so as to exclude unqualified persons.

(4) By elevation of 2.5 m (8 ft) or more above the floor or other working surface.

(B) Prevent Physical Damage. In locations where electric equipment is likely to be exposed to physical damage, enclosures or guards shall be so arranged and of such strength as to prevent such damage.

(C) Warning Signs. Entrances to rooms and other guarded locations that contain exposed live parts shall be marked with conspicuous warning signs forbidding unqualified persons to enter.

CHAPTER 3

TEMPORARY WIRING

INTRODUCTION

This chapter contains provisions from Article 527—Temporary Installations. Article 527 covers provisions pertaining to temporary electrical power and lighting installations. New construction generally requires the installation of temporary electrical equipment, which usually involves setting a pole and building a service. Although the feeders, branch circuits, and receptacles are specifically covered in Article 527, the service must be installed in accordance with Article 230 (see Chapter 5 of this book). For example, the minimum clearance of overhead service-drop conductors must comply with 230.24. Of course, not all electrical projects will require the installation of a temporary service. Remodeling and additions to existing installations typically already have available power through permanent wiring.

Because of the inherent hazards associated with construction sites, ground-fault protection of receptacles with certain ratings that are in use by personnel is required for all temporary wiring installations, not just for new construction sites with temporary services. If a receptacle(s) is installed or exists as part of the permanent wiring or the building or structure and is used for temporary electric power, ground-fault circuit-interrupter (GFCI) protection for personnel must be provided. This may require replacing the existing receptacle with a GFCI receptacle or the use of a portable GFCI device. All other outlets (besides 125-volt, single-phase, 15-, 20-, and 30-ampere receptacle outlets) must have GFCI protection for personnel, or an assured equipment grounding conductor program in accordance with 527.6(B)(2).

ARTICLE 527
Temporary Installations

527.1 Scope. The provisions of this article apply to temporary electrical power and lighting installations.

527.2 All Wiring Installations.

(A) Other Articles. Except as specifically modified in this article, all other requirements of this *Code* for permanent wiring shall apply to temporary wiring installations.

(B) Approval. Temporary wiring methods shall be acceptable only if approved based on the conditions of use and any special requirements of the temporary installation.

527.3 Time Constraints.

(A) During the Period of Construction. Temporary electrical power and lighting installations shall be permitted during the period of construction, remodeling, maintenance, repair, or demolition of buildings, structures, equipment, or similar activities.

(D) Removal. Temporary wiring shall be removed immediately upon completion of construction or purpose for which the wiring was installed.

527.4 General.

(A) Services. Services shall be installed in conformance with Article 230.

(B) Feeders. Feeders shall be protected as provided in Article 240. They shall originate in an approved distribution center. Conductors shall be permitted within cable assemblies or within multiconductor cords or cables of a type identified in Table 400.4 for hard usage or extra-hard usage. For the purpose of this section, Type NM and Type NMC cables shall be permitted to be used in any dwelling, building, or structure without any height limitation.

(C) Branch Circuits. All branch circuits shall originate in an approved power outlet or panelboard. Conductors shall be permitted within cable assemblies or within multiconductor cord or cable of a type identified in Table 400.4 for

Temporary Wiring

hard usage or extra-hard usage. All conductors shall be protected as provided in Article 240. For the purposes of this section, Type NM and Type NMC cables shall be permitted to be used in any dwelling, building, or structure without any height limitation.

(D) Receptacles. All receptacles shall be of the grounding type. Unless installed in a continuous grounded metal raceway or metal-covered cable, all branch circuits shall contain a separate equipment grounding conductor, and all receptacles shall be electrically connected to the equipment grounding conductors. Receptacles on construction sites shall not be installed on branch circuits that supply temporary lighting. Receptacles shall not be connected to the same ungrounded conductor of multiwire circuits that supply temporary lighting.

(E) Disconnecting Means. Suitable disconnecting switches or plug connectors shall be installed to permit the disconnection of all ungrounded conductors of each temporary circuit. Multiwire branch circuits shall be provided with a means to disconnect simultaneously all ungrounded conductors at the power outlet or panelboard where the branch circuit originated. Approved handle ties shall be permitted.

(F) Lamp Protection. All lamps for general illumination shall be protected from accidental contact or breakage by a suitable fixture or lampholder with a guard.

Brass shell, paper-lined sockets, or other metal-cased sockets shall not be used unless the shell is grounded.

(G) Splices. On construction sites, a box shall not be required for splices or junction connections where the circuit conductors are multiconductor cord or cable assemblies, provided that the equipment grounding continuity is maintained with or without the box. See 110.14(B) and 400.9. A box, conduit body, or terminal fitting having a separately bushed hole for each conductor shall be used wherever a change is made to a conduit or tubing system or a metal-sheathed cable system.

(H) Protection from Accidental Damage. Flexible cords and cables shall be protected from accidental damage.

Sharp corners and projections shall be avoided. Where passing through doorways or other pinch points, protection shall be provided to avoid damage.

(I) Termination(s) at Devices. Flexible cords and cables entering enclosures containing devices requiring termination shall be secured to the box with fittings designed for the purpose.

(J) Support. Cable assemblies and flexible cords and cables shall be supported in place at intervals that ensure that they will be protected from physical damage. Support shall be in the form of staples, cable ties, straps, or similar type fittings installed so as not to cause damage. Vegetation shall not be used for support of overhead spans of branch circuits or feeders.

527.6 Ground-Fault Protection for Personnel. Ground-fault protection for personnel for all temporary wiring installations shall be provided to comply with 527.6(A) and (B). This section shall apply only to temporary wiring installations used to supply temporary power to equipment used by personnel during construction, remodeling, maintenance, repair, or demolition of buildings, structures, equipment, or similar activities.

(A) Receptacle Outlets. All 125-volt, single-phase, 15-, 20-, and 30-ampere receptacle outlets that are not a part of the permanent wiring of the building or structure and that are in use by personnel shall have ground-fault circuit interrupter protection for personnel. If a receptacle(s) is installed or exists as part of the permanent wiring of the building or structure and is used for temporary electric power, ground-fault circuit-interrupter protection for personnel shall be provided. For the purposes of this section, cord sets or devices incorporating listed ground-fault circuit interrupter protection for personnel identified for portable use shall be permitted.

Exception: In industrial establishments only, where conditions of maintenance and supervision ensure that only qualified personnel are involved, an assured equipment grounding conductor program as specified in 527.6(B)(2)

Temporary Wiring

shall be permitted for only those receptacle outlets used to supply equipment that would create a greater hazard if power was interrupted or having a design that is not compatible with GFCI protection.

(B) Use of Other Outlets. Receptacles other than 125-volt, single-phase, 15-, 20-, and 30-ampere receptacles shall have protection in accordance with (1) or, the assured equipment grounding conductor program in accordance with (2).

(1) GFCI Protection. Ground-fault circuit interrupter protection for personnel.

(2) Assured Equipment Grounding Conductor Program. A written assured equipment grounding conductor program continuously enforced at the site by one or more designated persons to ensure that equipment grounding conductors for all cord sets, receptacles that are not a part of the permanent wiring of the building or structure, and equipment connected by cord and plug are installed and maintained in accordance with the applicable requirements of 250.114, 250.138, 406.3(C), and 527.4(D).

(a) The following tests shall be performed on all cord sets, receptacles that are not part of the permanent wiring of the building or structure, and cord- and plug-connected equipment required to be grounded:

(1) All equipment grounding conductors shall be tested for continuity and shall be electrically continuous.
(2) Each receptacle and attachment plug shall be tested for correct attachment of the equipment grounding conductor. The equipment grounding conductor shall be connected to its proper terminal.
(3) All required tests shall be performed as follows:

   a. Before first use on site
   b. When there is evidence of damage
   c. Before equipment is returned to service following any repairs
   d. At intervals not exceeding 3 months

(b) The tests required in item (2)(a) shall be recorded and made available to the authority having jurisdiction.

527.7 Guarding. For wiring over 600 volts, nominal, suitable fencing, barriers, or other effective means shall be provided to limit access only to authorized and qualified personnel.

CHAPTER 4

BRANCH-CIRCUIT, FEEDER, AND SERVICE CALCULATIONS

INTRODUCTION

This chapter contains provisions from Article 220—Branch-Circuit, Feeder, and Service Calculations. This article provides requirements for computing branch-circuit, feeder, and service loads. Contained in this pocket guide are two of the four parts that are included in Article 220. The two parts are I General and II Feeders and Services. Detailed instructions are provided for sizing branch-circuits, feeders, and services. These calculation provisions apply to various types of commercial and industrial occupancies. Article 220 specifies that motor loads must be calculated in accordance with Article 430 and that hermetic refrigerant motor compressors must be computed in accordance with Article 440.

ARTICLE 220
Branch-Circuit, Feeder, and Service Calculations

I. GENERAL

220.1 Scope. This article provides requirements for computing branch-circuit, feeder, and service loads.

Exception: Branch-circuit and feeder calculations for electrolytic cells as covered in 668.3(C)(1) and (4).

220.3 Computation of Branch Circuit Loads. Branch-circuit loads shall be computed as shown in 220.3(A) through (C).

(A) Lighting Load for Specified Occupancies. A unit load of not less than that specified in Table 220.3(A) for occupancies specified therein shall constitute the minimum lighting load. The floor area for each floor shall be computed from the outside dimensions of the building, dwelling unit, or other area involved. For dwelling units, the computed floor area shall not include open porches, garages, or unused or unfinished spaces not adaptable for future use.

(B) Other Loads—All Occupancies. In all occupancies, the minimum load for each outlet for general-use receptacles and outlets not used for general illumination shall not be less than that computed in 220.3(B)(1) through (11), the loads shown being based on nominal branch-circuit voltages.

(1) Specific Appliances or Loads. An outlet for a specific appliance or other load not covered in (2) through (11) shall be computed based on the ampere rating of the appliance or load served.

(2) Electric Dryers and Household Electric Cooking Appliances. Load computations shall be permitted as specified in 220.18 for electric dryers and in 220.19 for electric ranges and other cooking appliances.

(3) Motor Loads. Outlets for motor loads shall be computed in accordance with the requirements in 430.22, 430.24, and 440.6.

(4) Recessed Luminaires (Lighting Fixtures). An outlet supplying recessed luminaire(s) [lighting fixture(s)] shall be computed based on the maximum volt-ampere rating of the

Branch-Circuit, Feeder, and Service Calculations

Table 220.3(A) General Lighting Loads by Occupancy

Type of Occupancy	Unit Load Volt-Amperes per Square Meter	Volt-Amperes per Square Foot
Armories and auditoriums	11	1
Banks	39[b]	3-1/2[b]
Barber shops and beauty parlors	33	3
Churches	11	1
Clubs	22	2
Court rooms	22	2
Dwelling units[a]	33	3
Garages—commercial (storage)	6	1/2
Hospitals	22	2
Hotels and motels, including apartment houses without provision for cooking by tenants[a]	22	2
Industrial commercial (loft) buildings	22	2
Lodge rooms	17	1-1/2
Office buildings	39	3-1/2[b]
Restaurants	22	2
Schools	33	3
Stores	33	3
Warehouses (storage)	3	1/4
In any of the preceding occupancies except one-family dwellings and individual dwelling units of two-family and multifamily dwellings:		
Assembly halls and auditoriums	11	1
Halls, corridors, closets, stairways	6	1/2
Storage spaces	3	1/4

[a]See 220.3(B)(10).
[b]In addition, a unit load of 11 volt-amperes/m^2 or 1 volt-ampere/ft^2 shall be included for general-purpose receptacle outlets where the actual number of general-purpose receptacle outlets is unknown.

equipment and lamps for which the luminaire(s) [fixture(s)] is rated.

(5) Heavy-Duty Lampholders. Outlets for heavy-duty lampholders shall be computed at a minimum of 600 volt-amperes.

(6) Sign and Outline Lighting. Sign and outline lighting outlets shall be computed at a minimum of 1200 volt-amperes for each required branch circuit specified in 600.5(A).

(7) Show Windows. Show windows shall be computed in accordance with either of the following:

(1) The unit load per outlet as required in other provisions of this section
(2) At 200 volt-amperes per 300 mm (1 ft) of show window

(8) Fixed Multioutlet Assemblies. Fixed multioutlet assemblies used in other than dwelling units or the guest rooms of hotels or motels shall be computed in accordance with (1) or (2). For the purposes of this section, the computation shall be permitted to be based on the portion that contains receptacle outlets.

(1) Where appliances are unlikely to be used simultaneously, each 1.5 m (5 ft) or fraction thereof of each separate and continuous length shall be considered as one outlet of not less than 180 volt-amperes.
(2) Where appliances are likely to be used simultaneously, each 300 mm (1 ft) or fraction thereof shall be considered as an outlet of not less than 180 volt-amperes.

(9) Receptacle Outlets. Except as covered in 220.3(B)(10), receptacle outlets shall be computed at not less than 180 volt-amperes for each single or for each multiple receptacle on one yoke. A single piece of equipment consisting of a multiple receptacle comprised of four or more receptacles shall be computed at not less than 90 volt-amperes per receptacle.

This provision shall not be applicable to the receptacle outlets specified in 210.11(C)(1) and (2).

(11) Other Outlets. Other outlets not covered in 220.3(B)(1) through (10) shall be computed based on 180 volt-amperes per outlet.

II. FEEDERS AND SERVICES

220.10 General. The computed load of a feeder or service shall not be less than the sum of the loads on the branch circuits supplied, as determined by Part I of this article,

Branch-Circuit, Feeder, and Service Calculations

after any applicable demand factors permitted by Parts II, III, or IV have been applied.

FPN: See Examples D1(A) through D10 in Annex D. See 220.4(B) for the maximum load in amperes permitted for lighting units operating at less than 100 percent power factor.

220.11 General Lighting. The demand factors specified in Table 220.11 shall apply to that portion of the total branch-circuit load computed for general illumination. They shall not be applied in determining the number of branch circuits for general illumination.

220.12 Show-Window and Track Lighting.

(A) Show Windows. For show-window lighting, a load of not less than 660 volt-amperes/linear meter or 200 volt-amperes/linear foot shall be included for a show window, measured horizontally along its base.

FPN: See 220.3(B)(7) for branch circuits supplying show windows.

Table 220.11 Lighting Load Demand Factors

Type of Occupancy	Portion of Lighting Load to Which Demand Factor Applies (Volt-Amperes)	Demand Factor (Percent)
Hospitals*	First 50,000 or less at	40
	Remainder over 50,000 at	20
Hotels and motels, including apartment houses without provision for cooking by tenants*	First 20,000 or less at	50
	From 20,001 to 100,000 at	40
	Remainder over 100,000 at	30
Warehouses (storage)	First 12,500 or less at	100
	Remainder over 12,500 at	50
All others	Total volt-amperes	100

*The demand factors of this table shall not apply to the computed load of feeders or services supplying areas in hospitals, hotels, and motels where the entire lighting is likely to be used at one time, as in operating rooms, ballrooms, or dining rooms.

(B) Track Lighting. For track lighting in other than dwelling units or guest rooms of hotels or motels, an additional load of 150 volt-amperes shall be included for every 600 mm (2 ft) of lighting track or fraction thereof. Where multicircuit track is installed, the load shall be considered to be divided equally between the track circuits.

220.13 Receptacle Loads—Nondwelling Units. In other than dwelling units, receptacle loads computed at not more than 180 volt-amperes per outlet in accordance with 220.3(B)(9) and fixed multioutlet assemblies computed in accordance with 220.3(B)(8) shall be permitted to be added to the lighting loads and made subject to the demand factors given in Table 220.11, or they shall be permitted to be made subject to the demand factors given in Table 220.13.

220.14 Motors. Motor loads shall be computed in accordance with 430.24, 430.25, and 430.26 and with 440.6 for hermetic refrigerant motor compressors.

220.15 Fixed Electric Space Heating. Fixed electric space heating loads shall be computed at 100 percent of the total connected load; however, in no case shall a feeder or service load current rating be less than the rating of the largest branch circuit supplied.

220.20 Kitchen Equipment—Other Than Dwelling Unit(s). It shall be permissible to compute the load for commercial electric cooking equipment, dishwasher booster heaters, water heaters, and other kitchen equipment in accordance with Table 220.20. These demand factors shall be applied to all equipment that has either thermostatic control or intermit-

Table 220.13 Demand Factors for Nondwelling Receptacle Loads

Portion of Receptacle Load to Which Demand Factor Applies (Volt-Amperes)	Demand Factor (Percent)
First 10 kVA or less at	100
Remainder over 10 kVA at	50

Branch-Circuit, Feeder, and Service Calculations

tent use as kitchen equipment. They shall not apply to space-heating, ventilating, or air-conditioning equipment.

However, in no case shall the feeder or service demand be less than the sum of the largest two kitchen equipment loads.

220.21 Noncoincident Loads. Where it is unlikely that two or more noncoincident loads will be in use simultaneously, it shall be permissible to use only the largest load(s) that will be used at one time, in computing the total load of a feeder or service.

220.22 Feeder or Service Neutral Load. The feeder or service neutral load shall be the maximum unbalance of the load determined by this article. The maximum unbalanced load shall be the maximum net computed load between the neutral and any one ungrounded conductor, except that the load thus obtained shall be multiplied by 140 percent for 3-wire, 2-phase or 5-wire, 2-phase systems. For a feeder or service supplying household electric ranges, wall-mounted ovens, counter-mounted cooking units, and electric dryers, the maximum unbalanced load shall be considered as 70 percent of the load on the ungrounded conductors, as determined in accordance with Table 220.19 for ranges and Table 220.18 for dryers. For 3-wire dc or single-phase ac; 4-wire, 3-phase; 3-wire, 2-phase; or 5-wire, 2-phase systems, a further demand factor of 70 percent shall be permitted for that portion of the unbalanced load in excess of 200 amperes. There shall be no reduction of the neutral capacity

Table 220.20 Demand Factors for Kitchen Equipment—Other Than Dwelling Unit(s)

Number of Units of Equipment	Demand Factor (Percent)
1	100
2	100
3	90
4	80
5	70
6 and over	65

for that portion of the load that consists of nonlinear loads supplied from a 4-wire, wye-connected, 3-phase system. There shall be no reduction in the capacity of the grounded conductor of a 3-wire circuit consisting of two phase wires and the neutral of a 4-wire, 3-phase, wye-connected system.

FPN No. 1: See Examples D1(A), D1(B), D2(B), D4(A), and D5(A) in Annex D.

FPN No. 2: A 3-phase, 4-wire, wye-connected power system used to supply power to nonlinear loads may necessitate that the power system design allow for the possibility of high harmonic neutral currents.

CHAPTER 5

SERVICES

INTRODUCTION

This chapter contains provisions from Article 230—Services. Article 230 covers service conductors and equipment for control and protection of services and their installation requirements. Contained in this pocket guide are seven of the eight parts that make up Article 230. The seven parts are: I General; II Overhead Service-Drop Conductors; III Underground Service-Lateral Conductors; IV Service-Entrance Conductors; V Service Equipment—General; VI Service Equipment—Disconnecting Means; and VII Service Equipment—Overcurrent Protection. Reference the *2002 National Electrical Code* for services exceeding 600 volts, nominal.

While the service installation must comply with Parts I, IV, V, VI, and VII, compliance with Parts II and III depends on the type of service entrance. Part II contains requirements for overhead service-drop conductors, and Part III contains requirements for underground service-lateral conductors.

ARTICLE 230
Services

230.1 Scope. This article covers service conductors and equipment for control and protection of services and their installation requirements.

I. GENERAL

230.2 Number of Services. A building or other structure served shall be supplied by only one service unless permitted in 230.2(A) through (D). For the purpose of 230.40, Exception No. 2 only, underground sets of conductors, 1/0 AWG and larger, running to the same location and connected together at their supply end but not connected together at their load end shall be considered to be supplying one service.

(A) Special Conditions. Additional services shall be permitted to supply the following:

(1) Fire pumps
(2) Emergency systems
(3) Legally required standby systems
(4) Optional standby systems
(5) Parallel power production systems

(B) Special Occupancies. By special permission, additional services shall be permitted for the following:

(1) Multiple-occupancy buildings where there is no available space for service equipment accessible to all occupants, or
(2) A single building or other structure sufficiently large to make two or more services necessary

(C) Capacity Requirements. Additional services shall be permitted under any of the following:

(1) Where the capacity requirements are in excess of 2000 amperes at a supply voltage of 600 volts or less
(2) Where the load requirements of a single-phase installation are greater than the serving agency normally supplies through one service
(3) By special permission

Services

(D) Different Characteristics. Additional services shall be permitted for different voltages, frequencies, or phases, or for different uses, such as for different rate schedules.

(E) Identification. Where a building or structure is supplied by more than one service, or any combination of branch circuits, feeders, and services, a permanent plaque or directory shall be installed at each service disconnect location denoting all other services, feeders, and branch circuits supplying that building or structure and the area served by each. See 225.37.

230.3 One Building or Other Structure Not to Be Supplied Through Another. Service conductors supplying a building or other structure shall not pass through the interior of another building or other structure.

230.6 Conductors Considered Outside the Building. Conductors shall be considered outside of a building or other structure under any of the following conditions:

(1) Where installed under not less than 50 mm (2 in.) of concrete beneath a building or other structure
(2) Where installed within a building or other structure in a raceway that is encased in concrete or brick not less than 50 mm (2 in.) thick
(3) Where installed in any vault that meets the construction requirements of Article 450, Part III
(4) Where installed in conduit and under not less than 450 mm (18 in.) of earth beneath a building or other structure

230.7 Other Conductors in Raceway or Cable. Conductors other than service conductors shall not be installed in the same service raceway or service cable.

Exception No. 1: Grounding conductors and bonding jumpers.

Exception No. 2: Load management control conductors having overcurrent protection.

230.8 Raceway Seal. Where a service raceway enters a building or structure from an underground distribution system, it shall be sealed in accordance with 300.5(G). Spare or unused raceways shall also be sealed. Sealants shall be identified for use with the cable insulation, shield, or other components.

230.9 Clearance from Building Openings. Service conductors and final spans shall comply with 230.9(A), (B), and (C).

(A) Clearance from Windows. Service conductors installed as open conductors or multiconductor cable without an overall outer jacket shall have a clearance of not less than 900 mm (3 ft) from windows that are designed to be opened, doors, porches, balconies, ladders, stairs, fire escapes, or similar locations.

Exception: Conductors run above the top level of a window shall be permitted to be less than the 900-mm (3-ft) requirement.

(B) Vertical Clearance. The vertical clearance of final spans above, or within 900 mm (3 ft) measured horizontally of, platforms, projections, or surfaces from which they might be reached shall be maintained in accordance with 230.24(B).

(C) Building Openings. Overhead service conductors shall not be installed beneath openings through which materials may be moved, such as openings in farm and commercial buildings, and shall not be installed where they obstruct entrance to these building openings.

230.10 Vegetation as Support. Vegetation such as trees shall not be used for support of overhead service conductors.

II. OVERHEAD SERVICE-DROP CONDUCTORS

230.22 Insulation or Covering. Individual conductors shall be insulated or covered.

230.23 Size and Rating.

(A) General. Conductors shall have sufficient ampacity to carry the current for the load as computed in accordance with Article 220 and shall have adequate mechanical strength.

(B) Minimum Size. The conductors shall not be smaller than 8 AWG copper or 6 AWG aluminum or copper-clad aluminum.

Services

(C) Grounded Conductors. The grounded conductor shall not be less than the minimum size as required by 250.24(B).

230.24 Clearances. Service-drop conductors shall not be readily accessible and shall comply with 230.24(A) through (D) for services not over 600 volts, nominal.

(A) Above Roofs. Conductors shall have a vertical clearance of not less than 2.5 m (8 ft) above the roof surface. The vertical clearance above the roof level shall be maintained for a distance of not less than 900 mm (3 ft) in all directions from the edge of the roof.

Exception No. 1: The area above a roof surface subject to pedestrian or vehicular traffic shall have a vertical clearance from the roof surface in accordance with the clearance requirements of 230.24(B).

Exception No. 2: Where the voltage between conductors does not exceed 300 and the roof has a slope of 100 mm (4 in.) in 300 mm (12 in.), or greater, a reduction in clearance to 900 mm (3 ft) shall be permitted.

Exception No. 3: Where the voltage between conductors does not exceed 300, a reduction in clearance above only the overhanging portion of the roof to not less than 450 mm (18 in.) shall be permitted if (1) not more than 1.8 m (6 ft) of service-drop conductors, 1.2 m (4 ft) horizontally, pass above the roof overhang, and (2) they are terminated at a through-the-roof raceway or approved support.

Exception No. 4: The requirement for maintaining the vertical clearance 900 mm (3 ft) from the edge of the roof shall not apply to the final conductor span where the service drop is attached to the side of a building.

(B) Vertical Clearance from Ground. Service-drop conductors, where not in excess of 600 volts, nominal, shall have the following minimum clearance from final grade:

(1) 3.0 m (10 ft)—at the electric service entrance to buildings, also at the lowest point of the drip loop of the building electric entrance, and above areas or sidewalks accessible only to pedestrians, measured from final grade or other accessible surface only for service-drop cables supported on and cabled together with a

grounded bare messenger where the voltage does not exceed 150 volts to ground

(2) 3.7 m (12 ft)—over residential property and driveways, and those commercial areas not subject to truck traffic where the voltage does not exceed 300 volts to ground

(3) 4.5 m (15 ft)—for those areas listed in the 3.7 m (12 ft) classification where the voltage exceeds 300 volts to ground

(4) 5.5 m (18 ft)—over public streets, alleys, roads, parking areas subject to truck traffic, driveways on other than residential property, and other land such as cultivated, grazing, forest, and orchard

230.26 Point of Attachment. The point of attachment of the service-drop conductors to a building or other structure shall provide the minimum clearances as specified in 230.24. In no case shall this point of attachment be less than 3.0 m (10 ft) above finished grade.

230.27 Means of Attachment. Multiconductor cables used for service drops shall be attached to buildings or other structures by fittings identified for use with service conductors. Open conductors shall be attached to fittings identified for use with service conductors or to noncombustible, nonabsorbent insulators securely attached to the building or other structure.

230.28 Service Masts as Supports. Where a service mast is used for the support of service-drop conductors, it shall be of adequate strength or be supported by braces or guys to withstand safely the strain imposed by the service drop. Where raceway-type service masts are used, all raceway fittings shall be identified for use with service masts. Only power service-drop conductors shall be permitted to be attached to a service mast.

230.29 Supports over Buildings. Service-drop conductors passing over a roof shall be securely supported by substantial structures. Where practicable, such supports shall be independent of the building.

III. UNDERGROUND SERVICE-LATERAL CONDUCTORS

230.30 Insulation. Service-lateral conductors shall be insulated for the applied voltage.

230.31 Size and Rating.

(A) General. Service-lateral conductors shall have sufficient ampacity to carry the current for the load as computed in accordance with Article 220 and shall have adequate mechanical strength.

(B) Minimum Size. The conductors shall not be smaller than 8 AWG copper or 6 AWG aluminum or copper-clad aluminum.

(C) Grounded Conductors. The grounded conductor shall not be less than the minimum size required by 250.24(B).

230.32 Protection Against Damage.
Underground service-lateral conductors shall be protected against damage in accordance with 300.5. Service-lateral conductors entering a building shall be installed in accordance with 230.6 or protected by a raceway wiring method identified in 230.43.

230.33 Spliced Conductors.
Service-lateral conductors shall be permitted to be spliced or tapped in accordance with 110.14, 300.5(E), 300.13, and 300.15.

IV. SERVICE-ENTRANCE CONDUCTORS

230.40 Number of Service-Entrance Conductor Sets.
Each service drop or lateral shall supply only one set of service-entrance conductors.

Exception No. 1: A building with one or more than one occupancy shall be permitted to have one set of service-entrance conductors for each service of different characteristics, as defined in 230.2(D), run to each occupancy or group of occupancies.

Exception No. 2: Where two to six service disconnecting means in separate enclosures are grouped at one location and supply separate loads from one service drop or lateral, one set of service-entrance conductors shall be permitted to supply each or several such service equipment enclosures.

Exception No. 3: A single-family dwelling unit and a separate structure shall be permitted to have one set of service-entrance conductors run to each from a single service drop or lateral.

Exception No. 4: A two-family dwelling or a multifamily dwelling shall be permitted to have one set of service-entrance conductors installed to supply the circuits covered in 210.25.

Exception No. 5: One set of service-entrance conductors connected to the supply side of the normal service disconnecting means shall be permitted to supply each or several systems covered by 230.82(4) or (5).

230.41 Insulation of Service-Entrance Conductors. Service-entrance conductors entering or on the exterior of buildings or other structures shall be insulated.

230.42 Minimum Size and Rating.

(A) General. The ampacity of the service-entrance conductors before the application of any adjustment or correction factors shall not be less than either (1) or (2). Loads shall be determined in accordance with Article 220. Ampacity shall be determined from 310.15. The maximum allowable current of busways shall be that value for which the busway has been listed or labeled.

(1) The sum of the noncontinuous loads plus 125 percent of continuous loads
(2) The sum of the noncontinuous load plus the continuous load if the service-entrance conductors terminate in an overcurrent device where both the overcurrent device and its assembly are listed for operation at 100 percent of their rating

(B) Specific Installations. In addition to the requirements of 230.42(A), the minimum ampacity for ungrounded conductors for specific installations shall not be less than the rating of the service disconnecting means specified in 230.79(A) through (D).

(C) Grounded Conductors. The grounded conductor shall not be less than the minimum size as required by 250.24(B).

230.43 Wiring Methods for 600 Volts, Nominal, or Less. Service-entrance conductors shall be installed in accordance with the applicable requirements of this *Code* covering the type of wiring method used and shall be limited to the following methods:

(1) Open wiring on insulators
(2) Type IGS cable

Services

(3) Rigid metal conduit
(4) Intermediate metal conduit
(5) Electrical metallic tubing
(6) Electrical nonmetallic tubing (ENT)
(7) Service-entrance cables
(8) Wireways
(9) Busways
(10) Auxiliary gutters
(11) Rigid nonmetallic conduit
(12) Cablebus
(13) Type MC cable
(14) Mineral-insulated, metal-sheathed cable
(15) Flexible metal conduit not over 1.8 m (6 ft) long or liquidtight flexible metal conduit not over 1.8 m (6 ft) long between raceways, or between raceway and service equipment, with equipment bonding jumper routed with the flexible metal conduit or the liquidtight flexible metal conduit according to the provisions of 250.102(A), (B), (C), and (E)
(16) Liquidtight flexible nonmetallic conduit

230.44 Cable Trays. Cable tray systems shall be permitted to support cable used as service-entrance conductors.

230.46 Spliced Conductors. Service-entrance conductors shall be permitted to be spliced or tapped in accordance with 110.14, 300.5(E), 300.13, and 300.15.

230.49 Protection Against Physical Damage—Underground. Underground service-entrance conductors shall be protected against physical damage in accordance with 300.5.

230.50 Protection of Open Conductors and Cables Against Damage—Above Ground. Service-entrance conductors installed above ground shall be protected against physical damage as specified in 230.50(A) or (B).

(A) Service Cables. Service cables, where subject to physical damage, shall be protected by any of the following:

(1) Rigid metal conduit
(2) Intermediate metal conduit
(3) Schedule 80 rigid nonmetallic conduit
(4) Electrical metallic tubing
(5) Other approved means

(B) Other Than Service Cable. Individual open conductors and cables other than service cables shall not be installed within 3.0 m (10 ft) of grade level or where exposed to physical damage.

230.53 Raceways to Drain. Where exposed to the weather, raceways enclosing service-entrance conductors shall be raintight and arranged to drain. Where embedded in masonry, raceways shall be arranged to drain.

230.54 Overhead Service Locations.

(A) Raintight Service Head. Service raceways shall be equipped with a raintight service head at the point of connection to service-drop conductors.

(C) Service Heads Above Service-Drop Attachment. Service heads and goosenecks in service-entrance cables shall be located above the point of attachment of the service-drop conductors to the building or other structure.

Exception: Where it is impracticable to locate the service head above the point of attachment, the service head location shall be permitted not farther than 600 mm (24 in.) from the point of attachment.

(D) Secured. Service cables shall be held securely in place.

(E) Separately Bushed Openings. Service heads shall have conductors of different potential brought out through separately bushed openings.

(F) Drip Loops. Drip loops shall be formed on individual conductors. To prevent the entrance of moisture, service-entrance conductors shall be connected to the service-drop conductors either (1) below the level of the service head or (2) below the level of the termination of the service-entrance cable sheath.

(G) Arranged That Water Will Not Enter Service Raceway or Equipment. Service-drop conductors and service-entrance conductors shall be arranged so that water will not enter service raceway or equipment.

230.56 Service Conductor with the Higher Voltage to Ground. On a 4-wire, delta-connected service where the midpoint of one phase winding is grounded, the service conductor having the higher phase voltage to ground shall be durably and permanently marked by an outer finish that is orange in color, or by other effective means, at each termination or junction point.

V. SERVICE EQUIPMENT—GENERAL

230.62 Service Equipment—Enclosed or Guarded. Energized parts of service equipment shall be enclosed as specified in 230.62(A) or guarded as specified in 230.62(B).

(A) Enclosed. Energized parts shall be enclosed so that they will not be exposed to accidental contact or shall be guarded as in 230.62(B).

(B) Guarded. Energized parts that are not enclosed shall be installed on a switchboard, panelboard, or control board and guarded in accordance with 110.18 and 110.27. Where energized parts are guarded as provided in 110.27(A)(1) and (2), a means for locking or sealing doors providing access to energized parts shall be provided.

230.66 Marking. Service equipment rated at 600 volts or less shall be marked to identify it as being suitable for use as service equipment. Individual meter socket enclosures shall not be considered service equipment.

VI. SERVICE EQUIPMENT—DISCONNECTING MEANS

230.70 General. Means shall be provided to disconnect all conductors in a building or other structure from the service-entrance conductors.

(A) Location. The service disconnecting means shall be installed in accordance with 230.70(A)(1), (2), and (3).

(1) Readily Accessible Location. The service disconnecting means shall be installed at a readily accessible location either outside of a building or structure or inside nearest the point of entrance of the service conductors.

(2) Bathrooms. Service disconnecting means shall not be installed in bathrooms.

(3) Remote Control. Where a remote control device(s) is used to actuate the service disconnecting means, the service disconnecting means shall be located in accordance with 230.70(A)(1).

(B) Marking. Each service disconnect shall be permanently marked to identify it as a service disconnect.

(C) Suitable for Use. Each service disconnecting means shall be suitable for the prevailing conditions. Service equipment installed in hazardous (classified) locations shall comply with the requirements of Articles 500 through 517.

230.71 Maximum Number of Disconnects.

(A) General. The service disconnecting means for each service permitted by 230.2, or for each set of service-entrance conductors permitted by 230.40, Exception Nos. 1, 3, 4, or 5, shall consist of not more than six switches or sets of circuit breakers, or a combination of not more than six switches and sets of circuit breakers, mounted in a single enclosure, in a group of separate enclosures, or in or on a switchboard. There shall be no more than six sets of disconnects per service grouped in any one location. For the purpose of this section, disconnecting means used solely for power monitoring equipment, or the control circuit of the ground-fault protection system or power-operable service disconnecting means, installed as part of the listed equipment, shall not be considered a service disconnecting means.

(B) Single-Pole Units. Two or three single-pole switches or breakers, capable of individual operation, shall be permitted on multiwire circuits, one pole for each ungrounded conductor, as one multipole disconnect, provided they are equipped with handle ties or a master handle to disconnect all conductors of the service with no more than six operations of the hand.

230.72 Grouping of Disconnects.

(A) General. The two to six disconnects as permitted in 230.71 shall be grouped. Each disconnect shall be marked to indicate the load served.

Exception: One of the two to six service disconnecting means permitted in 230.71, where used only for a water

Services

43

pump also intended to provide fire protection, shall be permitted to be located remote from the other disconnecting means.

(B) Additional Service Disconnecting Means. The one or more additional service disconnecting means for fire pumps, for legally required standby, or for optional standby services permitted by 230.2 shall be installed remote from the one to six service disconnecting means for normal service to minimize the possibility of simultaneous interruption of supply.

(C) Access to Occupants. In a multiple-occupancy building, each occupant shall have access to the occupant's service disconnecting means.

230.74 Simultaneous Opening of Poles. Each service disconnect shall simultaneously disconnect all ungrounded service conductors that it controls from the premises wiring system.

230.75 Disconnection of Grounded Conductor. Where the service disconnecting means does not disconnect the grounded conductor from the premises wiring, other means shall be provided for this purpose in the service equipment. A terminal or bus to which all grounded conductors can be attached by means of pressure connectors shall be permitted for this purpose. In a multisection switchboard, disconnects for the grounded conductor shall be permitted to be in any section of the switchboard, provided any such switchboard section is marked.

230.76 Manually or Power Operable. The service disconnecting means for ungrounded service conductors shall consist of either (1) a manually operable switch or circuit breaker equipped with a handle or other suitable operating means or (2) a power-operated switch or circuit breaker, provided the switch or circuit breaker can be opened by hand in the event of a power supply failure.

230.77 Indicating. The service disconnecting means shall plainly indicate whether it is in the open or closed position.

230.79 Rating of Service Disconnecting Means. The service disconnecting means shall have a rating not less than the load to be carried, determined in accordance with Article 220. In no case shall the rating be lower than specified in 230.79(A), (B), (C), or (D).

(A) One-Circuit Installation. For installations to supply only limited loads of a single branch circuit, the service disconnecting means shall have a rating of not less than 15 amperes.

(B) Two-Circuit Installations. For installations consisting of not more than two 2-wire branch circuits, the service disconnecting means shall have a rating of not less than 30 amperes.

(D) All Others. For all other installations, the service disconnecting means shall have a rating of not less than 60 amperes.

230.82 Equipment Connected to the Supply Side of Service Disconnect. Only the following equipment shall be permitted to be connected to the supply side of the service disconnecting means:

(1) Cable limiters or other current-limiting devices.
(2) Meters, meter sockets, or meter disconnect switches nominally rated not in excess of 600 volts, provided all metal housings and service enclosures are grounded.
(3) Instrument transformers (current and voltage), high-impedance shunts, load management devices, and surge arresters.
(4) Taps used only to supply load management devices, circuits for standby power systems, fire pump equipment, and fire and sprinkler alarms, if provided with service equipment and installed in accordance with requirements for service-entrance conductors.
(5) Solar photovoltaic systems, fuel cell systems, or interconnected electric power production sources.
(6) Control circuits for power-operable service disconnecting means, if suitable overcurrent protection and disconnecting means are provided.
(7) Ground-fault protection systems where installed as part of listed equipment, if suitable overcurrent protection and disconnecting means are provided.

VII. SERVICE EQUIPMENT— OVERCURRENT PROTECTION

230.90 Where Required. Each ungrounded service conductor shall have overload protection.

(A) Ungrounded Conductor. Such protection shall be provided by an overcurrent device in series with each ungrounded service conductor that has a rating or setting not higher than the allowable ampacity of the conductor. A set of fuses shall be considered all the fuses required to protect all the ungrounded conductors of a circuit. Single-pole circuit breakers, grouped in accordance with 230.71(B), shall be considered as one protective device.

Exception No. 1: For motor-starting currents, ratings that conform with 430.52, 430.62, and 430.63 shall be permitted.

Exception No. 2: Fuses and circuit breakers with a rating or setting that conform with 240.4(B) or (C) and 240.6 shall be permitted.

Exception No. 3: Two to six circuit breakers or sets of fuses shall be permitted as the overcurrent device to provide the overload protection. The sum of the ratings of the circuit breakers or fuses shall be permitted to exceed the ampacity of the service conductors, provided the calculated load does not exceed the ampacity of the service conductors.

Exception No. 4: Overload protection for fire pump supply conductors shall conform with 695.4(B)(1).

Exception No. 5: Overload protection for 120/240-volt, 3-wire, single-phase dwelling services shall be permitted in accordance with the requirements of 310.15(B)(6).

(B) Not in Grounded Conductor. No overcurrent device shall be inserted in a grounded service conductor except a circuit breaker that simultaneously opens all conductors of the circuit.

230.91 Location. The service overcurrent device shall be an integral part of the service disconnecting means or shall be located immediately adjacent thereto.

230.95 Ground-Fault Protection of Equipment. Ground-fault protection of equipment shall be provided for solidly grounded wye electrical services of more than 150 volts to ground but not exceeding 600 volts phase-to-phase for each service disconnect rated 1000 amperes or more.

The rating of the service disconnect shall be considered to be the rating of the largest fuse that can be installed or the highest continuous current trip setting for which the actual overcurrent device installed in a circuit breaker is rated or can be adjusted.

Solidly Grounded—Definition. Connection of the grounded conductor to ground without inserting any resistor or impedance device.

Exception No. 1: The ground-fault protection provisions of this section shall not apply to a service disconnect for a continuous industrial process where a nonorderly shutdown will introduce additional or increased hazards.

Exception No. 2: The ground-fault protection provisions of this section shall not apply to fire pumps.

(A) Setting. The ground-fault protection system shall operate to cause the service disconnect to open all ungrounded conductors of the faulted circuit. The maximum setting of the ground-fault protection shall be 1200 amperes, and the maximum time delay shall be one second for ground-fault currents equal to or greater than 3000 amperes.

(B) Fuses. If a switch and fuse combination is used, the fuses employed shall be capable of interrupting any current higher than the interrupting capacity of the switch during a time that the ground-fault protective system will not cause the switch to open.

(C) Performance Testing. The ground-fault protection system shall be performance tested when first installed on site. The test shall be conducted in accordance with instructions that shall be provided with the equipment. A written record of this test shall be made and shall be available to the authority having jurisdiction.

CHAPTER 6

FEEDERS

INTRODUCTION

This chapter contains provisions from Article 215—Feeders. This article covers installation, overcurrent protection, minimum size, and ampacity of conductors for feeders that supply branch-circuit loads. These branch-circuit loads must be calculated in accordance with Article 220. A feeder consists of all circuit conductors located between the service equipment, the source of a separately derived system, or other power supply source and the final branch-circuit overcurrent device. Feeder conductors supply power to remote panelboards, which are sometimes referred to as subpanels. Feeder overcurrent protection must meet the provisions stipulated in Article 240, Part I.

ARTICLE 215

Feeders

215.1 Scope. This article covers the installation requirements, overcurrent protection requirements, minimum size, and ampacity of conductors for feeders supplying branch-circuit loads as computed in accordance with Article 220.

215.2 Minimum Rating and Size.

(A) Feeders Not More Than 600 Volts.

(1) General. Feeder conductors shall have an ampacity not less than required to supply the load as computed in Parts II, III, and IV of Article 220. The minimum feeder-circuit conductor size, before the application of any adjustment or correction factors, shall have an allowable ampacity not less than the noncontinuous load plus 125 percent of the continuous load.

Additional minimum sizes shall be as specified in (2), (3), and (4) under the conditions stipulated.

(2) For Specified Circuits. The ampacity of feeder conductors shall not be less than 30 amperes where the load supplied consists of any of the following number and types of circuits:

(1) Two or more 2-wire branch circuits supplied by a 2-wire feeder
(2) More than two 2-wire branch circuits supplied by a 3-wire feeder
(3) Two or more 3-wire branch circuits supplied by a 3-wire feeder
(4) Two or more 4-wire branch circuits supplied by a 3-phase, 4-wire feeder

(3) Ampacity Relative to Service-Entrance Conductors. The feeder conductor ampacity shall not be less than that of the service-entrance conductors where the feeder conductors carry the total load supplied by service-entrance conductors with an ampacity of 55 amperes or less.

> FPN No. 2: Conductors for feeders as defined in Article 100, sized to prevent a voltage drop exceeding 3 percent at the farthest outlet of power, heat-

ing, and lighting loads, or combinations of such loads, and where the maximum total voltage drop on both feeders and branch circuits to the farthest outlet does not exceed 5 percent, will provide reasonable efficiency of operation.

215.3 Overcurrent Protection. Feeders shall be protected against overcurrent in accordance with the provisions of Part I of Article 240. Where a feeder supplies continuous loads or any combination of continuous and noncontinuous loads, the rating of the overcurrent device shall not be less than the noncontinuous load plus 125 percent of the continuous load.

215.4 Feeders with Common Neutral.

(A) Feeders with Common Neutral. Two or three sets of 3-wire feeders or two sets of 4-wire or 5-wire feeders shall be permitted to utilize a common neutral.

(B) In Metal Raceway or Enclosure. Where installed in a metal raceway or other metal enclosure, all conductors of all feeders using a common neutral shall be enclosed within the same raceway or other enclosure as required in 300.20.

215.5 Diagrams of Feeders. If required by the authority having jurisdiction, a diagram showing feeder details shall be provided prior to the installation of the feeders. Such a diagram shall show the area in square feet of the building or other structure supplied by each feeder, the total computed load before applying demand factors, the demand factors used, the computed load after applying demand factors, and the size and type of conductors to be used.

215.6 Feeder Conductor Grounding Means. Where a feeder supplies branch circuits in which equipment grounding conductors are required, the feeder shall include or provide a grounding means, in accordance with the provisions of 250.134, to which the equipment grounding conductors of the branch circuits shall be connected.

215.8 Means of Identifying Conductor with the Higher Voltage to Ground. On a 4-wire, delta-connected secondary where the midpoint of one phase winding is grounded to supply lighting and similar loads, the phase conductor having the higher voltage to ground shall be identified by an outer finish that is orange in color or by tagging or other effective means. Such identification shall be placed at each point where a connection is made if the grounded conductor is also present.

215.9 Ground-Fault Circuit-Interrupter Protection for Personnel. Feeders supplying 15- and 20-ampere receptacle branch circuits shall be permitted to be protected by a ground-fault circuit interrupter in lieu of the provisions for such interrupters as specified in 210.8 and Article 527.

215.10 Ground-Fault Protection of Equipment. Each feeder disconnect rated 1000 amperes or more and installed on solidly grounded wye electrical systems of more than 150 volts to ground, but not exceeding 600 volts phase-to-phase, shall be provided with ground-fault protection of equipment in accordance with the provisions of 230.95.

Exception No. 1: The provisions of this section shall not apply to a disconnecting means for a continuous industrial process where a nonorderly shutdown will introduce additional or increased hazards.

Exception No. 2: The provisions of this section shall not apply to fire pumps.

Exception No. 3: The provisions of this section shall not apply if ground-fault protection of equipment is provided on the supply side of the feeder.

215.11 Circuits Derived from Autotransformers. Feeders shall not be derived from autotransformers unless the system supplied has a grounded conductor that is electrically connected to a grounded conductor of the system supplying the autotransformer.

Exception No. 1: An autotransformer shall be permitted without the connection to a grounded conductor where transforming from a nominal 208 volts to a nominal 240-volt supply or similarly from 240 volts to 208 volts.

Exception No. 2: In industrial occupancies, where conditions of maintenance and supervision ensure that only qualified persons service the installation, autotransformers shall be permitted to supply nominal 600-volt loads from nominal 480-volt systems, and 480-volt loads from nominal 600-volt systems, without the connection to a similar grounded conductor.

CHAPTER 7

SWITCHBOARDS, PANELBOARDS, AND DISCONNECTS

INTRODUCTION

This chapter contains provisions from Article 408—Switchboards and Panelboards and Article 404—Switches. Article 408 covers switchboards, panelboards, and distribution boards installed for the control of light and power circuits. A switchboard is a large single panel, frame, or assembly of panels on which are mounted, on the front or back (or both), switches, overcurrent and other protective devices, buses, and usually instruments. Switchboards are generally accessible from both the front and the back and are not intended to be installed in cabinets. A panelboard consists of one (or more) panel units designed for assembly in the form of a single panel. It includes buses and automatic overcurrent devices and may contain switches for the control of light, heat, or power circuits. Panelboards are accessible only from the front and are designed to be placed in cabinets or cutout boxes. Switchboards and panelboards can be used as service equipment where identified for such use.

Provisions from Article 404 are located in two chapters of this *Pocket Guide,* Chapters 7 and 17. This chapter covers switches and circuit breakers where used as switches. The switches discussed here are more commonly called disconnect switches or safety switches. Disconnects may contain overcurrent protective devices (fuses or circuit breakers), or they may be nonfusible or molded-case switches.

ARTICLE 408
Switchboards and Panelboards

I. GENERAL

408.1 Scope. This article covers the following:

(1) All switchboards, panelboards, and distribution boards installed for the control of light and power circuits
(2) Battery-charging panels supplied from light or power circuits

408.3 Support and Arrangement of Busbars and Conductors.

(A) Conductors and Busbars on a Switchboard or Panelboard. Conductors and busbars on a switchboard or panelboard shall comply with 408.3(A)(1), (2), and (3) as applicable.

(1) Location. Conductors and busbars shall be located so as to be free from physical damage and shall be held firmly in place.

(2) Service Switchboards. Barriers shall be placed in all service switchboards such that no uninsulated, ungrounded service busbar or service terminal is exposed to inadvertent contact by persons or maintenance equipment while servicing load terminations.

(3) Same Vertical Section. Other than the required interconnections and control wiring, only those conductors that are intended for termination in a vertical section of a switchboard shall be located in that section.

(B) Overheating and Inductive Effects. The arrangement of busbars and conductors shall be such as to avoid overheating due to inductive effects.

(C) Used as Service Equipment. Each switchboard or panelboard, if used as service equipment, shall be provided with a main bonding jumper sized in accordance with 250.28(D) or the equivalent placed within the panelboard or one of the sections of the switchboard for connecting the

Switchboards, Panelboards, and Disconnects 55

grounded service conductor on its supply side to the switchboard or panelboard frame. All sections of a switchboard shall be bonded together using an equipment grounding conductor sized in accordance with Table 250.122.

(E) Phase Arrangement. The phase arrangement on 3-phase buses shall be A, B, C from front to back, top to bottom, or left to right, as viewed from the front of the switchboard or panelboard. The B phase shall be that phase having the higher voltage to ground on 3-phase, 4-wire, delta-connected systems. Other busbar arrangements shall be permitted for additions to existing installations and shall be marked.

(F) Minimum Wire-Bending Space. The minimum wire-bending space at terminals and minimum gutter space provided in panelboards and switchboards shall be as required in 312.6.

408.4 Circuit Directory. All circuits and circuit modifications shall be legibly identified as to purpose or use on a circuit directory located on the face or inside of the panel door in the case of a panelboard, and at each switch on a switchboard.

II. SWITCHBOARDS

408.5 Location of Switchboards. Switchboards that have any exposed live parts shall be located in permanently dry locations and then only where under competent supervision and accessible only to qualified persons. Switchboards shall be located so that the probability of damage from equipment or processes is reduced to a minimum.

408.7 Location Relative to Easily Ignitible Material. Switchboards shall be placed so as to reduce to a minimum the probability of communicating fire to adjacent combustible materials. Where installed over a combustible floor, suitable protection thereto shall be provided.

408.8 Clearances.

(A) From Ceiling. For other than a totally enclosed switchboard, a space not less than 900 mm (3 ft) shall be provided between the top of the switchboard and any combustible ceiling, unless a noncombustible shield is provided between the switchboard and the ceiling.

(B) Around Switchboards. Clearances around switchboards shall comply with the provisions of 110.26.

408.10 Clearance for Conductors Entering Bus Enclosures. Where conduits or other raceways enter a switchboard, floor-standing panelboard, or similar enclosure at the bottom, sufficient space shall be provided to permit installation of conductors in the enclosure. The wiring space shall not be less than shown in Table 408.10 where the conduit or raceways enter or leave the enclosure below the busbars, their supports, or other obstructions. The conduit or raceways, including their end fittings, shall not rise more than 75 mm (3 in.) above the bottom of the enclosure.

III. PANELBOARDS

408.13 General. All panelboards shall have a rating not less than the minimum feeder capacity required for the load computed in accordance with Article 220. Panelboards shall be durably marked by the manufacturer with the voltage and the current rating and the number of phases for which they are designed and with the manufacturer's name or trademark in such a manner so as to be visible after installation, without disturbing the interior parts or wiring.

Table 408.10 Clearance for Conductors Entering Bus Enclosures

Conductor	Minimum Spacing Between Bottom of Enclosure and Busbars, Their Supports, or Other Obstructions	
	mm	in.
Insulated busbars, their supports, or other obstructions	200	8
Noninsulated busbars	250	10

Switchboards, Panelboards, and Disconnects

408.14 Classification of Panelboards. Panelboards shall be classified for the purposes of this article as either lighting and appliance branch-circuit panelboards or power panelboards, based on their content. A lighting and appliance branch circuit is a branch circuit that has a connection to the neutral of the panelboard and that has overcurrent protection of 30 amperes or less in one or more conductors.

(A) Lighting and Appliance Branch-Circuit Panelboard. A lighting and appliance branch-circuit panelboard is one having more than 10 percent of its overcurrent devices protecting lighting and appliance branch circuits.

(B) Power Panelboard. A power panelboard is one having 10 percent or fewer of its overcurrent devices protecting lighting and appliance branch circuits.

408.15 Number of Overcurrent Devices on One Panelboard. Not more than 42 overcurrent devices (other than those provided for in the mains) of a lighting and appliance branch-circuit panelboard shall be installed in any one cabinet or cutout box.

A lighting and appliance branch-circuit panelboard shall be provided with physical means to prevent the installation of more overcurrent devices than that number for which the panelboard was designed, rated, and approved.

For the purposes of this article, a 2-pole circuit breaker shall be considered two overcurrent devices; a 3-pole circuit breaker shall be considered three overcurrent devices.

408.16 Overcurrent Protection.

(A) Lighting and Appliance Branch-Circuit Panelboard Individually Protected. Each lighting and appliance branch-circuit panelboard shall be individually protected on the supply side by not more than two main circuit breakers or two sets of fuses having a combined rating not greater than that of the panelboard.

Exception No. 1: Individual protection for a lighting and appliance panelboard shall not be required if the panelboard feeder has overcurrent protection not greater than the rating of the panelboard.

(B) Power Panelboard Protection. In addition to the requirements of 408.13, a power panelboard with supply conductors that include a neutral and having more than 10 percent of its overcurrent devices protecting branch circuits rated 30 amperes or less shall be protected by an overcurrent protective device having a rating not greater than that of the panelboard. The overcurrent protective device shall be located within or at any point on the supply side of the panelboard.

Exception: This individual protection shall not be required for a power panelboard used as service equipment with multiple disconnecting means in accordance with 230.71.

(C) Snap Switches Rated at 30 Amperes or Less. Panelboards equipped with snap switches rated at 30 amperes or less shall have overcurrent protection not in excess of 200 amperes.

(D) Supplied Through a Transformer. Where a panelboard is supplied through a transformer, the overcurrent protection in 408.16(A), (B), and (C) shall be located on the secondary side of the transformer.

Exception: A panelboard supplied by the secondary side of a transformer shall be considered as protected by the overcurrent protection provided on the primary side of the transformer where that protection is in accordance with 240.21(C)(1).

(E) Delta Breakers. A 3-phase disconnect or overcurrent device shall not be connected to the bus of any panelboard that has less than 3-phase buses. Delta breakers shall not be installed in panelboards.

(F) Back-Fed Devices. Plug-in-type overcurrent protection devices or plug-in type-main lug assemblies that are back-fed and used to terminate field-installed ungrounded supply conductors shall be secured in place by an additional fastener that requires other than a pull to release the device from the mounting means on the panel.

408.20 Grounding of Panelboards. Panelboard cabinets and panelboard frames, if of metal, shall be in physical con-

Switchboards, Panelboards, and Disconnects

tact with each other and shall be grounded. Where the panelboard is used with nonmetallic raceway or cable or where separate grounding conductors are provided, a terminal bar for the grounding conductors shall be secured inside the cabinet. The terminal bar shall be bonded to the cabinet and panelboard frame, if of metal; otherwise it shall be connected to the grounding conductor that is run with the conductors feeding the panelboard.

Exception: Where an isolated equipment grounding conductor is provided as permitted by 250.146(D), the insulated equipment grounding conductor that is run with the circuit conductors shall be permitted to pass through the panelboard without being connected to the panelboard's equipment grounding terminal bar.

Grounding conductors shall not be connected to a terminal bar provided for grounded conductors (may be a neutral) unless the bar is identified for the purpose and is located where interconnection between equipment grounding conductors and grounded circuit conductors is permitted or required by Article 250.

408.21 Grounded Conductor Terminations. Each grounded conductor shall terminate within the panelboard in an individual terminal that is not also used for another conductor.

Exception: Grounded conductors of circuits with parallel conductors shall be permitted to terminate in a single terminal if the terminal is identified for connection of more than one conductor.

ARTICLE 404
Switches

404.3 Enclosure.

(A) General. Switches and circuit breakers shall be of the externally operable type mounted in an enclosure listed for the intended use. The minimum wire-bending space at terminals and minimum gutter space provided in switch enclosures shall be as required in 312.6.

(B) Used as a Raceway. Enclosures shall not be used as junction boxes, auxiliary gutters, or raceways for conductors feeding through or tapping off to other switches or overcurrent devices, unless the enclosure complies with 312.8.

404.4 Wet Locations. A switch or circuit breaker in a wet location or outside of a building shall be enclosed in a weatherproof enclosure or cabinet that shall comply with 312.2(A). Switches shall not be installed within wet locations in tub or shower spaces unless installed as part of a listed tub or shower assembly.

404.5 Time Switches, Flashers, and Similar Devices. Time switches, flashers, and similar devices shall be of the enclosed type or shall be mounted in cabinets or boxes or equipment enclosures. Energized parts shall be barriered to prevent operator exposure when making manual adjustments or switching.

Exception: Devices mounted so they are accessible only to qualified persons shall be permitted without barriers, provided they are located within an enclosure such that any energized parts within 152 mm (6.0 in.) of the manual adjustment or switch are covered by suitable barriers.

404.6 Position and Connection of Switches.

(A) Single-Throw Knife Switches. Single-throw knife switches shall be placed so that gravity will not tend to close them. Single-throw knife switches, approved for use in the inverted position, shall be provided with a locking device that ensures that the blades remain in the open position when so set.

404.7 Indicating. General-use and motor-circuit switches, circuit breakers, and molded case switches, where mounted in an enclosure as described in 404.3, shall clearly indicate whether they are in the open (off) or closed (on) position.

Where these switch or circuit breaker handles are operated vertically rather than rotationally or horizontally, the up position of the handle shall be the (on) position.

404.8 Accessibility and Grouping.

(A) Location. All switches and circuit breakers used as switches shall be located so that they may be operated from

Switchboards, Panelboards, and Disconnects

a readily accessible place. They shall be installed so that the center of the grip of the operating handle of the switch or circuit breaker, when in its highest position, is not more than 2.0 m (6 ft 7 in.) above the floor or working platform.

Exception No. 1: On busway installations, fused switches and circuit breakers shall be permitted to be located at the same level as the busway. Suitable means shall be provided to operate the handle of the device from the floor.

Exception No. 2: Switches and circuit breakers installed adjacent to motors, appliances, or other equipment that they supply shall be permitted to be located higher than specified in the foregoing and to be accessible by portable means.

Exception No. 3: Hookstick operable isolating switches shall be permitted at greater heights.

404.11 Circuit Breakers as Switches. A hand-operable circuit breaker equipped with a lever or handle, or a power-operated circuit breaker capable of being opened by hand in the event of a power failure, shall be permitted to serve as a switch if it has the required number of poles.

CHAPTER 8

OVERCURRENT PROTECTION

INTRODUCTION

This chapter contains provisions from Article 240—Overcurrent Protection. Article 240 provides general requirements for overcurrent protection and overcurrent protective devices. Generally, overcurrent protective devices are plug fuses, cartridge fuses, and circuit breakers. Overcurrent protection for conductors and equipment is provided to open the circuit if the current reaches a value that will cause an excessive or dangerous temperature in conductors or conductor insulation.

Contained in this *Pocket Guide* are six of the nine parts that are included in Article 240. The six parts are: I General; II Location; III Enclosures; IV Disconnecting and Guarding; VII Circuit Breakers; and VIII Supervised Industrial Installations.

ARTICLE 240
Overcurrent Protection

I. GENERAL

240.1 Scope. Parts I through VII of this article provide the general requirements for overcurrent protection and overcurrent protective devices not more than 600 volts, nominal. Part VIII covers overcurrent protection for those portions of supervised industrial installations operating at voltages of not more than 600 volts, nominal.

240.2 Definitions.

Supervised Industrial Installation. For the purposes of Part VIII, the industrial portions of a facility where all of the following conditions are met:

(1) Conditions of maintenance and engineering supervision ensure that only qualified persons monitor and service the system.
(2) The premises wiring system has 2500 kVA or greater of load used in industrial process(es), manufacturing activities, or both, as calculated in accordance with Article 220.
(3) The premises has at least one service that is more than 150 volts to ground and more than 300 volts phase-to-phase.

This definition excludes installations in buildings used by the industrial facility for offices, warehouses, garages, machine shops, and recreational facilities that are not an integral part of the industrial plant, substation, or control center.

Tap Conductors. As used in this article, a tap conductor is defined as a conductor, other than a service conductor, that has overcurrent protection ahead of its point of supply that exceeds the value permitted for similar conductors that are protected as described elsewhere in 240.4.

240.4 Protection of Conductors. Conductors, other than flexible cords, flexible cables, and fixture wires, shall be protected against overcurrent in accordance with their am-

Overcurrent Protection

pacities specified in 310.15, unless otherwise permitted or required in 240.4(A) through (G).

(A) Power Loss Hazard. Conductor overload protection shall not be required where the interruption of the circuit would create a hazard, such as in a material-handling magnet circuit or fire pump circuit. Short-circuit protection shall be provided.

(B) Devices Rated 800 Amperes or Less. The next higher standard overcurrent device rating (above the ampacity of the conductors being protected) shall be permitted to be used, provided all of the following conditions are met:

(1) The conductors being protected are not part of a multioutlet branch circuit supplying receptacles for cord-and-plug-connected portable loads.
(2) The ampacity of the conductors does not correspond with the standard ampere rating of a fuse or a circuit breaker without overload trip adjustments above its rating (but that shall be permitted to have other trip or rating adjustments).
(3) The next higher standard rating selected does not exceed 800 amperes.

(C) Devices Rated Over 800 Amperes. Where the overcurrent device is rated over 800 amperes, the ampacity of the conductors it protects shall be equal to or greater than the rating of the overcurrent device defined in 240.6.

(D) Small Conductors. Unless specifically permitted in 240.4(E) through (G), the overcurrent protection shall not exceed 15 amperes for 14 AWG, 20 amperes for 12 AWG, and 30 amperes for 10 AWG copper; or 15 amperes for 12 AWG and 25 amperes for 10 AWG aluminum and copperclad aluminum after any correction factors for ambient temperature and number of conductors have been applied.

(E) Tap Conductors. Tap conductors shall be permitted to be protected against overcurrent in accordance with 210.19(A)(3) and (A)(4), 240.5(B)(2), 240.21, 368.11, 368.12, and 430.53(D).

(F) Transformer Secondary Conductors. Single-phase (other than 2-wire) and multiphase (other than delta-delta,

3-wire) transformer secondary conductors shall not be considered to be protected by the primary overcurrent protective device. Conductors supplied by the secondary side of a single-phase transformer having a 2-wire (single-voltage) secondary, or a three-phase, delta-delta connected transformer having a 3-wire (single-voltage) secondary, shall be permitted to be protected by overcurrent protection provided on the primary (supply) side of the transformer, provided this protection is in accordance with 450.3 and does not exceed the value determined by multiplying the secondary conductor ampacity by the secondary to primary transformer voltage ratio.

240.6 Standard Ampere Ratings.

(A) Fuses and Fixed-Trip Circuit Breakers. The standard ampere ratings for fuses and inverse time circuit breakers shall be considered 15, 20, 25, 30, 35, 40, 45, 50, 60, 70, 80, 90, 100, 110, 125, 150, 175, 200, 225, 250, 300, 350, 400, 450, 500, 600, 700, 800, 1000, 1200, 1600, 2000, 2500, 3000, 4000, 5000, and 6000 amperes. Additional standard ampere ratings for fuses shall be 1, 3, 6, 10, and 601. The use of fuses and inverse time circuit breakers with nonstandard ampere ratings shall be permitted.

(B) Adjustable-Trip Circuit Breakers. The rating of adjustable-trip circuit breakers having external means for adjusting the current setting (long-time pickup setting), not meeting the requirements of 240.6(C), shall be the maximum setting possible.

(C) Restricted Access Adjustable-Trip Circuit Breakers. A circuit breaker(s) that has restricted access to the adjusting means shall be permitted to have an ampere rating(s) that is equal to the adjusted current setting (long-time pickup setting). Restricted access shall be defined as located behind one of the following:

(1) Removable and sealable covers over the adjusting means
(2) Bolted equipment enclosure doors
(3) Locked doors accessible only to qualified personnel

240.8 Fuses or Circuit Breakers in Parallel.
Fuses and circuit breakers shall be permitted to be connected in par-

/ Overcurrent Protection

allel where they are factory assembled in parallel and listed as a unit. Individual fuses, circuit breakers, or combinations thereof shall not otherwise be connected in parallel.

240.9 Thermal Devices. Thermal relays and other devices not designed to open short circuits or ground faults shall not be used for the protection of conductors against overcurrent due to short circuits or ground faults, but the use of such devices shall be permitted to protect motor branch-circuit conductors from overload if protected in accordance with 430.40.

240.10 Supplementary Overcurrent Protection. Where supplementary overcurrent protection is used for luminaires (lighting fixtures), appliances, and other equipment or for internal circuits and components of equipment, it shall not be used as a substitute for branch-circuit overcurrent devices or in place of the branch-circuit protection specified in Article 210. Supplementary overcurrent devices shall not be required to be readily accessible.

240.12 Electrical System Coordination. Where an orderly shutdown is required to minimize the hazard(s) to personnel and equipment, a system of coordination based on the following two conditions shall be permitted:

(1) Coordinated short-circuit protection
(2) Overload indication based on monitoring systems or devices

240.13 Ground-Fault Protection of Equipment. Ground-fault protection of equipment shall be provided in accordance with the provisions of 230.95 for solidly grounded wye electrical systems of more than 150 volts to ground but not exceeding 600 volts phase-to-phase for each individual device used as a building or structure main disconnecting means rated 1000 amperes or more.

The provisions of this section shall not apply to the disconnecting means for the following:

(1) Continuous industrial processes where a nonorderly shutdown will introduce additional or increased hazards
(2) Installations where ground-fault protection is provided by other requirements for services or feeders
(3) Fire pumps installed in accordance with Article 695

II. LOCATION

240.20 Ungrounded Conductors.

(A) Overcurrent Device Required. A fuse or an overcurrent trip unit of a circuit breaker shall be connected in series with each ungrounded conductor. A combination of a current transformer and overcurrent relay shall be considered equivalent to an overcurrent trip unit.

(B) Circuit Breaker as Overcurrent Device. Circuit breakers shall open all ungrounded conductors of the circuit unless otherwise permitted in 240.20(B)(1), (B)(2), and (B)(3).

(1) Multiwire Branch Circuit. Except where limited by 210.4(B), individual single-pole circuit breakers, with or without approved handle ties, shall be permitted as the protection for each ungrounded conductor of multiwire branch circuits that serve only single-phase line-to-neutral loads.

(2) Grounded Single-Phase and 3-wire dc Circuits. In grounded systems, individual single-pole circuit breakers with approved handle ties shall be permitted as the protection for each ungrounded conductor for line-to-line connected loads for single-phase circuits or 3-wire, direct-current circuits.

(3) 3-Phase and 2-Phase Systems. For line-to-line loads in 4-wire, 3-phase systems or 5-wire, 2-phase systems having a grounded neutral and no conductor operating at a voltage greater than permitted in 210.6, individual single-pole circuit breakers with approved handle ties shall be permitted as the protection for each ungrounded conductor.

240.21 Location in Circuit. Overcurrent protection shall be provided in each ungrounded circuit conductor and shall be located at the point where the conductors receive their supply except as specified in 240.21(A) through (G). No conductor supplied under the provisions of 240.21(A) through (G) shall supply another conductor under those provisions, except through an overcurrent protective device meeting the requirements of 240.4.

(A) Branch-Circuit Conductors. Branch-circuit tap conductors meeting the requirements specified in 210.19 shall

Overcurrent Protection

be permitted to have overcurrent protection located as specified in that section.

(B) Feeder Taps. Conductors shall be permitted to be tapped, without overcurrent protection at the tap, to a feeder as specified in 240.21(B)(1) through (5).

(1) Taps Not Over 3 m (10 ft) Long. Where the length of the tap conductors does not exceed 3 m (10 ft) and the tap conductors comply with all of the following:

(1) The ampacity of the tap conductors is

   a. Not less than the combined computed loads on the circuits supplied by the tap conductors, and
   b. Not less than the rating of the device supplied by the tap conductors or not less than the rating of the overcurrent-protective device at the termination of the tap conductors.

(2) The tap conductors do not extend beyond the switchboard, panelboard, disconnecting means, or control devices they supply.
(3) Except at the point of connection to the feeder, the tap conductors are enclosed in a raceway, which shall extend from the tap to the enclosure of an enclosed switchboard, panelboard, or control devices, or to the back of an open switchboard.
(4) For field installations where the tap conductors leave the enclosure or vault in which the tap is made, the rating of the overcurrent device on the line side of the tap conductors shall not exceed 10 times the ampacity of the tap conductor.

(2) Taps Not Over 7.5 m (25 ft) Long. Where the length of the tap conductors does not exceed 7.5 m (25 ft) and the tap conductors comply with all of the following:

(1) The ampacity of the tap conductors is not less than one-third of the rating of the overcurrent device protecting the feeder conductors.
(2) The tap conductors terminate in a single circuit breaker or a single set of fuses that will limit the load to the ampacity of the tap conductors. This device shall

be permitted to supply any number of additional overcurrent devices on its load side.

(3) The tap conductors are suitably protected from physical damage or are enclosed in a raceway.

(3) Taps Supplying a Transformer [Primary Plus Secondary Not Over 7.5 m (25 ft) Long]. Where the tap conductors supply a transformer and comply with all the following conditions:

(1) The conductors supplying the primary of a transformer have an ampacity at least one-third the rating of the overcurrent device protecting the feeder conductors.

(2) The conductors supplied by the secondary of the transformer shall have an ampacity that, when multiplied by the ratio of the secondary-to-primary voltage, is at least one-third of the rating of the overcurrent device protecting the feeder conductors.

(3) The total length of one primary plus one secondary conductor, excluding any portion of the primary conductor that is protected at its ampacity, is not over 7.5 m (25 ft).

(4) The primary and secondary conductors are suitably protected from physical damage.

(5) The secondary conductors terminate in a single circuit breaker or set of fuses that limit the load current to not more than the conductor ampacity that is permitted by 310.15.

(4) Taps Over 7.5 m (25 ft) Long. Where the feeder is in a high bay manufacturing building over 11 m (35 ft) high at walls and the installation complies with all the following conditions:

(1) Conditions of maintenance and supervision ensure that only qualified persons service the systems.

(2) The tap conductors are not over 7.5 m (25 ft) long horizontally and not over 30 m (100 ft) total length.

(3) The ampacity of the tap conductors is not less than one-third the rating of the overcurrent device protecting the feeder conductors.

(4) The tap conductors terminate at a single circuit breaker or a single set of fuses that limit the load to the ampacity of the tap conductors. This single overcurrent device shall be permitted to supply any number of additional overcurrent devices on its load side.

Overcurrent Protection

(5) The tap conductors are suitably protected from physical damage or are enclosed in a raceway.

(6) The tap conductors are continuous from end-to-end and contain no splices.

(7) The tap conductors are sized 6 AWG copper or 4 AWG aluminum or larger.

(8) The tap conductors do not penetrate walls, floors, or ceilings.

(9) The tap is made no less than 9 m (30 ft) from the floor.

(5) Outside Taps of Unlimited Length. Where the conductors are located outdoors of a building or structure, except at the point of load termination, and comply with all of the following conditions:

(1) The conductors are suitably protected from physical damage.

(2) The conductors terminate at a single circuit breaker or a single set of fuses that limit the load to the ampacity of the conductors. This single overcurrent device shall be permitted to supply any number of additional overcurrent devices on its load side.

(3) The overcurrent device for the conductors is an integral part of a disconnecting means or shall be located immediately adjacent thereto.

(4) The disconnecting means for the conductors is installed at a readily accessible location complying with one of the following:

    a. Outside of a building or structure
    b. Inside, nearest the point of entrance of the conductors
    c. Where installed in accordance with 230.6, nearest the point of entrance of the conductors

(C) Transformer Secondary Conductors. Conductors shall be permitted to be connected to a transformer secondary, without overcurrent protection at the secondary, as specified in 240.21(C)(1) through (6).

(1) Protection by Primary Overcurrent Device. Conductors supplied by the secondary side of a single-phase transformer having a 2-wire (single-voltage) secondary, or a three-phase, delta-delta connected transformer having a 3-wire (single-voltage) secondary, shall be permitted to be protected by overcurrent protection provided on the primary (supply) side of the transformer, provided this protection is in accordance with 450.3 and does not exceed the value determined by multiplying the secondary conductor ampacity by the secondary to primary transformer voltage ratio.

Single-phase (other than 2-wire) and multiphase (other than delta-delta, 3-wire) transformer secondary conductors are not considered to be protected by the primary overcurrent protective device.

(2) Transformer Secondary Conductors Not Over 3 m (10 ft) Long. Where the length of secondary conductor does not exceed 3 m (10 ft) and complies with all of the following:

(1) The ampacity of the secondary conductors is
   a. Not less than the combined computed loads on the circuits supplied by the secondary conductors, and
   b. Not less than the rating of the device supplied by the secondary conductors or not less than the rating of the overcurrent-protective device at the termination of the secondary conductors

(2) The secondary conductors do not extend beyond the switchboard, panelboard, disconnecting means, or control devices they supply.

(3) The secondary conductors are enclosed in a raceway, which shall extend from the transformer to the enclosure of an enclosed switchboard, panelboard, or control devices or to the back of an open switchboard.

(3) Industrial Installation Secondary Conductors Not Over 7.5 m (25 ft) Long. For industrial installations only, where the length of the secondary conductors does not exceed 7.5 m (25 ft) and complies with all of the following:

(1) The ampacity of the secondary conductors is not less than the secondary current rating of the transformer, and the sum of the ratings of the overcurrent devices does not exceed the ampacity of the secondary conductors.
(2) All overcurrent devices are grouped.
(3) The secondary conductors are suitably protected from physical damage.

Overcurrent Protection

(4) Outside Secondary of Building or Structure Conductors. Where the conductors are located outdoors of a building or structure, except at the point of load termination, and comply with all of the following conditions:

(1) The conductors are suitably protected from physical damage.
(2) The conductors terminate at a single circuit breaker or a single set of fuses that limit the load to the ampacity of the conductors. This single overcurrent device shall be permitted to supply any number of additional overcurrent devices on its load side.
(3) The overcurrent device for the conductors is an integral part of a disconnecting means or shall be located immediately adjacent thereto.
(4) The disconnecting means for the conductors is installed at a readily accessible location complying with one of the following:
   a. Outside of a building or structure
   b. Inside, nearest the point of entrance of the conductors
   c. Where installed in accordance with 230.6, nearest the point of entrance of the conductors

(5) Secondary Conductors from a Feeder Tapped Transformer. Transformer secondary conductors installed in accordance with 240.21(B)(3) shall be permitted to have overcurrent protection as specified in that section.

(6) Secondary Conductors Not Over 7.5 m (25 ft) Long. Where the length of secondary conductor does not exceed 7.5 m (25 ft) and complies with all of the following:

(1) The secondary conductors shall have an ampacity that, when multiplied by the ratio of the secondary-to-primary voltage, is at least one-third of the rating of the overcurrent device protecting the primary of the transformer.
(2) The secondary conductors terminate in a single circuit breaker or set of fuses that limit the load current to not more than the conductor ampacity that is permitted by 310.15.
(3) The secondary conductors are suitably protected from physical damage.

(D) Service Conductors. Service-entrance conductors shall be permitted to be protected by overcurrent devices in accordance with 230.91.

(E) Busway Taps. Busways and busway taps shall be permitted to be protected against overcurrent in accordance with 368.10 through 368.13.

(F) Motor Circuit Taps. Motor-feeder and branch-circuit conductors shall be permitted to be protected against overcurrent in accordance with 430.28 and 430.53, respectively.

(G) Conductors from Generator Terminals. Conductors from generator terminals that meet the size requirement in 445.13 shall be permitted to be protected against overload by the generator overload protective device(s) required by 445.12.

240.23 Change in Size of Grounded Conductor. Where a change occurs in the size of the ungrounded conductor, a similar change shall be permitted to be made in the size of the grounded conductor.

240.24 Location in or on Premises.

(A) Accessibility. Overcurrent devices shall be readily accessible unless one of the following applies:

(1) For busways, as provided in 368.12.
(2) For supplementary overcurrent protection, as described in 240.10.
(3) For overcurrent devices, as described in 225.40 and 230.92.
(4) For overcurrent devices adjacent to utilization equipment that they supply, access shall be permitted to be by portable means.

(B) Occupancy. Each occupant shall have ready access to all overcurrent devices protecting the conductors supplying that occupancy.

Exception No. 2: Where electric service and electrical maintenance are provided by the building management and where these are under continuous building management supervision, the branch circuit overcurrent devices supplying any guest rooms shall be permitted to be accessible to only authorized management personnel for guest

Overcurrent Protection 75

rooms of hotels and motels that are intended for transient occupancy.

(C) Not Exposed to Physical Damage. Overcurrent devices shall be located where they will not be exposed to physical damage.

(D) Not in Vicinity of Easily Ignitible Material. Overcurrent devices shall not be located in the vicinity of easily ignitible material, such as in clothes closets.

III. ENCLOSURES

240.30 General.

(A) Protection from Physical Damage. Overcurrent devices shall be protected from physical damage by one of the following:

(1) Installation in enclosures, cabinets, cutout boxes, or equipment assemblies
(2) Mounting on open-type switchboards, panelboards, or control boards that are in rooms or enclosures free from dampness and easily ignitible material and are accessible only to qualified personnel

(B) Operating Handle. The operating handle of a circuit breaker shall be permitted to be accessible without opening a door or cover.

240.33 Vertical Position. Enclosures for overcurrent devices shall be mounted in a vertical position unless that is shown to be impracticable. Circuit breaker enclosures shall be permitted to be installed horizontally where the circuit breaker is installed in accordance with 240.81. Listed busway plug-in units shall be permitted to be mounted in orientations corresponding to the busway mounting position.

IV. DISCONNECTING AND GUARDING

240.40 Disconnecting Means for Fuses. A disconnecting means shall be provided on the supply side of all fuses in circuits over 150 volts to ground and cartridge fuses in circuits of any voltage where accessible to other than qualified persons so that each individual circuit containing fuses can be independently disconnected from the source of power. A current-limiting device without a disconnecting means shall be permitted on the supply side of the service disconnecting means as permitted by 230.82. A single disconnecting means shall be permitted on the supply side of more than one set of fuses as permitted by 430.112, Exception, for group operation of motors and 424.22(C) for fixed electric space-heating equipment.

240.41 Arcing or Suddenly Moving Parts. Arcing or suddenly moving parts shall comply with 240.41(A) and (B).

(A) Location. Fuses and circuit breakers shall be located or shielded so that persons will not be burned or otherwise injured by their operation.

(B) Suddenly Moving Parts. Handles or levers of circuit breakers, and similar parts that may move suddenly in such a way that persons in the vicinity are likely to be injured by being struck by them, shall be guarded or isolated.

VI. CARTRIDGE FUSES AND FUSEHOLDERS

240.60 General.

(A) Maximum Voltage—300-Volt Type. Cartridge fuses and fuseholders of the 300-volt type shall be permitted to be used in the following circuits:

(1) Circuits not exceeding 300 volts between conductors
(2) Single-phase line-to-neutral circuits supplied from a 3-phase, 4-wire, solidly grounded neutral source where the line-to-neutral voltage does not exceed 300 volts

(B) Noninterchangeable—0–6000-Ampere Cartridge Fuseholders. Fuseholders shall be designed so that it will be difficult to put a fuse of any given class into a fuseholder that is designed for a current lower, or voltage higher, than that of the class to which the fuse belongs. Fuseholders for current-limiting fuses shall not permit insertion of fuses that are not current-limiting.

(C) Marking. Fuses shall be plainly marked, either by printing on the fuse barrel or by a label attached to the barrel showing the following:

Overcurrent Protection

(1) Ampere rating
(2) Voltage rating
(3) Interrupting rating where other than 10,000 amperes
(4) Current limiting where applicable
(5) The name or trademark of the manufacturer

The interrupting rating shall not be required to be marked on fuses used for supplementary protection.

VII. CIRCUIT BREAKERS

240.80 Method of Operation. Circuit breakers shall be trip free and capable of being closed and opened by manual operation. Their normal method of operation by other than manual means, such as electrical or pneumatic, shall be permitted if means for manual operation are also provided.

240.81 Indicating. Circuit breakers shall clearly indicate whether they are in the open "off" or closed "on" position.

Where circuit breaker handles are operated vertically rather than rotationally or horizontally, the "up" position of the handle shall be the "on" position.

240.83 Marking.

(D) Used as Switches. Circuit breakers used as switches in 120-volt and 277-volt fluorescent lighting circuits shall be listed and shall be marked SWD or HID. Circuit breakers used as switches in high-intensity discharge lighting circuits shall be listed and shall be marked as HID.

(E) Voltage Marking. Circuit breakers shall be marked with a voltage rating not less than the nominal system voltage that is indicative of their capability to interrupt fault currents between phases or phase to ground.

240.85 Applications. A circuit breaker with a straight voltage rating, such as 240V or 480V, shall be permitted to be applied in a circuit in which the nominal voltage between any two conductors does not exceed the circuit breaker's voltage rating. A two-pole circuit breaker shall not be used for protecting a 3-phase, corner-grounded delta circuit unless the circuit breaker is marked 1ϕ—3ϕ to indicate such suitability.

A circuit breaker with a slash rating, such as 120/240V or 480Y/277V, shall be permitted to be applied in a solidly grounded circuit where the nominal voltage of any conductor to ground does not exceed the lower of the two values of the circuit breaker's voltage rating and the nominal voltage between any two conductors does not exceed the higher value of the circuit breaker's voltage rating.

240.86 Series Ratings. Where a circuit breaker is used on a circuit having an available fault current higher than its marked interrupting rating by being connected on the load side of an acceptable overcurrent protective device having the higher rating, 240.86(A) and (B) shall apply.

(A) Marking. The additional series combination interrupting rating shall be marked on the end use equipment, such as switchboards and panelboards.

(B) Motor Contribution. Series ratings shall not be used where

(1) Motors are connected on the load side of the higher-rated overcurrent device and on the line side of the lower-rated overcurrent device, and
(2) The sum of the motor full-load currents exceeds 1 percent of the interrupting rating of the lower-rated circuit breaker.

VIII. SUPERVISED INDUSTRIAL INSTALLATIONS

240.90 General. Overcurrent protection in areas of supervised industrial installations shall comply with all of the other applicable provisions of this article, except as provided in Part VIII. The provisions of Part VIII shall only be permitted to apply to those portions of the electrical system in the supervised industrial installation used exclusively for manufacturing or process control activities.

240.92 Location in Circuit. An overcurrent device shall be connected in each ungrounded circuit conductor as required in 240.92(A) through (D).

(A) Feeder and Branch-Circuit Conductors. Feeder and branch-circuit conductors shall be protected at the point the conductors receive their supply as permitted in 240.21 or as otherwise permitted in 240.92(B), (C), or (D).

Overcurrent Protection

(B) Transformer Secondary Conductors of Separately Derived Systems. Conductors shall be permitted to be connected to a transformer secondary of a separately derived system, without overcurrent protection at the connection, where the conditions of 240.92(B)(1), (2), and (3) are met.

(1) Short-Circuit and Ground-Fault Protection. The conductors shall be protected from short-circuit and ground-fault conditions by complying with one of the following conditions:

(1) The length of the secondary conductors does not exceed 30 m (100 ft) and the transformer primary overcurrent device has a rating or setting that does not exceed 150 percent of the value determined by multiplying the secondary conductor ampacity by the secondary-to-primary transformer voltage ratio.
(2) The conductors are protected by a differential relay with a trip setting equal to or less than the conductor ampacity.
(3) The conductors shall be considered to be protected if calculations, made under engineering supervision, determine that the system overcurrent devices will protect the conductors within recognized time vs. current limits for all short-circuit and ground-fault conditions.

(2) Overload Protection. The conductors shall be protected against overload conditions by complying with one of the following:

(1) The conductors terminate in a single overcurrent device that will limit the load to the conductor ampacity.
(2) The sum of the overcurrent devices at the conductor termination limits the load to the conductor ampacity. The overcurrent devices shall consist of not more than six circuit breakers or sets of fuses, mounted in a single enclosure, in a group of separate enclosures, or in or on a switchboard. There shall be no more than six overcurrent devices grouped in any one location.
(3) Overcurrent relaying is connected [with a current transformer(s), if needed] to sense all of the secondary conductor current and limit the load to the conductor ampacity by opening upstream or downstream devices.
(4) Conductors shall be considered to be protected if calculations, made under engineering supervision,

determine that the system overcurrent devices will protect the conductors from overload conditions.

(3) Physical Protection. The secondary conductors shall be suitably protected from physical damage.

(C) Outside Feeder Taps. Outside conductors shall be permitted to be tapped to a feeder or to be connected at a transformer secondary, without overcurrent protection at the tap or connection, where all the following conditions are met:

(1) The conductors are suitably protected from physical damage.
(2) The sum of the overcurrent devices at the conductor termination limits the load to the conductor ampacity. The overcurrent devices shall consist of not more than six circuit breakers or sets of fuses mounted in a single enclosure, in a group of separate enclosures, or in or on a switchboard. There shall be no more than six overcurrent devices grouped in any one location.
(3) The tap conductors are installed outdoors of a building or structure except at the point of load termination.
(4) The overcurrent device for the conductors is an integral part of a disconnecting means or shall be located immediately adjacent thereto.
(5) The disconnecting means for the conductors is installed at a readily accessible location complying with one of the following:

   a. Outside of a building or structure
   b. Inside, nearest the point of entrance of the conductors
   c. Where installed in accordance with 230.6, nearest the point of entrance of the conductors

(D) Protection by Primary Overcurrent Device. Conductors supplied by the secondary side of a transformer shall be permitted to be protected by overcurrent protection provided on the primary (supply) side of the transformer, provided the primary device time—current protection characteristic, multiplied by the maximum effective primary-to-secondary transformer voltage ratio, effectively protects the secondary conductors.

CHAPTER 9
GROUNDED (NEUTRAL) CONDUCTORS

INTRODUCTION

This chapter contains provisions from Article 200—Use and Identification of Grounded Conductors. Article 200 covers grounded conductors in premises wiring systems and identification of grounded conductors and their terminals. A grounded conductor is a system or circuit conductor that is intentionally grounded. Grounded conductors are generally identified by a continuous white or gray outer finish or by three continuous white stripes on other than green insulation along the entire length. Although Article 200 is a short article, do not overlook the importance of proper use and identification of grounded conductors.

ARTICLE 200
Use and Identification of Grounded Conductors

200.1 Scope. This article provides requirements for the following:

(1) Identification of terminals
(2) Grounded conductors in premises wiring systems
(3) Identification of grounded conductors

200.2 General. All premises wiring systems, other than circuits and systems exempted or prohibited by 210.10, 215.7, 250.21, 250.22, 250.162, 503.13, 517.63, 668.11, 668.21, and 690.41, Exception, shall have a grounded conductor that is identified in accordance with 200.6.

The grounded conductor, where insulated, shall have insulation that is (1) suitable, other than color, for any ungrounded conductor of the same circuit on circuits of less than 1000 volts or impedance grounded neutral systems of 1 kV and over, or (2) rated not less than 600 volts for solidly grounded neutral systems of 1 kV and over as described in 250.184(A).

200.3 Connection to Grounded System. Premises wiring shall not be electrically connected to a supply system unless the latter contains, for any grounded conductor of the interior system, a corresponding conductor that is grounded. For the purpose of this section, *electrically connected* shall mean connected so as to be capable of carrying current, as distinguished from connection through electromagnetic induction.

200.6 Means of Identifying Grounded Conductors.

(A) Sizes 6 AWG or Smaller. An insulated grounded conductor of 6 AWG or smaller shall be identified by a continuous white or gray outer finish or by three continuous white stripes on other than green insulation along its entire length. Wires that have their outer covering finished to show a white or gray color but have colored tracer threads in the braid identifying the source of manufacture shall be considered as meeting the provisions of this section.

(B) Sizes Larger Than 6 AWG. An insulated grounded conductor larger than 6 AWG shall be identified either by a continuous white or gray outer finish or by three continuous

Grounded (Neutral) Conductors

white stripes on other than green insulation along its entire length or at the time of installation by a distinctive white marking at its terminations. This marking shall encircle the conductor or insulation.

(C) Flexible Cords. An insulated conductor that is intended for use as a grounded conductor, where contained within a flexible cord, shall be identified by a white or gray outer finish or by three continuous white stripes on other than green insulation or by methods permitted by 400.22.

(D) Grounded Conductors of Different Systems. Where conductors of different systems are installed in the same raceway, cable, box, auxiliary gutter, or other type of enclosure, one system grounded conductor, if required, shall have an outer covering conforming to 200.6(A) or 200.6(B). Each other system grounded conductor shall have an outer covering of white with a readily distinguishable, different colored stripe other than green running along the insulation, or shall have other and different means of identification as allowed by 200.6(A) or (B) that will distinguish each system grounded conductor.

(E) Grounded Conductors of Multiconductor Cables. The insulated grounded conductors in a multiconductor cable shall be identified by a continuous white or gray outer finish or by three continuous white stripes on other than green insulation along its entire length. Multiconductor flat 4 AWG or larger shall be permitted to employ an external ridge on the grounded conductor.

Exception No. 1: Where the conditions of maintenance and supervision ensure that only qualified persons service the installation, grounded conductors in multiconductor cables shall be permitted to be permanently identified at their terminations at the time of installation by a distinctive white marking or other equally effective means.

Exception No. 2: The grounded conductor of a multiconductor varnished-cloth-insulated cable shall be permitted to be identified at its terminations at the time of installation by a distinctive white marking or other equally effective means.

> FPN: The color gray may have been used in the past as an ungrounded conductor. Care should be taken when working on existing systems.

200.7 Use of Insulation of a White or Gray Color or with Three Continuous White Stripes.

(A) General. The following shall be used only for the grounded circuit conductor, unless otherwise permitted in 200.7(B) and (C):

(1) A conductor with continuous white or gray covering
(2) A conductor with three continuous white stripes on other than green insulation
(3) A marking of white or gray color at the termination

(B) Circuits of Less Than 50 Volts. A conductor with white or gray color insulation or three continuous white stripes or having a marking of white or gray at the termination for circuits of less than 50 volts shall be required to be grounded only as required by 250.20(A).

(C) Circuits of 50 Volts or More. The use of insulation that is white or gray or that has three continuous white stripes for other than a grounded conductor for circuits of 50 volts or more shall be permitted only as in (1) through (3).

(1) If part of a cable assembly and where the insulation is permanently reidentified to indicate its use as an ungrounded conductor, by painting or other effective means at its termination, and at each location where the conductor is visible and accessible.
(2) Where a cable assembly contains an insulated conductor for single-pole, 3-way or 4-way switch loops and the conductor with white or gray insulation or a marking of three continuous white stripes is used for the supply to the switch but not as a return conductor from the switch to the switched outlet. In these applications, the conductor with white or gray insulation or with three continuous white stripes shall be permanently reidentified to indicate its use by painting or other effective means at its terminations and at each location where the conductor is visible and accessible.
(3) Where a flexible cord, having one conductor identified by a white or gray outer finish or three continuous white stripes or by any other means permitted by 400.22, is used for connecting an appliance or equipment permitted by 400.7. This shall apply to flexible cords connected to outlets whether or not the outlet is supplied by a circuit that has a grounded conductor.

CHAPTER 10

GROUNDING

INTRODUCTION

This chapter contains provisions from Article 250—Grounding. Article 250 covers general requirements for grounding and bonding of electrical systems, circuits, equipment, and conductors. A conductor or equipment is effectively grounded when it is intentionally connected to earth through a ground connection(s) of sufficiently low impedance and having sufficient current-carrying capacity to prevent the buildup of voltages that may result in undue hazards to connected equipment or to persons. Bonding (bonded) is the permanent joining of metallic parts to form an electrically conductive path that will ensure electrical continuity and the capacity to conduct safely any current likely to be imposed.

Contained in this *Pocket Guide* are seven of the ten parts that are included in Article 250. Those seven parts are: I General; II Circuit System Grounding; III Grounding Electrode System and Grounding Electrode Conductor; IV Enclosure, Raceway, and Service Cable Grounding; V Bonding; VI Equipment Grounding and Equipment Grounding Conductors; and VII Methods of Equipment Grounding.

I. GENERAL

250.1 Scope. This article covers general requirements for grounding and bonding of electrical installations, and specific requirements in (1) through (6).

(1) Systems, circuits, and equipment required, permitted, or not permitted to be grounded
(2) Circuit conductor to be grounded on grounded systems
(3) Location of grounding connections
(4) Types and sizes of grounding and bonding conductors and electrodes
(5) Methods of grounding and bonding
(6) Conditions under which guards, isolation, or insulation may be substituted for grounding

250.2 Definitions.

Effective Ground-Fault Current Path. An intentionally constructed, permanent, low-impedance electrically conductive path designed and intended to carry current under ground-fault conditions from the point of a ground fault on a wiring system to the electrical supply source.

Ground Fault. An unintentional, electrically conducting connection between an ungrounded conductor of an electrical circuit and the normally non–current-carrying conductors, metallic enclosures, metallic raceways, metallic equipment, or earth.

Ground-Fault Current Path. An electrically conductive path from the point of a ground fault on a wiring system through normally non–current-carrying conductors, equipment, or the earth to the electrical supply source.

> FPN: Examples of ground-fault current paths could consist of any combination of equipment grounding conductors, metallic raceways, metallic cable sheaths, electrical equipment, and any other electrically conductive material such as metal water and gas piping, steel framing members, stucco mesh, metal ducting, reinforcing steel, shields of communications cables, and the earth itself.

250.4 General Requirements for Grounding and Bonding. The following general requirements identify what grounding and bonding of electrical systems are required

Grounding

to accomplish. The prescriptive methods contained in Article 250 shall be followed to comply with the performance requirements of this section.

(A) Grounded Systems.

(1) Electrical System Grounding. Electrical systems that are grounded shall be connected to earth in a manner that will limit the voltage imposed by lightning, line surges, or unintentional contact with higher-voltage lines and that will stabilize the voltage to earth during normal operation.

(2) Grounding of Electrical Equipment. Non–current-carrying conductive materials enclosing electrical conductors or equipment, or forming part of such equipment, shall be connected to earth so as to limit the voltage to ground on these materials.

(3) Bonding of Electrical Equipment. Non–current-carrying conductive materials enclosing electrical conductors or equipment, or forming part of such equipment, shall be connected together and to the electrical supply source in a manner that establishes an effective ground-fault current path.

(4) Bonding of Electrically Conductive Materials and Other Equipment. Electrically conductive materials that are likely to become energized shall be connected together and to the electrical supply source in a manner that establishes an effective ground-fault current path.

(5) Effective Ground-Fault Current Path. Electrical equipment and wiring and other electrically conductive material likely to become energized shall be installed in a manner that creates a permanent, low-impedance circuit capable of safely carrying the maximum ground-fault current likely to be imposed on it from any point on the wiring system where a ground fault may occur to the electrical supply source. The earth shall not be used as the sole equipment grounding conductor or effective ground-fault current path.

250.6 Objectionable Current over Grounding Conductors.

(A) Arrangement to Prevent Objectionable Current. The grounding of electrical systems, circuit conductors, surge arresters, and conductive non–current-carrying materials and equipment shall be installed and arranged in a manner that will prevent objectionable current over the grounding conductors or grounding paths.

(B) Alterations to Stop Objectionable Current. If the use of multiple grounding connections results in objectionable current, one or more of the following alterations shall be permitted to be made, provided that the requirements of 250.4(A)(5) or 250.4(B)(4) are met:

(1) Discontinue one or more but not all of such grounding connections.
(2) Change the locations of the grounding connections.
(3) Interrupt the continuity of the conductor or conductive path interconnecting the grounding connections.
(4) Take other suitable remedial and approved action.

(C) Temporary Currents Not Classified as Objectionable Currents. Temporary currents resulting from accidental conditions, such as ground-fault currents, that occur only while the grounding conductors are performing their intended protective functions shall not be classified as objectionable current for the purposes specified in 250.6(A) and (B).

(D) Limitations to Permissible Alterations. The provisions of this section shall not be considered as permitting electronic equipment from being operated on ac systems or branch circuits that are not grounded as required by this article. Currents that introduce noise or data errors in electronic equipment shall not be considered the objectionable currents addressed in this section.

250.8 Connection of Grounding and Bonding Equipment. Grounding conductors and bonding jumpers shall be connected by exothermic welding, listed pressure connectors, listed clamps, or other listed means. Connection devices or fittings that depend solely on solder shall not be used. Sheet metal screws shall not be used to connect grounding conductors to enclosures.

250.10 Protection of Ground Clamps and Fittings. Ground clamps or other fittings shall be approved for general use without protection or shall be protected from physical damage as indicated in (1) or (2).

(1) In installations where they are not likely to be damaged
(2) Where enclosed in metal, wood, or equivalent protective covering

250.12 Clean Surfaces. Nonconductive coatings (such as paint, lacquer, and enamel) on equipment to be grounded shall be removed from threads and other contact surfaces to ensure good electrical continuity or be connected by means of fittings designed so as to make such removal unnecessary.

II. CIRCUIT AND SYSTEM GROUNDING

250.20 Alternating-Current Circuits and Systems to Be Grounded. Alternating-current circuits and systems shall be grounded as provided for in 250.20(A), (B), (C), or (D). Other circuits and systems shall be permitted to be grounded. If such systems are grounded, they shall comply with the applicable provisions of this article.

(A) Alternating-Current Circuits of Less Than 50 Volts. Alternating-current circuits of less than 50 volts shall be grounded under any of the following conditions:

(1) Where supplied by transformers, if the transformer supply system exceeds 150 volts to ground
(2) Where supplied by transformers, if the transformer supply system is ungrounded
(3) Where installed as overhead conductors outside of buildings

(B) Alternating-Current Systems of 50 Volts to 1000 Volts. Alternating-current systems of 50 volts to 1000 volts that supply premises wiring and premises wiring systems shall be grounded under any of the following conditions:

(1) Where the system can be grounded so that the maximum voltage to ground on the ungrounded conductors does not exceed 150 volts
(2) Where the system is 3-phase, 4-wire, wye connected in which the neutral is used as a circuit conductor
(3) Where the system is 3-phase, 4-wire, delta connected in which the midpoint of one phase winding is used as a circuit conductor

(D) Separately Derived Systems. Separately derived systems, as covered in 250.20(A) or (B), shall be grounded as specified in 250.30.

250.22 Circuits Not to Be Grounded. The following circuits shall not be grounded:

(1) Cranes (circuits for electric cranes operating over combustible fibers in Class III locations, as provided in 503.13)
(2) Health care facilities (circuits as provided in Article 517)
(3) Electrolytic cells (circuits as provided in Article 668)
(4) Lighting systems [secondary circuits as provided in 411.5(A)]

250.24 Grounding Service-Supplied Alternating-Current Systems.

(A) System Grounding Connections. A premises wiring system supplied by a grounded ac service shall have a grounding electrode conductor connected to the grounded service conductor, at each service, in accordance with 250.24(A)(1) through (A)(5).

(1) General. The connection shall be made at any accessible point from the load end of the service drop or service lateral to and including the terminal or bus to which the grounded service conductor is connected at the service disconnecting means.

(2) Outdoor Transformer. Where the transformer supplying the service is located outside the building, at least one additional grounding connection shall be made from the grounded service conductor to a grounding electrode, either at the transformer or elsewhere outside the building.

(3) Dual Fed Services. For services that are dual fed (double ended) in a common enclosure or grouped together in separate enclosures and employing a secondary tie, a single grounding electrode connection to the tie point of the grounded circuit conductors from each power source shall be permitted.

(4) Main Bonding Jumper as Wire or Busbar. Where the main bonding jumper specified in 250.28 is a wire or busbar and is installed from the neutral bar or bus to the equipment grounding terminal bar or bus in the service equipment, the

Grounding

grounding electrode conductor shall be permitted to be connected to the equipment grounding terminal bar or bus to which the main bonding jumper is connected.

(5) Load-Side Grounding Connections. A grounding connection shall not be made to any grounded circuit conductor on the load side of the service disconnecting means except as otherwise permitted in this article.

(B) Grounded Conductor Brought to Service Equipment. Where an ac system operating at less than 1000 volts is grounded at any point, the grounded conductor(s) shall be run to each service disconnecting means and shall be bonded to each disconnecting means enclosure. The grounded conductor(s) shall be installed in accordance with 250.24(B)(1) through (B)(3).

(1) Routing and Sizing. This conductor shall be routed with the phase conductors and shall not be smaller than the required grounding electrode conductor specified in Table 250.66 but shall not be required to be larger than the largest ungrounded service-entrance phase conductor. In addition, for service-entrance phase conductors larger than 1100 kcmil copper or 1750 kcmil aluminum, the grounded conductor shall not be smaller than 12-1/2 percent of the area of the largest service-entrance phase conductor. The grounded service entrance conductor of a 3-phase, 3-wire delta service shall have an ampacity not less than the ungrounded conductors.

(2) Parallel Conductors. Where the service-entrance phase conductors are installed in parallel, the size of the grounded conductor shall be based on the total circular mil area of the parallel conductors as indicated in this section. Where installed in two or more raceways, the size of the grounded conductor in each raceway shall be based on the size of the ungrounded service-entrance conductor in the raceway but not smaller than 1/0 AWG.

(C) Grounding Electrode Conductor. A grounding electrode conductor shall be used to connect the equipment grounding conductors, the service-equipment enclosures, and, where the system is grounded, the grounded service conductor to the grounding electrode(s) required by Part III of this article.

(D) Ungrounded System Grounding Connections. A premises wiring system that is supplied by an ac service that is ungrounded shall have, at each service, a grounding electrode conductor connected to the grounding electrode(s) required by Part III of this article. The grounding electrode conductor shall be connected to a metal enclosure of the service conductors at any accessible point from the load end of the service drop or service lateral to the service disconnecting means.

250.26 Conductor to Be Grounded—Alternating-Current Systems. For ac premises wiring systems, the conductor to be grounded shall be as specified in the following:

(1) Single-phase, 2-wire—one conductor
(2) Single-phase, 3-wire—the neutral conductor
(3) Multiphase systems having one wire common to all phases—the common conductor
(4) Multiphase systems where one phase is grounded—one phase conductor
(5) Multiphase systems in which one phase is used as in (2)—the neutral conductor

250.28 Main Bonding Jumper. For a grounded system, an unspliced main bonding jumper shall be used to connect the equipment grounding conductor(s) and the service-disconnect enclosure to the grounded conductor of the system within the enclosure for each service disconnect.

(A) Material. Main bonding jumpers shall be of copper or other corrosion-resistant material. A main bonding jumper shall be a wire, bus, screw, or similar suitable conductor.

(B) Construction. Where a main bonding jumper is a screw only, the screw shall be identified with a green finish that shall be visible with the screw installed.

(C) Attachment. Main bonding jumpers shall be attached in the manner specified by the applicable provisions of 250.8.

(D) Size. The main bonding jumper shall not be smaller than the sizes shown in Table 250.66 for grounding electrode conductors. Where the service-entrance phase conductors are larger than 1100 kcmil copper or 1750 kcmil aluminum, the bonding jumper shall have an area that is

Grounding

not less than 12-1/2 percent of the area of the largest phase conductor except that, where the phase conductors and the bonding jumper are of different materials (copper or aluminum), the minimum size of the bonding jumper shall be based on the assumed use of phase conductors of the same material as the bonding jumper and with an ampacity equivalent to that of the installed phase conductors.

250.30 Grounding Separately Derived Alternating-Current Systems.

(A) Grounded Systems. A separately derived ac system that is grounded shall comply with 250.30(A)(1) through (6).

(1) Bonding Jumper. A bonding jumper in compliance with 250.28(A) through (D) that is sized for the derived phase conductors shall be used to connect the equipment grounding conductors of the separately derived system to the grounded conductor. Except as permitted by 250.24(A)(3), this connection shall be made at any point on the separately derived system from the source to the first system disconnecting means or overcurrent device, or it shall be made at the source of a separately derived system that has no disconnecting means or overcurrent devices. The point of connection shall be the same as the grounding electrode conductor as required in 250.30(A)(2).

Exception No. 1: A bonding jumper at both the source and the first disconnecting means shall be permitted where doing so does not establish a parallel path for the grounded circuit conductor. Where a grounded conductor is used in this manner, it shall not be smaller than the size specified for the bonding jumper but shall not be required to be larger than the ungrounded conductor(s). For the purposes of this exception, connection through the earth shall not be considered as providing a parallel path.

(2) Grounding Electrode Conductor. The grounding electrode conductor shall be installed in accordance with (a) or (b). Where taps are connected to a common grounding electrode conductor, the installation shall comply with 250.30(A)(3).

(a) Single Separately Derived System. A grounding electrode conductor for a single separately derived system shall be sized in accordance with 250.66 for the derived phase conductors and shall be used to connect the grounded conductor of the derived system to the grounding electrode as specified in 250.30(A)(4). Except as permitted by 250.24(A)(3) or (A)(4), this connection shall be made at the same point on the separately derived system where the bonding jumper is installed.

(b) Multiple Separately Derived Systems. Where more than one separately derived system is connected to a common grounding electrode conductor as provided in 250.30(A)(3), the common grounding electrode conductor shall be sized in accordance with 250.66, based on the total area of the largest derived phase conductor from each separately derived system.

(3) Grounding Electrode Conductor Taps. It shall be permissible to connect taps from a separately derived system to a common grounding electrode conductor. Each tap conductor shall connect the grounded conductor of the separately derived system to the common grounding electrode conductor.

(a) Tap Conductor Size. Each tap conductor shall be sized in accordance with 250.66 for the derived phase conductors of the separately derived system it serves.

(b) Connections. All connections shall be made at an accessible location by an irreversible compression connector listed for the purpose, listed connections to copper busbars not less than 6 mm x 50 mm (1/4 in. x 2 in.), or by the exothermic welding process. The tap conductors shall be connected to the common grounding electrode conductor as specified in 250.30(A)(2)(b) in such a manner that the common grounding electrode conductor remains without a splice or joint.

(c) Installation. The common grounding electrode conductor and the taps to each separately derived system shall comply with 250.64(A), (B), (C), and (E).

(d) Bonding. Where exposed structural steel that is interconnected to form the building frame or interior metal piping exists in the area served by the separately derived system, it shall be bonded to the grounding electrode conductor in accordance with 250.104.

Grounding

(4) Grounding Electrode. The grounding electrode shall be as near as practicable to and preferably in the same area as the grounding electrode conductor connection to the system. The grounding electrode shall be the nearest one of the following:

(1) An effectively grounded structural metal member of the structure
(2) An effectively grounded metal water pipe within 1.5 m (5 ft) from the point of entrance into the building

Exception: In industrial and commercial buildings where conditions of maintenance and supervision ensure that only qualified persons service the installation and the entire length of the interior metal water pipe that is being used for the grounding electrode is exposed, the connection shall be permitted at any point on the water pipe system.

(3) Other electrodes as specified by 250.52 where the electrodes specified by 250.30(A)(4)(1) or (A)(4)(2) are not available

Exception to (1), (2), and (3): Where a separately derived system originates in listed equipment suitable for use as service equipment, the grounding electrode used for the service or feeder shall be permitted as the grounding electrode for the separately derived system, provided the grounding electrode conductor from the service or feeder to the grounding electrode is of sufficient size for the separation derived system. Where the equipment ground bus internal to the service equipment is not smaller than the required grounding electrode conductor, the grounding electrode connection for the separately derived system shall be permitted to be made to the bus.

(5) Equipment Bonding Jumper Size. Where a bonding jumper is run with the derived phase conductors from the source of a separately derived system to the first disconnecting means, it shall be sized in accordance with 250.28(A) through (D), based on the size of the derived phase conductors.

(6) Grounded Conductor. Where a grounded conductor is installed and the bonding jumper is not located at the source of the separately derived system, the following shall apply:

(a) Routing and Sizing. This conductor shall be routed with the derived phase conductors and shall not be smaller than the required grounding electrode conductor specified in Table 250.66, but shall not be required to be larger than the largest ungrounded derived phase conductor. In addition, for phase conductors larger than 1100 kcmil copper or 1750 kcmil aluminum, the grounded conductor shall not be smaller than 12-1/2 percent of the area of the largest derived phase conductor. The grounded conductor of a 3-phase, 3-wire delta system shall have an ampacity not less than the ungrounded conductors.

(b) Parallel Conductors. Where the derived phase conductors are installed in parallel, the size of the grounded conductor shall be based on the total circular mil area of the parallel conductors as indicated in this section. Where installed in two or more raceways, the size of the grounded conductor in each raceway shall be based on the size of the ungrounded conductors in the raceway but not smaller than 1/0 AWG.

250.32 Two or More Buildings or Structures Supplied from a Common Service.

(A) Grounding Electrode. Where two or more buildings or structures are supplied from a common ac service by a feeder(s) or branch circuit(s), the grounding electrode(s) required in Part III of this article at each building or structure shall be connected in the manner specified in 250.32(B) or (C). Where there are no existing grounding electrodes, the grounding electrode(s) required in Part III of this article shall be installed.

Exception: A grounding electrode at separate buildings or structures shall not be required where only one branch circuit supplies the building or structure and the branch circuit includes an equipment grounding conductor for grounding the conductive non–current-carrying parts of all equipment.

(B) Grounded Systems. For a grounded system at the separate building or structure, the connection to the grounding electrode and grounding or bonding of equipment, structures, or frames required to be grounded or bonded shall comply with either 250.32(B)(1) or (2).

Grounding

(1) Equipment Grounding Conductor. An equipment grounding conductor as described in 250.118 shall be run with the supply conductors and connected to the building or structure disconnecting means and to the grounding electrode(s). The equipment grounding conductor shall be used for grounding or bonding of equipment, structures, or frames required to be grounded or bonded. The equipment grounding conductor shall be sized in accordance with 250.122. Any installed grounded conductor shall not be connected to the equipment grounding conductor or to the grounding electrode(s).

(2) Grounded Conductor. Where (1) an equipment grounding conductor is not run with the supply to the building or structure, (2) there are no continuous metallic paths bonded to the grounding system in both buildings or structures involved, and (3) ground-fault protection of equipment has not been installed on the common ac service, the grounded circuit conductor run with the supply to the building or structure shall be connected to the building or structure disconnecting means and to the grounding electrode(s) and shall be used for grounding or bonding of equipment, structures, or frames required to be grounded or bonded. The size of the grounded conductor shall not be smaller than the larger of

(1) That required by 220.22
(2) That required by 250.122

(D) Disconnecting Means Located in Separate Building or Structure on the Same Premises. Where one or more disconnecting means supply one or more additional buildings or structures under single management, and where these disconnecting means are located remote from those buildings or structures in accordance with the provisions of 225.32, Exception Nos. 1 and 2, all of the following conditions shall be met:

(1) The connection of the grounded circuit conductor to the grounding electrode at a separate building or structure shall not be made.
(2) An equipment grounding conductor for grounding any non–current-carrying equipment, interior metal piping

systems, and building or structural metal frames is run with the circuit conductors to a separate building or structure and bonded to existing grounding electrode(s) required in Part III of this article, or, where there are no existing electrodes, the grounding electrode(s) required in Part III of this article shall be installed where a separate building or structure is supplied by more than one branch circuit.

(3) Bonding the equipment grounding conductor to the grounding electrode at a separate building or structure shall be made in a junction box, panelboard, or similar enclosure located immediately inside or outside the separate building or structure.

(E) Grounding Electrode Conductor. The size of the grounding electrode conductor to the grounding electrode(s) shall not be smaller than given in 250.66, based on the largest ungrounded supply conductor. The installation shall comply with Part III of this article.

III. GROUNDING ELECTRODE SYSTEM AND GROUNDING ELECTRODE CONDUCTOR

250.50 Grounding Electrode System. If available on the premises at each building or structure served, each item in 250.52(A)(1) through (A)(6) shall be bonded together to form the grounding electrode system. Where none of these electrodes are available, one or more of the electrodes specified in 250.52(A)(4) through (A)(7) shall be installed and used.

250.52 Grounding Electrodes.

(A) Electrodes Permitted for Grounding.

(1) Metal Underground Water Pipe. A metal underground water pipe in direct contact with the earth for 3.0 m (10 ft) or more (including any metal well casing effectively bonded to the pipe) and electrically continuous (or made electrically continuous by bonding around insulating joints or insulating pipe) to the points of connection of the grounding electrode conductor and the bonding conductors. Interior metal water piping located more than 1.52 m (5 ft) from the point of en-

Grounding

trance to the building shall not be used as a part of the grounding electrode system or as a conductor to interconnect electrodes that are part of the grounding electrode system.

Exception: In industrial and commercial buildings or structures where conditions of maintenance and supervision ensure that only qualified persons service the installation, interior metal water piping located more than 1.52 m (5 ft) from the point of entrance to the building shall be permitted as a part of the grounding electrode system or as a conductor to interconnect electrodes that are part of the grounding electrode system, provided that the entire length, other than short sections passing perpendicular through walls, floors, or ceilings, of the interior metal water pipe that is being used for the conductor is exposed.

(2) Metal Frame of the Building or Structure. The metal frame of the building or structure, where effectively grounded.

(3) Concrete-Encased Electrode. An electrode encased by at least 50 mm (2 in.) of concrete, located within and near the bottom of a concrete foundation or footing that is in direct contact with the earth, consisting of at least 6.0 m (20 ft) of one or more bare or zinc galvanized or other electrically conductive coated steel reinforcing bars or rods of not less than 13 mm (1/2 in.) in diameter, or consisting of at least 6.0 m (20 ft) of bare copper conductor not smaller than 4 AWG. Reinforcing bars shall be permitted to be bonded together by the usual steel tie wires or other effective means.

(4) Ground Ring. A ground ring encircling the building or structure, in direct contact with the earth, consisting of at least 6.0 m (20 ft) of bare copper conductor not smaller than 2 AWG.

(5) Rod and Pipe Electrodes. Rod and pipe electrodes shall not be less than 2.5 m (8 ft) in length and shall consist of the following materials.

(a) Electrodes of pipe or conduit shall not be smaller than metric designator 21 (trade size 3/4) and, where of iron or steel, shall have the outer surface galvanized or otherwise metal-coated for corrosion protection.

(b) Electrodes of rods of iron or steel shall be at least 15.87 mm (5/8 in.) in diameter. Stainless steel rods less than 16 mm (5/8 in.) in diameter, nonferrous rods, or their equivalent shall be listed and shall not be less than 13 mm (1/2 in.) in diameter.

(6) Plate Electrodes. Each plate electrode shall expose not less than 0.186 m^2 (2 ft^2) of surface to exterior soil. Electrodes of iron or steel plates shall be at least 6.4 mm (1/4 in.) in thickness. Electrodes of nonferrous metal shall be at least 1.5 mm (0.06 in.) in thickness.

(7) Other Local Metal Underground Systems or Structures. Other local metal underground systems or structures such as piping systems and underground tanks.

(B) Electrodes Not Permitted for Grounding. The following shall not be used as grounding electrodes:

(1) Metal underground gas piping system
(2) Aluminum electrodes

250.53 Grounding Electrode System Installation.

(A) Rod, Pipe, and Plate Electrodes. Where practicable, rod, pipe, and plate electrodes shall be embedded below permanent moisture level. Rod, pipe, and plate electrodes shall be free from nonconductive coatings such as paint or enamel.

(B) Electrode Spacing. Where more than one of the electrodes of the type specified in 250.52(A)(5) or (A)(6) are used, each electrode of one grounding system (including that used for air terminals) shall not be less than 1.83 m (6 ft) from any other electrode of another grounding system. Two or more grounding electrodes that are effectively bonded together shall be considered a single grounding electrode system.

(C) Bonding Jumper. The bonding jumper(s) used to connect the grounding electrodes together to form the grounding electrode system shall be installed in accordance with 250.64(A), (B), and (E), shall be sized in accordance with 250.66, and shall be connected in the manner specified in 250.70.

Grounding

(D) Metal Underground Water Pipe. Where used as a grounding electrode, metal underground water pipe shall meet the requirements of 250.53(D)(1) and (D)(2).

(1) Continuity. Continuity of the grounding path or the bonding connection to interior piping shall not rely on water meters or filtering devices and similar equipment.

(2) Supplemental Electrode Required. A metal underground water pipe shall be supplemented by an additional electrode of a type specified in 250.52(A)(2) through (A)(7). Where the supplemental electrode is a rod, pipe, or plate type, it shall comply with 250.56. The supplemental electrode shall be permitted to be bonded to the grounding electrode conductor, the grounded service-entrance conductor, the nonflexible grounded service raceway, or any grounded service enclosure.

Exception: The supplemental electrode shall be permitted to be bonded to the interior metal water piping at any convenient point as covered in 250.52(A)(1), Exception.

(E) Supplemental Electrode Bonding Connection Size. Where the supplemental electrode is a rod, pipe, or plate electrode, that portion of the bonding jumper that is the sole connection to the supplemental grounding electrode shall not be required to be larger than 6 AWG copper wire or 4 AWG aluminum wire.

(F) Ground Ring. The ground ring shall be buried at a depth below the earth's surface of not less than 750 mm (30 in.).

(G) Rod and Pipe Electrodes. The electrode shall be installed such that at least 2.44 m (8 ft) of length is in contact with the soil. It shall be driven to a depth of not less than 2.44 m (8 ft) except that, where rock bottom is encountered, the electrode shall be driven at an oblique angle not to exceed 45 degrees from the vertical or, where rock bottom is encountered at an angle up to 45 degrees, the electrode shall be permitted to be buried in a trench that is at least 750 mm (30 in.) deep. The upper end of the electrode shall be flush with or below ground level unless the aboveground end and the grounding electrode conductor attachment are protected against physical damage as specified in 250.10.

(H) Plate Electrode. Plate electrodes shall be installed not less than 750 mm (30 in.) below the surface of the earth.

250.54 Supplementary Grounding Electrodes. Supplementary grounding electrodes shall be permitted to be connected to the equipment grounding conductors specified in 250.118 and shall not be required to comply with the electrode bonding requirements of 250.50 or 250.53(C) or the resistance requirements of 250.56, but the earth shall not be used as the sole equipment grounding conductor.

250.56 Resistance of Rod, Pipe, and Plate Electrodes. A single electrode consisting of a rod, pipe, or plate that does not have a resistance to ground of 25 ohms or less shall be augmented by one additional electrode of any of the types specified by 250.52(A)(2) through (A)(7). Where multiple rod, pipe, or plate electrodes are installed to meet the requirements of this section, they shall not be less than 1.8 m (6 ft) apart.

250.58 Common Grounding Electrode. Where an ac system is connected to a grounding electrode in or at a building as specified in 250.24 and 250.32, the same electrode shall be used to ground conductor enclosures and equipment in or on that building. Where separate services supply a building and are required to be connected to a grounding electrode, the same grounding electrode shall be used.

Two or more grounding electrodes that are effectively bonded together shall be considered as a single grounding electrode system in this sense.

250.60 Use of Air Terminals. Air terminal conductors and driven pipes, rods, or plate electrodes used for grounding air terminals shall not be used in lieu of the grounding electrodes required by 250.50 for grounding wiring systems and equipment. This provision shall not prohibit the required bonding together of grounding electrodes of different systems.

> FPN No. 1: See 250.106 for spacing from air terminals. See 800.40(D), 810.21(J), and 820.40(D) for bonding of electrodes.
>
> FPN No. 2: Bonding together of all separate grounding electrodes will limit potential differences

Grounding

between them and between their associated wiring systems.

250.62 Grounding Electrode Conductor Material. The grounding electrode conductor shall be of copper, aluminum, or copper-clad aluminum. The material selected shall be resistant to any corrosive condition existing at the installation or shall be suitably protected against corrosion. The conductor shall be solid or stranded, insulated, covered, or bare.

250.64 Grounding Electrode Conductor Installation. Grounding electrode conductors shall be installed as specified in 250.64(A) through (F).

(A) Aluminum or Copper-Clad Aluminum Conductors. Bare aluminum or copper-clad aluminum grounding conductors shall not be used where in direct contact with masonry or the earth or where subject to corrosive conditions. Where used outside, aluminum or copper-clad aluminum grounding conductors shall not be terminated within 450 mm (18 in.) of the earth.

(B) Securing and Protection from Physical Damage. A grounding electrode conductor or its enclosure shall be securely fastened to the surface on which it is carried. A 4 AWG copper or aluminum or larger conductor shall be protected if exposed to severe physical damage. A 6 AWG grounding conductor that is free from exposure to physical damage shall be permitted to be run along the surface of the building construction without metal covering or protection where it is securely fastened to the construction; otherwise, it shall be in rigid metal conduit, intermediate metal conduit, rigid nonmetallic conduit, electrical metallic tubing, or cable armor. Grounding conductors smaller than 6 AWG shall be in rigid metal conduit, intermediate metal conduit, rigid nonmetallic conduit, electrical metallic tubing, or cable armor.

(C) Continuous. The grounding electrode conductor shall be installed in one continuous length without a splice or joint, unless spliced only by irreversible compression-type connectors listed for the purpose or by the exothermic welding process.

Exception: Sections of busbars shall be permitted to be connected together to form a grounding electrode conductor.

(D) Grounding Electrode Conductor Taps. Where a service consists of more than a single enclosure as permitted in 230.40, Exception No. 2, it shall be permitted to connect taps to the grounding electrode conductor. Each such tap conductor shall extend to the inside of each such enclosure. The grounding electrode conductor shall be sized in accordance with 250.66, but the tap conductors shall be permitted to be sized in accordance with the grounding electrode conductors specified in 250.66 for the largest conductor serving the respective enclosures. The tap conductors shall be connected to the grounding electrode conductor in such a manner that the grounding electrode conductor remains without a splice.

(E) Enclosures for Grounding Electrode Conductors. Metal enclosures for grounding electrode conductors shall be electrically continuous from the point of attachment to cabinets or equipment to the grounding electrode and shall be securely fastened to the ground clamp or fitting. Metal enclosures that are not physically continuous from cabinet or equipment to the grounding electrode shall be made electrically continuous by bonding each end to the grounding electrode conductor. Where a raceway is used as protection for a grounding electrode conductor, the installation shall comply with the requirements of the appropriate raceway article.

(F) To Electrode(s). A grounding electrode conductor shall be permitted to be run to any convenient grounding electrode available in the grounding electrode system or to one or more grounding electrode(s) individually. The grounding electrode conductor shall be sized for the largest grounding electrode conductor required among all the electrodes connected to it.

250.66 Size of Alternating-Current Grounding Electrode Conductor. The size of the grounding electrode conductor of a grounded or ungrounded ac system shall not be less than given in Table 250.66, except as permitted in 250.66(A) through (C).

(A) Connections to Rod, Pipe, or Plate Electrodes. Where the grounding electrode conductor is connected to rod, pipe, or plate electrodes as permitted in 250.52(A)(5)

Grounding

Table 250.66 Grounding Electrode Conductor for Alternating-Current Systems

Size of Largest Ungrounded Service-Entrance Conductor or Equivalent Area for Parallel Conductors[a] (AWG/kcmil)		Size of Grounding Electrode Conductor (AWG/kcmil)	
Copper	Aluminum or Copper-Clad Aluminum	Copper	Aluminum or Copper-Clad Aluminum[b]
2 or smaller	1/0 or smaller	8	6
1 or 1/0	2/0 or 3/0	6	4
2/0 or 3/0	4/0 or 250	4	2
Over 3/0 through 350	Over 250 through 500	2	1/0
Over 350 through 600	Over 500 through 900	1/0	3/0
Over 600 through 1100	Over 900 through 1750	2/0	4/0
Over 1100	Over 1750	3/0	250

Notes:

1. Where multiple sets of service-entrance conductors are used as permitted in 230.40, Exception No. 2, the equivalent size of the largest service-entrance conductor shall be determined by the largest sum of the areas of the corresponding conductors of each set.

2. Where there are no service-entrance conductors, the grounding electrode conductor size shall be determined by the equivalent size of the largest service-entrance conductor required for the load to be served.

[a]This table also applies to the derived conductors of separately derived ac systems

[b]See installation restrictions in 250.64(A).

or 250.52(A)(6), that portion of the conductor that is the sole connection to the grounding electrode shall not be required to be larger than 6 AWG copper wire or 4 AWG aluminum wire.

(B) Connections to Concrete-Encased Electrodes. Where the grounding electrode conductor is connected to a concrete-encased electrode as permitted in 250.52(A)(3), that portion of the conductor that is the sole connection to the grounding electrode shall not be required to be larger than 4 AWG copper wire.

(C) Connections to Ground Rings. Where the grounding electrode conductor is connected to a ground ring as permitted in 250.52(4), that portion of the conductor that is the sole connection to the grounding electrode shall not be required to be larger than the conductor used for the ground ring.

250.68 Grounding Electrode Conductor and Bonding Jumper Connection to Grounding Electrodes.

(A) Accessibility. The connection of a grounding electrode conductor or bonding jumper to a grounding electrode shall be accessible.

Exception: An encased or buried connection to a concrete-encased, driven, or buried grounding electrode shall not be required to be accessible.

(B) Effective Grounding Path. The connection of a grounding electrode conductor or bonding jumper to a grounding electrode shall be made in a manner that will ensure a permanent and effective grounding path. Where necessary to ensure the grounding path for a metal piping system used as a grounding electrode, effective bonding shall be provided around insulated joints and around any equipment likely to be disconnected for repairs or replacement. Bonding conductors shall be of sufficient length to permit removal of such equipment while retaining the integrity of the bond.

250.70 Methods of Grounding and Bonding Conductor Connection to Electrodes.

The grounding or bonding conductor shall be connected to the grounding electrode by exothermic welding, listed lugs, listed pressure connectors, listed clamps, or other listed means. Connections depending on solder shall not be used. Ground clamps shall be listed for the materials of the grounding electrode and the grounding electrode conductor and, where used on pipe, rod, or other buried electrodes, shall also be listed for direct soil burial or concrete encasement. Not more than one conductor shall be connected to the grounding elec-

Grounding

trode by a single clamp or fitting unless the clamp or fitting is listed for multiple conductors. One of the following methods shall be used:

(1) A pipe fitting, pipe plug, or other approved device screwed into a pipe or pipe fitting
(2) A listed bolted clamp of cast bronze or brass, or plain or malleable iron
(3) For indoor telecommunications purposes only, a listed sheet metal strap-type ground clamp having a rigid metal base that seats on the electrode and having a strap of such material and dimensions that it is not likely to stretch during or after installation
(4) An equally substantial approved means

IV. ENCLOSURE, RACEWAY, AND SERVICE CABLE GROUNDING

250.80 Service Raceways and Enclosures. Metal enclosures and raceways for service conductors and equipment shall be grounded.

Exception: A metal elbow that is installed in an underground installation of rigid nonmetallic conduit and is isolated from possible contact by a minimum cover of 450 mm (18 in.) to any part of the elbow shall not be required to be grounded.

250.86 Other Conductor Enclosures and Raceways. Except as permitted by 250.112(l), metal enclosures and raceways for other than service conductors shall be grounded.

V. BONDING

250.90 General. Bonding shall be provided where necessary to ensure electrical continuity and the capacity to conduct safely any fault current likely to be imposed.

250.92 Services.

(A) Bonding of Services. The non–current-carrying metal parts of equipment indicated in 250.92(A)(1), (2), and (3) shall be effectively bonded together.

(1) The service raceways, cable trays, cablebus framework, auxiliary gutters, or service cable armor or sheath except as permitted in 250.84.
(2) All service enclosures containing service conductors, including meter fittings, boxes, or the like, interposed in the service raceway or armor.
(3) Any metallic raceway or armor enclosing a grounding electrode conductor as specified in 250.64(B). Bonding shall apply at each end and to all intervening raceways, boxes, and enclosures between the service equipment and the grounding electrode.

(B) Method of Bonding at the Service. Electrical continuity at service equipment, service raceways, and service conductor enclosures shall be ensured by one of the following methods:

(1) Bonding equipment to the grounded service conductor in a manner provided in 250.8
(2) Connections utilizing threaded couplings or threaded bosses on enclosures where made up wrenchtight
(3) Threadless couplings and connectors where made up tight for metal raceways and metal-clad cables
(4) Other approved devices, such as bonding-type locknuts and bushings

Bonding jumpers meeting the other requirements of this article shall be used around concentric or eccentric knockouts that are punched or otherwise formed so as to impair the electrical connection to ground. Standard locknuts or bushings shall not be the sole means for the bonding required by this section.

250.94 Bonding for Other Systems. An accessible means external to enclosures for connecting intersystem bonding and grounding conductors shall be provided at the service equipment and at the disconnecting means for any additional buildings or structures by at least one of the following means:

(1) Exposed nonflexible metallic raceways
(2) Exposed grounding electrode conductor

Grounding

(3) Approved means for the external connection of a copper or other corrosion-resistant bonding or grounding conductor to the grounded raceway or equipment

250.96 Bonding Other Enclosures.

(A) General. Metal raceways, cable trays, cable armor, cable sheath, enclosures, frames, fittings, and other metal non–current-carrying parts that are to serve as grounding conductors, with or without the use of supplementary equipment grounding conductors, shall be effectively bonded where necessary to ensure electrical continuity and the capacity to conduct safely any fault current likely to be imposed on them. Any nonconductive paint, enamel, or similar coating shall be removed at threads, contact points, and contact surfaces or be connected by means of fittings designed so as to make such removal unnecessary.

(B) Isolated Grounding Circuits. Where required for the reduction of electrical noise (electromagnetic interference) on the grounding circuit, an equipment enclosure supplied by a branch circuit shall be permitted to be isolated from a raceway containing circuits supplying only that equipment by one or more listed nonmetallic raceway fittings located at the point of attachment of the raceway to the equipment enclosure. The metal raceway shall comply with provisions of this article and shall be supplemented by an internal insulated equipment grounding conductor installed in accordance with 250.146(D) to ground the equipment enclosure.

250.97 Bonding for Over 250 Volts.

For circuits of over 250 volts to ground, the electrical continuity of metal raceways and cables with metal sheaths that contain any conductor other than service conductors shall be ensured by one or more of the methods specified for services in 250.92(B), except for (1).

Exception: Where oversized, concentric, or eccentric knockouts are not encountered, or where a box or enclosure with concentric or eccentric knockouts is listed for the purpose, the following methods shall be permitted:

(a) Threadless couplings and connectors for cables with metal sheaths

(b) Two locknuts, on rigid metal conduit or intermediate metal conduit, one inside and one outside of boxes and cabinets

(c) Fittings with shoulders that seat firmly against the box or cabinet, such as electrical metallic tubing connectors, flexible metal conduit connectors, and cable connectors, with one locknut on the inside of boxes and cabinets

(d) Listed fittings that are identified for the purpose

250.98 Bonding Loosely Jointed Metal Raceways. Expansion fittings and telescoping sections of metal raceways shall be made electrically continuous by equipment bonding jumpers or other means.

250.100 Bonding in Hazardous (Classified) Locations. Regardless of the voltage of the electrical system, the electrical continuity of non–current-carrying metal parts of equipment, raceways, and other enclosures in any hazardous (classified) location as defined in Article 500 shall be ensured by any of the methods specified for services in 250.92(B) that are approved for the wiring method used.

250.102 Equipment Bonding Jumpers.

(A) Material. Equipment bonding jumpers shall be of copper or other corrosion-resistant material. A bonding jumper shall be a wire, bus, screw, or similar suitable conductor.

(B) Attachment. Equipment bonding jumpers shall be attached in the manner specified by the applicable provisions of 250.8 for circuits and equipment and by 250.70 for grounding electrodes.

(C) Size—Equipment Bonding Jumper on Supply Side of Service. The bonding jumper shall not be smaller than the sizes shown in Table 250.66 for grounding electrode conductors. Where the service-entrance phase conductors are larger than 1100 kcmil copper or 1750 kcmil aluminum, the bonding jumper shall have an area not less than 12-1/2 percent of the area of the largest phase conductor except that, where the phase conductors and the bonding jumper are of different materials (copper or aluminum), the mini-

Grounding

mum size of the bonding jumper shall be based on the assumed use of phase conductors of the same material as the bonding jumper and with an ampacity equivalent to that of the installed phase conductors. Where the service-entrance conductors are paralleled in two or more raceways or cables, the equipment bonding jumper, where routed with the raceways or cables, shall be run in parallel. The size of the bonding jumper for each raceway or cable shall be based on the size of the service-entrance conductors in each raceway or cable.

The bonding jumper for a grounding electrode conductor raceway or cable armor as covered in 250.64(E) shall be the same size or larger than the required enclosed grounding electrode conductor.

(D) Size—Equipment Bonding Jumper on Load Side of Service. The equipment bonding jumper on the load side of the service overcurrent devices shall be sized, as a minimum, in accordance with the sizes listed in Table 250.122, but shall not be required to be larger than the largest ungrounded circuit conductors supplying the equipment and shall not be smaller than 14 AWG.

A single common continuous equipment bonding jumper shall be permitted to bond two or more raceways or cables where the bonding jumper is sized in accordance with Table 250.122 for the largest overcurrent device supplying circuits therein.

(E) Installation. The equipment bonding jumper shall be permitted to be installed inside or outside of a raceway or enclosure. Where installed on the outside, the length of the equipment bonding jumper shall not exceed 1.8 m (6 ft) and shall be routed with the raceway or enclosure. Where installed inside of a raceway, the equipment bonding jumper shall comply with the requirements of 250.119 and 250.148.

Exception: An equipment bonding jumper longer than 1.8 m (6 ft) shall be permitted at outside pole locations for the purpose of bonding or grounding isolated sections of metal raceways or elbows installed in exposed risers of metal conduit or other metal raceway.

250.104 Bonding of Piping Systems and Exposed Structural Steel.

(A) Metal Water Piping. The metal water piping system shall be bonded as required in (1), (2), (3), or (4) of this section. The bonding jumper(s) shall be installed in accordance with 250.64(A), (B), and (E). The points of attachment of the bonding jumper(s) shall be accessible.

(1) General. Metal water piping system(s) installed in or attached to a building or structure shall be bonded to the service equipment enclosure, the grounded conductor at the service, the grounding electrode conductor where of sufficient size, or to the one or more grounding electrodes used. The bonding jumper(s) shall be sized in accordance with Table 250.66 except as permitted in 250.104(A)(2) and (A)(3).

(2) Buildings of Multiple Occupancy. In buildings of multiple occupancy where the metal water piping system(s) installed in or attached to a building or structure for the individual occupancies is metallically isolated from all other occupancies by use of nonmetallic water piping, the metal water piping system(s) for each occupancy shall be permitted to be bonded to the equipment grounding terminal of the panelboard or switchboard enclosure (other than service equipment) supplying that occupancy. The bonding jumper shall be sized in accordance with Table 250.122.

(3) Multiple Buildings or Structures Supplied from a Common Service. The metal water piping system(s) installed in or attached to a building or structure shall be bonded to the building or structure disconnecting means enclosure where located at the building or structure, to the equipment grounding conductor run with the supply conductors, or to the one or more grounding electrodes used. The bonding jumper(s) shall be sized in accordance with 250.66, based on the size of the feeder or branch circuit conductors that supply the building. The bonding jumper shall not be required to be larger than the largest ungrounded feeder or branch circuit conductor supplying the building.

(4) Separately Derived Systems. The grounded conductor of each separately derived system shall be bonded to

Grounding

the nearest available point of the interior metal water piping system(s) in the area served by each separately derived system. This connection shall be made at the same point on the separately derived system where the grounding electrode conductor is connected. Each bonding jumper shall be sized in accordance with Table 250.66.

Exception: A separate water piping bonding jumper shall not be required where the effectively grounded metal frame of a building or structure is used as the grounding electrode for a separately derived system and is bonded to the metallic water piping in the area served by the separately derived system.

(B) Other Metal Piping. Where installed in or attached to a building or structure, metal piping system(s), including gas piping, that may become energized shall be bonded to the service equipment enclosure, the grounded conductor at the service, the grounding electrode conductor where of sufficient size, or to the one or more grounding electrodes used. The bonding jumper(s) shall be sized in accordance with 250.122 using the rating of the circuit that may energize the piping system(s). The equipment grounding conductor for the circuit that may energize the piping shall be permitted to serve as the bonding means. The points of attachment of the bonding jumper(s) shall be accessible.

> FPN: Bonding all piping and metal air ducts within the premises will provide additional safety.

(C) Structural Steel. Exposed structural steel that is interconnected to form a steel building frame and is not intentionally grounded and may become energized shall be bonded to the service equipment enclosure, the grounded conductor at the service, the grounding electrode conductor where of sufficient size, or the one or more grounding electrodes used. The bonding jumper(s) shall be sized in accordance with Table 250.66 and installed in accordance with 250.64(A), (B), and (E). The points of attachment of the bonding jumper(s) shall be accessible.

250.106 Lightning Protection Systems. The lightning protection system ground terminals shall be bonded to the building or structure grounding electrode system.

VI. EQUIPMENT GROUNDING AND EQUIPMENT GROUNDING CONDUCTORS

250.110 Equipment Fastened in Place or Connected by Permanent Wiring Methods (Fixed). Exposed non–current-carrying metal parts of fixed equipment likely to become energized shall be grounded under any of the following conditions:

(1) Where within 2.5 m (8 ft) vertically or 1.5 m (5 ft) horizontally of ground or grounded metal objects and subject to contact by persons
(2) Where located in a wet or damp location and not isolated
(3) Where in electrical contact with metal
(4) Where in a hazardous (classified) location as covered by Articles 500 through 517
(5) Where supplied by a metal-clad, metal-sheathed, metal-raceway, or other wiring method that provides an equipment ground, except as permitted by 250.86, Exception No. 2, for short sections of metal enclosures
(6) Where equipment operates with any terminal at over 150 volts to ground

250.112 Fastened in Place or Connected by Permanent Wiring Methods (Fixed)—Specific. Exposed, non–current-carrying metal parts of the kinds of equipment described in 250.112(A) through (K), and non–current-carrying metal parts of equipment and enclosures described in 250.112(L) and (M), shall be grounded regardless of voltage.

(A) Switchboard Frames and Structures. Switchboard frames and structures supporting switching equipment, except frames of 2-wire dc switchboards where effectively insulated from ground.

(D) Enclosures for Motor Controllers. Enclosures for motor controllers unless attached to ungrounded portable equipment.

(E) Elevators and Cranes. Electric equipment for elevators and cranes.

Grounding 115

(F) Garages, Theaters, and Motion Picture Studios. Electric equipment in commercial garages, theaters, and motion picture studios, except pendant lampholders supplied by circuits not over 150 volts to ground.

(G) Electric Signs. Electric signs, outline lighting, and associated equipment as provided in Article 600.

(H) Motion Picture Projection Equipment. Motion picture projection equipment.

(I) Power-Limited Remote-Control, Signaling, and Fire Alarm Circuits. Equipment supplied by Class 1 power-limited circuits and Class 1, Class 2, and Class 3 remote-control and signaling circuits, and by fire alarm circuits, shall be grounded where system grounding is required by Part II or Part VIII of this article.

(J) Luminaires (Lighting Fixtures). Luminaires (lighting fixtures) as provided in Part V of Article 410.

(K) Skid Mounted Equipment. Permanently mounted electrical equipment and skids shall be grounded with an equipment bonding jumper sized as required by 250.122.

(L) Motor-Operated Water Pumps. Motor-operated water pumps, including the submersible type.

(M) Metal Well Casings. Where a submersible pump is used in a metal well casing, the well casing shall be bonded to the pump circuit equipment grounding conductor.

250.114 Equipment Connected by Cord and Plug. Under any of the conditions described in (1) through (4), exposed non–current-carrying metal parts of cord-and-plug-connected equipment likely to become energized shall be grounded.

(4) In other than residential occupancies:

    a. Refrigerators, freezers, and air conditioners
    b. Clothes-washing, clothes-drying, dish-washing machines; information technology equipment; sump pumps and electrical aquarium equipment

c. Hand-held motor-operated tools, stationary and fixed motor-operated tools, light industrial motor-operated tools
d. Motor-operated appliances of the following types: hedge clippers, lawn mowers, snow blowers, and wet scrubbers
e. Portable handlamps
f. Cord-and-plug-connected appliances used in damp or wet locations or by persons standing on the ground or on metal floors or working inside of metal tanks or boilers
g. Tools likely to be used in wet or conductive locations

250.118 Types of Equipment Grounding Conductors.

The equipment grounding conductor run with or enclosing the circuit conductors shall be one or more or a combination of the following:

(1) A copper, aluminum, or copper-clad aluminum conductor. This conductor shall be solid or stranded; insulated, covered, or bare; and in the form of a wire or a busbar of any shape.
(2) Rigid metal conduit.
(3) Intermediate metal conduit.
(4) Electrical metallic tubing.
(5) Flexible metal conduit where both the conduit and fittings are listed for grounding.
(6) Listed flexible metal conduit that is not listed for grounding, meeting all the following conditions:

   a. The conduit is terminated in fittings listed for grounding.
   b. The circuit conductors contained in the conduit are protected by overcurrent devices rated at 20 amperes or less.
   c. The combined length of flexible metal conduit and flexible metallic tubing and liquidtight flexible metal conduit in the same ground return path does not exceed 1.8 m (6 ft).
   d. The conduit is not installed for flexibility.

(7) Listed liquidtight flexible metal conduit meeting all the following conditions:

Grounding

a. The conduit is terminated in fittings listed for grounding.
b. For metric designators 12 through 16 (trade sizes 3/8 through 1/2), the circuit conductors contained in the conduit are protected by overcurrent devices rated at 20 amperes or less.
c. For metric designators 21 through 35 (trade sizes 3/4 through 1-1/4), the circuit conductors contained in the conduit are protected by overcurrent devices rated not more than 60 amperes and there is no flexible metal conduit, flexible metallic tubing, or liquidtight flexible metal conduit in trade sizes metric designators 12 through 16 (trade sizes 3/8 through 1/2) in the grounding path.
d. The combined length of flexible metal conduit and flexible metallic tubing and liquidtight flexible metal conduit in the same ground return path does not exceed 1.8 m (6 ft).
e. The conduit is not installed for flexibility.

(8) Flexible metallic tubing where the tubing is terminated in fittings listed for grounding and meeting the following conditions:

a. The circuit conductors contained in the tubing are protected by overcurrent devices rated at 20 amperes or less.
b. The combined length of flexible metal conduit and flexible metallic tubing and liquidtight flexible metal conduit in the same ground return path does not exceed 1.8 m (6 ft).

(9) Armor of Type AC cable as provided in 320.108.
(10) The copper sheath of mineral-insulated, metal-sheathed cable.
(11) Type MC cable where listed and identified for grounding in accordance with the following:

a. The combined metallic sheath and grounding conductor of interlocked metal tape—type MC cable
b. The metallic sheath or the combined metallic sheath and grounding conductors of the smooth or corrugated tube type MC cable

(12) Cable trays as permitted in 392.3(C) and 392.7.
(13) Cablebus framework as permitted in 370.3.
(14) Other electrically continuous metal raceways and auxiliary gutters listed for grounding.

250.119 Identification of Equipment Grounding Conductors. Unless required elsewhere in this *Code*, equipment grounding conductors shall be permitted to be bare, covered, or insulated. Individually covered or insulated equipment grounding conductors shall have a continuous outer finish that is either green or green with one or more yellow stripes except as permitted in this section.

(A) Conductors Larger Than 6 AWG. An insulated or covered conductor larger than 6 AWG copper or aluminum shall be permitted, at the time of installation, to be permanently identified as an equipment grounding conductor at each end and at every point where the conductor is accessible. Identification shall encircle the conductor and shall be accomplished by one of the following:

(1) Stripping the insulation or covering from the entire exposed length
(2) Coloring the exposed insulation or covering green
(3) Marking the exposed insulation or covering with green tape or green adhesive labels

(B) Multiconductor Cable. Where the conditions of maintenance and supervision ensure that only qualified persons service the installation, one or more insulated conductors in a multiconductor cable, at the time of installation, shall be permitted to be permanently identified as equipment grounding conductors at each end and at every point where the conductors are accessible by one of the following means:

(1) Stripping the insulation from the entire exposed length
(2) Coloring the exposed insulation green
(3) Marking the exposed insulation with green tape or green adhesive labels

(C) Flexible Cord. An uninsulated equipment grounding conductor shall be permitted, but, if individually covered, the covering shall have a continuous outer finish that is either green or green with one or more yellow stripes.

Grounding

250.120 Equipment Grounding Conductor Installation.
An equipment grounding conductor shall be installed in accordance with 250.120(A), (B), and (C).

(A) Raceway, Cable Trays, Cable Armor, Cablebus, or Cable Sheaths. Where it consists of a raceway, cable tray, cable armor, cablebus framework, or cable sheath or where it is a wire within a raceway or cable, it shall be installed in accordance with the applicable provisions in this *Code* using fittings for joints and terminations approved for use with the type raceway or cable used. All connections, joints, and fittings shall be made tight using suitable tools.

(B) Aluminum and Copper-Clad Aluminum Conductors. Equipment grounding conductors of bare or insulated aluminum or copper-clad aluminum shall be permitted. Bare conductors shall not come in direct contact with masonry or the earth or where subject to corrosive conditions. Aluminum or copper-clad aluminum conductors shall not be terminated within 450 mm (18 in.) of the earth.

(C) Equipment Grounding Conductors Smaller Than 6 AWG. Equipment grounding conductors smaller than 6 AWG shall be protected from physical damage by a raceway or cable armor except where run in hollow spaces of walls or partitions, where not subject to physical damage, or where protected from physical damage.

250.122 Size of Equipment Grounding Conductors.

(A) General. Copper, aluminum, or copper-clad aluminum equipment grounding conductors of the wire type shall not be smaller than shown in Table 250.122 but shall not be required to be larger than the circuit conductors supplying the equipment. Where a raceway or a cable armor or sheath is used as the equipment grounding conductor, as provided in 250.118 and 250.134(A), it shall comply with 250.4(A)(5) or 250.4(B)(4).

(B) Increased in Size. Where ungrounded conductors are increased in size, equipment grounding conductors, where installed, shall be increased in size proportionately according to circular mil area of the ungrounded conductors.

(C) Multiple Circuits. Where a single equipment grounding conductor is run with multiple circuits in the same raceway or cable, it shall be sized for the largest overcurrent device protecting conductors in the raceway or cable.

(D) Motor Circuits. Where the overcurrent device consists of an instantaneous trip circuit breaker or a motor short-

Table 250.122 Minimum Size Equipment Grounding Conductors for Grounding Raceway and Equipment

Rating or Setting of Automatic Overcurrent Device in Circuit Ahead of Equipment, Conduit, etc., Not Exceeding (Amperes)	Size (AWG or kcmil)	
	Copper	Aluminum or Copper-Clad Aluminum*
15	14	12
20	12	10
30	10	8
40	10	8
60	10	8
100	8	6
200	6	4
300	4	2
400	3	1
500	2	1/0
600	1	2/0
800	1/0	3/0
1000	2/0	4/0
1200	3/0	250
1600	4/0	350
2000	250	400
2500	350	600
3000	400	600
4000	500	800
5000	700	1200
6000	800	1200

Note: Where necessary to comply with 250.4(A)(5) or 240.4(B)(4), the equipment grounding conductor shall be sized larger than given in this table.

*See installation restrictions in 250.120.

Grounding

circuit protector, as allowed in 430.52, the equipment grounding conductor size shall be permitted to be based on the rating of the motor overload protective device but not less than the size shown in Table 250.122.

(E) Flexible Cord and Fixture Wire. Equipment grounding conductors that are part of flexible cords or used with fixture wires in accordance with 240.5 shall be not smaller than 18 AWG copper and not smaller than the circuit conductors.

(F) Conductors in Parallel. Where conductors are run in parallel in multiple raceways or cables as permitted in 310.4, the equipment grounding conductors, where used, shall be run in parallel in each raceway or cable. One of the methods in 250.122(F)(1) or (2) shall be used to ensure the equipment grounding conductors are protected.

(1) Each parallel equipment grounding conductor shall be sized on the basis of the ampere rating of the overcurrent device protecting the circuit conductors in the raceway or cable in accordance with Table 250.122.

250.126 Identification of Wiring Device Terminals. The terminal for the connection of the equipment grounding conductor shall be identified by one of the following:

(1) A green, not readily removable terminal screw with a hexagonal head.
(2) A green, hexagonal, not readily removable terminal nut.
(3) A green pressure wire connector. If the terminal for the grounding conductor is not visible, the conductor entrance hole shall be marked with the word *green* or *ground,* the letters *G* or *GR* or the grounding symbol shown in Figure 250.126, or otherwise identified by a distinctive green color. If the terminal for the equipment grounding conductor is readily removable, the area adjacent to the terminal shall be similarly marked.

Figure 250.126 Grounding symbol.

VII. METHODS OF EQUIPMENT GROUNDING

250.130 Equipment Grounding Conductor Connections. Equipment grounding conductor connections at the source of separately derived systems shall be made in accordance with 250.30(A)(1). Equipment grounding conductor connections at service equipment shall be made as indicated in 250.130(A) or (B). For replacement of non–grounding-type receptacles with grounding-type receptacles and for branch-circuit extensions only in existing installations that do not have an equipment grounding conductor in the branch circuit, connections shall be permitted as indicated in 250.130(C).

(A) For Grounded Systems. The connection shall be made by bonding the equipment grounding conductor to the grounded service conductor and the grounding electrode conductor.

(B) For Ungrounded Systems. The connection shall be made by bonding the equipment grounding conductor to the grounding electrode conductor.

(C) Nongrounding Receptacle Replacement or Branch Circuit Extensions. The equipment grounding conductor of a grounding-type receptacle or a branch-circuit extension shall be permitted to be connected to any of the following:

(1) Any accessible point on the grounding electrode system as described in 250.50
(2) Any accessible point on the grounding electrode conductor
(3) The equipment grounding terminal bar within the enclosure where the branch circuit for the receptacle or branch circuit originates
(4) For grounded systems, the grounded service conductor within the service equipment enclosure
(5) For ungrounded systems, the grounding terminal bar within the service equipment enclosure

250.132 Short Sections of Raceway. Isolated sections of metal raceway of cable armor, where required to be grounded, shall be grounded in accordance with 250.134.

250.134 Equipment Fastened in Place or Connected by Permanent Wiring Methods (Fixed)—Grounding. Un-

Grounding

less grounded by connection to the grounded circuit conductor as permitted by 250.32, 250.140, and 250.142, non–current-carrying metal parts of equipment, raceways, and other enclosures, if grounded, shall be grounded by one of the following methods.

(A) Equipment Grounding Conductor Types. By any of the equipment grounding conductors permitted by 250.118.

(B) With Circuit Conductors. By an equipment grounding conductor contained within the same raceway, cable, or otherwise run with the circuit conductors.

250.142. Use of Grounded Circuit Conductor for Grounding Equipment.

(A) Supply-Side Equipment. A grounded circuit conductor shall be permitted to ground non–current-carrying metal parts of equipment, raceways, and other enclosures at any of the following locations:

(1) On the supply side or within the enclosure of the ac service-disconnecting means
(2) On the supply side or within the enclosure of the main disconnecting means for separate buildings as provided in 250.32(B)
(3) On the supply side or within the enclosure of the main disconnecting means or overcurrent devices of a separately derived system where permitted by 250.30(A)(1)

(B) Load-Side Equipment. Except as permitted in 250.30(A)(1) and 250.32(B), a grounded circuit conductor shall not be used for grounding non–current-carrying metal parts of equipment on the load side of the service disconnecting means or on the load side of a separately derived system disconnecting means or the overcurrent devices for a separately derived system not having a main disconnecting means.

250.146 Connecting Receptacle Grounding Terminal to Box.
An equipment bonding jumper shall be used to connect the grounding terminal of a grounding-type receptacle to a grounded box unless grounded as in 250.146(A) through (D).

(A) Surface Mounted Box. Where the box is mounted on the surface, direct metal-to-metal contact between the device yoke and the box shall be permitted to ground the receptacle to the box. This provision shall not apply to cover-mounted receptacles unless the box and cover combination are listed as providing satisfactory ground continuity between the box and the receptacle.

(B) Contact Devices or Yokes. Contact devices or yokes designed and listed for the purpose shall be permitted in conjunction with the supporting screws to establish the grounding circuit between the device yoke and flush-type boxes.

(C) Floor Boxes. Floor boxes designed for and listed as providing satisfactory ground continuity between the box and the device shall be permitted.

(D) Isolated Receptacles. Where required for the reduction of electrical noise (electromagnetic interference) on the grounding circuit, a receptacle in which the grounding terminal is purposely insulated from the receptacle mounting means shall be permitted. The receptacle grounding terminal shall be grounded by an insulated equipment grounding conductor run with the circuit conductors. This grounding conductor shall be permitted to pass through one or more panelboards without connection to the panelboard grounding terminal as permitted in 408.20, Exception, so as to terminate within the same building or structure directly at an equipment grounding conductor terminal of the applicable derived system or service.

250.148 Continuity and Attachment of Equipment Grounding Conductors to Boxes. Where circuit conductors are spliced within a box, or terminated on equipment within or supported by a box, any separate equipment grounding conductors associated with those circuit conductors shall be spliced or joined within the box or to the box with devices suitable for the use. Connections depending solely on solder shall not be used. Splices shall be made in accordance with 110.14(B) except that insulation shall not be required. The arrangement of grounding connections shall be such that the disconnection or the re-

Grounding

moval of a receptacle, luminaire (fixture), or other device fed from the box will not interfere with or interrupt the grounding continuity.

Exception: The equipment grounding conductor permitted in 250.146(D) shall not be required to be connected to the other equipment grounding conductors or to the box.

(A) Metal Boxes. A connection shall be made between the one or more equipment grounding conductors and a metal box by means of a grounding screw that shall be used for no other purpose or a listed grounding device.

(B) Nonmetallic Boxes. One or more equipment grounding conductors brought into a nonmetallic outlet box shall be arranged so that a connection can be made to any fitting or device in that box requiring grounding.

CHAPTER 11

BOXES AND ENCLOSURES

INTRODUCTION

This chapter contains provisions from Article 314—Outlet, Device, Pull and Junction Boxes; Conduit Bodies; and Fittings. This chapter also contains requirements from Article 312—Cabinets, Cutout Boxes, and Meter Socket Enclosures. Article 314 covers the installation and use of all enclosures (and conduit bodies) used as outlet, device, pull, or junction boxes, depending on their use, and the installation requirements for fittings used to join raceways and to connect raceways (and cables) to boxes and conduit bodies. Article 312 covers the installation of cabinets, cutout boxes, and meter sockets.

Article 314's provisions range from selecting the size and type of box to installing the cover or canopy. Provisions for boxes enclosing 18 through 6 AWG conductors are located in 314.16. Enclosures (outlet, device, and junction boxes) covered by 314.16 are calculated from the sizes and numbers of conductors. Requirements for boxes containing larger conductors (4 AWG and larger) are in 314.28. Pull and junction boxes covered by 314.28 are calculated from the sizes and numbers of raceways.

ARTICLE 314

Outlet, Device, Pull, and Junction Boxes; Conduit Bodies; Fittings; and Manholes

I. SCOPE AND GENERAL

314.1 Scope. This article covers the installation and use of all boxes and conduit bodies used as outlet, device, junction, or pull boxes, depending on their use, and manholes and other electric enclosures intended for personnel entry. Cast, sheet metal, nonmetallic, and other boxes such as FS, FD, and larger boxes are not classified as conduit bodies. This article also includes installation requirements for fittings used to join raceways and to connect raceways and cables to boxes and conduit bodies.

314.2 Round Boxes. Round boxes shall not be used where conduits or connectors requiring the use of locknuts or bushings are to be connected to the side of the box.

314.4 Metal Boxes. All metal boxes shall be grounded in accordance with the provisions of Article 250.

II. INSTALLATION

314.15 Damp, Wet, or Hazardous (Classified) Locations.

(A) Damp or Wet Locations. In damp or wet locations, boxes, conduit bodies, and fittings shall be placed or equipped so as to prevent moisture from entering or accumulating within the box, conduit body, or fitting. Boxes, conduit bodies, and fittings installed in wet locations shall be listed for use in wet locations.

314.16 Number of Conductors in Outlet, Device, and Junction Boxes, and Conduit Bodies. Boxes and conduit bodies shall be of sufficient size to provide free space for all enclosed conductors. In no case shall the volume of the box, as calculated in 314.16(A), be less than the fill calculation as calculated in 314.16(B). The minimum volume for conduit bodies shall be as calculated in 314.16(C).

The provisions of this section shall not apply to terminal housings supplied with motors.

Boxes and Enclosures

Boxes and conduit bodies enclosing conductors 4 AWG or larger shall also comply with the provisions of 314.28.

(A) Box Volume Calculations. The volume of a wiring enclosure (box) shall be the total volume of the assembled sections, and, where used, the space provided by plaster rings, domed covers, extension rings, and so forth, that are marked with their volume or are made from boxes the dimensions of which are listed in Table 314.16(A).

(1) Standard Boxes. The volumes of standard boxes that are not marked with their volume shall be as given in Table 314.16(A).

(2) Other Boxes. Boxes 1650 cm^3 (100 in.^{3}) or less, other than those described in Table 314.16(A), and nonmetallic boxes shall be durably and legibly marked by the manufacturer with their volume. Boxes described in Table 314.16(A) that have a volume larger than is designated in the table shall be permitted to have their volume marked as required by this section.

(B) Box Fill Calculations. The volumes in paragraphs 314.16(B)(1) through (5), as applicable, shall be added together. No allowance shall be required for small fittings such as locknuts and bushings.

(1) Conductor Fill. Each conductor that originates outside the box and terminates or is spliced within the box shall be counted once, and each conductor that passes through the box without splice or termination shall be counted once. The conductor fill shall be computed using Table 314.16(B). A conductor, no part of which leaves the box, shall not be counted.

Exception: An equipment grounding conductor or conductors or not over four luminaire (fixture) wires smaller than 14 AWG, or both, shall be permitted to be omitted from the calculations where they enter a box from a domed luminaire (fixture) or similar canopy and terminate within that box.

(2) Clamp Fill. Where one or more internal cable clamps, whether factory or field supplied, are present in the box, a single volume allowance in accordance with Table 314.16(B)

Table 314.16(A) Metal Boxes

Box Trade Size			Minimum Volume		Maximum Number of Conductors*						
mm	in.		cm³	in.³	18	16	14	12	10	8	6
100 × 32	(4 × 1-1/4)	round/octagonal	205	12.5	8	7	6	5	5	5	2
100 × 38	(4 × 1-1/2)	round/octagonal	254	15.5	10	8	7	6	6	5	3
100 × 54	(4 × 2-1/8)	round/octagonal	353	21.5	14	12	10	9	8	7	4
100 × 32	(4 × 1-1/4)	square	295	18.0	12	10	9	8	7	6	3
100 × 38	(4 × 1-1/2)	square	344	21.0	14	12	10	9	8	7	4
100 × 54	(4 × 2-1/8)	square	497	30.3	20	17	15	13	12	10	6
120 × 32	(4-11/16 × 1-1/4)	square	418	25.5	17	14	12	11	10	8	5
120 × 38	(4-11/16 × 1-1/2)	square	484	29.5	19	16	14	13	11	9	5
120 × 54	(4-11/16 × 2-1/8)	square	689	42.0	28	24	21	18	16	14	8
75 × 50 × 38	(3 × 2 × 1-1/2)	device	123	7.5	5	4	3	3	3	2	1
75 × 50 × 50	(3 × 2 × 2)	device	164	10.0	6	5	5	4	4	3	2
75 × 50 × 57	(3 × 2 × 2-1/4)	device	172	10.5	7	6	5	4	4	3	2
75 × 50 × 65	(3 × 2 × 2-1/2)	device	205	12.5	8	7	6	5	5	4	2

Boxes and Enclosures

75 × 50 × 70	(3 × 2 × 2-3/4)		230	14.0	9	8	7	6	5	4	2	
75 × 50 × 90	(3 × 2 × 3-1/2)		295	18.0	12	10	9	8	7	6	3	
100 × 54 × 38	(4 × 2-1/8 × 1-1/2)	device	169	10.3	6	5	5	4	4	3	2	
100 × 54 × 48	(4 × 2-1/8 × 1-7/8)	device	213	13.0	8	7	6	5	5	4	2	
100 × 54 × 54	(4 × 2-1/8 × 2-1/8)	device	238	14.5	9	8	7	6	5	4	2	
95 × 50 × 65	(3-3/4 × 2 × 2-1/2)	masonry box/gang	230	14.0	9	8	7	6	5	4	2	
95 × 50 × 90	(3-3/4 × 2 × 3-1/2)	masonry box/gang	344	21.0	14	12	10	9	8	7	2	
min. 44.5 depth	FS—single cover/gang (1-3/4)		221	13.5	9	7	6	6	5	4	2	
min. 60.3 depth	FD—single cover/gang (2-3/8)		295	18.0	12	10	9	8	7	6	3	
min. 44.5 depth	FS—multiple cover/gang (1-3/4)		295	18.0	12	10	9	8	7	6	3	
min. 60.3 depth	FD—multiple cover/gang (2-3/8)		395	24.0	16	13	12	10	9	8	4	

*Where no volume allowances are required by 314.16(B)(2) through 314.16(B)(5).

Table 314.16(B) Volume Allowance Required per Conductor

Size of Conductor (AWG)	Free Space Within Box for Each Conductor	
	cm^3	in.^{3}
18	24.6	1.50
16	28.7	1.75
14	32.8	2.00
12	36.9	2.25
10	41.0	2.50
8	49.2	3.00
6	81.9	5.00

shall be made based on the largest conductor present in the box. No allowance shall be required for a cable connector with its clamping mechanism outside the box.

(3) Support Fittings Fill. Where one or more luminaire (fixture) studs or hickeys are present in the box, a single volume allowance in accordance with Table 314.16(B) shall be made for each type of fitting based on the largest conductor present in the box.

(4) Device or Equipment Fill. For each yoke or strap containing one or more devices or equipment, a double volume allowance in accordance with Table 314.16(B) shall be made for each yoke or strap based on the largest conductor connected to a device(s) or equipment supported by that yoke or strap.

(5) Equipment Grounding Conductor Fill. Where one or more equipment grounding conductors or equipment bonding jumpers enter a box, a single volume allowance in accordance with Table 314.16(B) shall be made based on the largest equipment grounding conductor or equipment bonding jumper present in the box. Where an additional set of equipment grounding conductors, as permitted by 250.146(D), is present in the box, an additional volume allowance shall be made based on the largest equipment grounding conductor in the additional set.

(C) Conduit Bodies.

(2) With Splices, Taps, or Devices. Only those conduit bodies that are durably and legibly marked by the manufacturer with their volume shall be permitted to contain splices, taps, or devices. The maximum number of conductors shall be computed in accordance with 314.16(B). Conduit bodies shall be supported in a rigid and secure manner.

314.17 Conductors Entering Boxes, Conduit Bodies, or Fittings.
Conductors entering boxes, conduit bodies, or fittings shall be protected from abrasion and shall comply with 314.17(A) through (D).

(A) Openings to Be Closed. Openings through which conductors enter shall be adequately closed.

(B) Metal Boxes and Conduit Bodies. Where metal boxes or conduit bodies are installed with open wiring or concealed knob-and-tube wiring, conductors shall enter through insulating bushings or, in dry locations, through flexible tubing extending from the last insulating support to not less than 6 mm (1/4 in.) inside the box and beyond any cable clamps. Except as provided in 300.15(C), the wiring shall be firmly secured to the box or conduit body. Where raceway or cable is installed with metal boxes or conduit bodies, the raceway or cable shall be secured to such boxes and conduit bodies.

314.19 Boxes Enclosing Flush Devices.
Boxes used to enclose flush devices shall be of such design that the devices will be completely enclosed on back and sides and substantial support for the devices will be provided. Screws for supporting the box shall not be used in attachment of the device contained therein.

314.20 In Wall or Ceiling.
In walls or ceilings with a surface of concrete, tile, gypsum, plaster, or other noncombustible material, boxes shall be installed so that the front edge of the box will not be set back of the finished surface more than 6 mm (1/4 in.).

In walls and ceilings constructed of wood or other combustible surface material, boxes shall be flush with the finished surface or project therefrom.

314.21 Repairing Plaster and Drywall or Plasterboard. Plaster, drywall, or plasterboard surfaces that are broken or incomplete shall be repaired so there will be no gaps or open spaces greater than 3 mm (1/8 in.) at the edge of the box or fitting.

314.22 Exposed Surface Extensions. Surface extensions from a flush-mounted box shall be made by mounting and mechanically securing an extension ring over the flush box. Equipment grounding and bonding shall be in accordance with Article 250.

Exception: A surface extension shall be permitted to be made from the cover of a flush-mounted box where the cover is designed so it is unlikely to fall off or be removed if its securing means becomes loose. The wiring method shall be flexible for a length sufficient to permit removal of the cover and provide access to the box interior, and arranged so that any bonding or grounding continuity is independent of the connection between the box and cover.

314.23 Supports. Enclosures within the scope of this article shall be supported in accordance with one or more of the provisions in 314.23(A) through (H).

(A) Surface Mounting. An enclosure mounted on a building or other surface shall be rigidly and securely fastened in place. If the surface does not provide rigid and secure support, additional support in accordance with other provisions of this section shall be provided.

(B) Structural Mounting. An enclosure supported from a structural member of a building or from grade shall be rigidly supported either directly or by using a metal, polymeric, or wood brace.

(1) Nails and Screws. Nails and screws, where used as a fastening means, shall be attached by using brackets on the outside of the enclosure, or they shall pass through the interior within 6 mm (1/4 in.) of the back or ends of the enclosure.

(2) Braces. Metal braces shall be protected against corrosion and formed from metal that is not less than 0.51 mm (0.020 in.) thick uncoated. Wood braces shall have a cross section not less than nominal 25 mm × 50 mm (1 in. × 2 in.). Wood braces in wet locations shall be treated for the conditions. Polymeric braces shall be identified as being suitable for the use.

(C) Mounting in Finished Surfaces. An enclosure mounted in a finished surface shall be rigidly secured thereto by clamps, anchors, or fittings identified for the application.

(D) Suspended Ceilings. An enclosure mounted to structural or supporting elements of a suspended ceiling shall be not more than 1650 cm^3 (100 in.^{3}) in size and shall be securely fastened in place in accordance with either (D)(1) or (D)(2).

(1) Framing Members. An enclosure shall be fastened to the framing members by mechanical means such as bolts, screws, or rivets, or by the use of clips or other securing means identified for use with the type of ceiling framing member(s) and enclosure(s) employed. The framing members shall be adequately supported and securely fastened to each other and to the building structure.

(2) Support Wires. The installation shall comply with the provisions of 300.11(A). The enclosure shall be secured, using methods identified for the purpose, to ceiling support wire(s), including any additional support wire(s) installed for that purpose. Support wire(s) used for enclosure support shall be fastened at each end so as to be taut within the ceiling cavity.

(E) Raceway Supported Enclosure, Without Devices, Luminaires (Fixtures), or Lampholders. An enclosure that does not contain a device(s) other than splicing devices or support a luminaire(s) [fixture(s)], lampholder, or other equipment and is supported by entering raceways shall not exceed 1650 cm^3 (100 in.^{3}) in size. It shall have threaded entries or have hubs identified for the purpose. It shall be supported by two or more conduits threaded wrenchtight into the enclosure or hubs. Each conduit shall be secured within 900 mm (3 ft) of the enclosure, or within 450 mm (18 in.) of the enclosure if all conduit entries are on the same side.

Exception: Rigid metal, intermediate metal, or rigid nonmetallic conduit or electrical metallic tubing shall be permitted to support a conduit body of any size, including a conduit body constructed with only one conduit entry, provided the trade size of the conduit body is not larger than the largest trade size of the conduit or electrical metallic tubing.

(F) Raceway Supported Enclosures, with Devices, Luminaires (Fixtures), or Lampholders. An enclosure that contains a device(s) or supports a luminaire(s) [fixture(s)], lampholder, or other equipment and is supported by entering raceways shall not exceed 1650 cm^3 (100 in.^{3}) in size. It shall have threaded entries or have hubs identified for the purpose. It shall be supported by two or more conduits threaded wrenchtight into the enclosure or hubs. Each conduit shall be secured within 450 mm (18 in.) of the enclosure.

Exception No. 1: Rigid metal or intermediate metal conduit shall be permitted to support a conduit body of any size, including a conduit body constructed with only one conduit entry, provided the trade size of the conduit body is not larger than the largest trade size of the conduit.

Exception No. 2: An unbroken length(s) of rigid or intermediate metal conduit shall be permitted to support a box used for luminaire (fixture) or lampholder support, or to support a wiring enclosure that is an integral part of a luminaire (fixture) and used in lieu of a box in accordance with 300.15(B), where all of the following conditions are met.

(a) The conduit is securely fastened at a point so that the length of conduit beyond the last point of conduit support does not exceed 900 mm (3 ft).

(b) The unbroken conduit length before the last point of conduit support is 300 mm (12 in.) or greater, and that portion of the conduit is securely fastened at some point not less than 300 mm (12 in.) from its last point of support.

(c) Where accessible to unqualified persons, the luminaire (fixture) or lampholder, measured to its lowest point, is at least 2.5 m (8 ft) above grade or standing area and at least 900 mm (3 ft) measured horizontally to the 2.5 m

(8 ft) elevation from windows, doors, porches, fire escapes, or similar locations.

(d) A luminaire (fixture) supported by a single conduit does not exceed 300 mm (12 in.) in any direction from the point of conduit entry.

(e) The weight supported by any single conduit does not exceed 9 kg (20 lb).

(f) At the luminaire (fixture) or lampholder end, the conduit(s) is threaded wrenchtight into the box, conduit body, or integral wiring enclosure, or into hubs identified for the purpose. Where a box or conduit body is used for support, the luminaire (fixture) shall be secured directly to the box or conduit body, or through a threaded conduit nipple not over 75 mm (3 in.) long.

(G) Enclosures in Concrete or Masonry. An enclosure supported by embedment shall be identified as suitably protected from corrosion and securely embedded in concrete or masonry.

(H) Pendant Boxes. An enclosure supported by a pendant shall comply with 314.23(H)(1) or (2).

(1) Flexible Cord. A box shall be supported from a multiconductor cord or cable in an approved manner that protects the conductors against strain, such as a strain-relief connector threaded into a box with a hub.

(2) Conduit. A box supporting lampholders or luminaires (lighting fixtures), or wiring enclosures within luminaires (fixtures) used in lieu of boxes in accordance with 300.15(B), shall be supported by rigid or intermediate metal conduit stems. For stems longer than 450 mm (18 in.), the stems shall be connected to the wiring system with flexible fittings suitable for the location. At the luminaire (fixture) end, the conduit(s) shall be threaded wrenchtight into the box or wiring enclosure, or into hubs identified for the purpose.

Where supported by only a single conduit, the threaded joints shall be prevented from loosening by the use of setscrews or other effective means, or the luminaire (fixture), at any point, shall be at least 2.5 m (8 ft) above grade or standing area and at least 900 mm (3 ft) measured horizontally to the 2.5 m (8 ft) elevation from windows, doors, porches, fire escapes, or similar locations. A luminaire (fixture) supported by a single conduit shall not exceed 300 mm (12 in.) in any horizontal direction from the point of conduit entry.

314.24 Depth of Outlet Boxes. No box shall have an internal depth of less than 12.7 mm (1/2 in.). Boxes intended to enclose flush devices shall have an internal depth of not less than 23.8 mm (15/16 in.).

314.25 Covers and Canopies. In completed installations, each box shall have a cover, faceplate, lampholder, or luminaire (fixture) canopy, except where the installation complies with 410.14(B).

(A) Nonmetallic or Metal Covers and Plates. Nonmetallic or metal covers and plates shall be permitted. Where metal covers or plates are used, they shall comply with the grounding requirements of 250.110.

(B) Exposed Combustible Wall or Ceiling Finish. Where a luminaire (fixture) canopy or pan is used, any combustible wall or ceiling finish exposed between the edge of the canopy or pan and the outlet box shall be covered with noncombustible material.

(C) Flexible Cord Pendants. Covers of outlet boxes and conduit bodies having holes through which flexible cord pendants pass shall be provided with bushings designed for the purpose or shall have smooth, well-rounded surfaces on which the cords may bear. So-called hard rubber or composition bushings shall not be used.

314.27 Outlet Boxes.

(A) Boxes at Luminaire (Lighting Fixture) Outlets. Boxes used at luminaire (lighting fixture) or lampholder outlets shall be designed for the purpose. At every outlet used exclusively for lighting, the box shall be designed or installed so that a luminaire (lighting fixture) may be attached.

Exception: A wall-mounted luminaire (fixture) weighing not more than 3 kg (6 lb) shall be permitted to be supported on other boxes or plaster rings that are secured to other boxes, provided the luminaire (fixture) or its supporting yoke is secured to the box with no fewer than two No. 6 or larger screws.

(B) Maximum Luminaire (Fixture) Weight. Outlet boxes or fittings installed as required by 314.23 shall be permitted to support luminaires (lighting fixtures) weighing 23 kg (50 lb) or less. A luminaire (lighting fixture) that weighs more than 23 kg (50 lb) shall be supported independently of the outlet box unless the outlet box is listed for the weight to be supported.

(C) Floor Boxes. Boxes listed specifically for this application shall be used for receptacles located in the floor.

Exception: Where the authority having jurisdiction judges them free from likely exposure to physical damage, moisture, and dirt, boxes located in elevated floors of show windows and similar locations shall be permitted to be other than those listed for floor applications. Receptacles and covers shall be listed as an assembly for this type of location.

(D) Boxes at Ceiling-Suspended (Paddle) Fan Outlets. Where a box is used as the sole support of a ceiling-suspended (paddle) fan, the box shall be listed for the application and for the weight of the fan to be supported. The installation shall comply with 422.18.

314.28 Pull and Junction Boxes and Conduit Bodies. Boxes and conduit bodies used as pull or junction boxes shall comply with 314.28(A) through (D).

(A) Minimum Size. For raceways containing conductors of 4 AWG or larger, and for cables containing conductors of 4 AWG or larger, the minimum dimensions of pull or junction boxes installed in a raceway or cable run shall comply with the following. Where an enclosure dimension is to be calculated based on the diameter of entering raceways, the diameter shall be the metric designator (trade size) expressed in the units of measurement employed.

(1) Straight Pulls. In straight pulls, the length of the box shall not be less than eight times the metric designator (trade size) of the largest raceway.

(2) Angle or U Pulls. Where splices or where angle or U pulls are made, the distance between each raceway entry inside the box and the opposite wall of the box shall not be less than six times the metric designator (trade size) of the largest raceway in a row. This distance shall be increased for additional entries by the amount of the sum of the diameters of all other raceway entries in the same row on the same wall of the box. Each row shall be calculated individually, and the single row that provides the maximum distance shall be used.

Exception: Where a raceway or cable entry is in the wall of a box or conduit body opposite a removable cover, the distance from that wall to the cover shall be permitted to comply with the distance required for one wire per terminal in Table 312.6(A).

The distance between raceway entries enclosing the same conductor shall not be less than six times the metric designator (trade size) of the larger raceway.

When transposing cable size into raceway size in 314.28(A)(1) and (A)(2), the minimum metric designator (trade size) raceway required for the number and size of conductors in the cable shall be used.

(3) Smaller Dimensions. Boxes or conduit bodies of dimensions less than those required in 314.28(A)(1) and (A)(2) shall be permitted for installations of combinations of conductors that are less than the maximum conduit or tubing fill (of conduits or tubing being used) permitted by Table 1 of Chapter 9, provided the box or conduit body has been listed for and is permanently marked with the maximum number and maximum size of conductors permitted.

(B) Conductors in Pull or Junction Boxes. In pull boxes or junction boxes having any dimension over 1.8 m (6 ft), all conductors shall be cabled or racked up in an approved manner.

(C) Covers. All pull boxes, junction boxes, and conduit bodies shall be provided with covers compatible with the box or conduit body construction and suitable for the conditions of use. Where metal covers are used, they shall comply with the grounding requirements of 250.110. An extension from the cover of an exposed box shall comply with 314.22, Exception.

(D) Permanent Barriers. Where permanent barriers are installed in a box, each section shall be considered as a separate box.

314.29 Boxes and Conduit Bodies to Be Accessible. Boxes and conduit bodies shall be installed so that the wiring contained in them can be rendered accessible without removing any part of the building or, in underground circuits, without excavating sidewalks, paving, earth, or other substance that is to be used to establish the finished grade.

ARTICLE 312
Cabinets, Cutout Boxes, and Meter Socket Enclosures

312.1 Scope. This article covers the installation and construction specifications of cabinets, cutout boxes, and meter socket enclosures.

I. INSTALLATION

312.2 Damp, Wet, or Hazardous (Classified) Locations.

(A) Damp and Wet Locations. In damp or wet locations, surface-type enclosures within the scope of this article shall be placed or equipped so as to prevent moisture or water from entering and accumulating within the cabinet or cutout box, and shall be mounted so there is at least 6 mm (1/4 in.) airspace between the enclosure and the wall or other supporting surface. Enclosures installed in wet locations shall be weatherproof.

Exception: Nonmetallic enclosures shall be permitted to be installed without the airspace on a concrete, masonry, tile, or similar surface.

FPN: For protection against corrosion, see 300.6.

312.3 Position in Wall. In walls of concrete, tile, or other noncombustible material, cabinets shall be installed so that the front edge of the cabinet is not set back of the finished surface more than 6 mm (1/4 in.). In walls constructed of wood or other combustible material, cabinets shall be flush with the finished surface or project therefrom.

312.6 Deflection of Conductors. Conductors at terminals or conductors entering or leaving cabinets or cutout boxes and the like shall comply with 312.6(A) through (C).

(A) Width of Wiring Gutters. Conductors shall not be deflected within a cabinet or cutout box unless a gutter having a width in accordance with Table 312.6(A) is provided. Conductors in parallel in accordance with 310.4 shall be judged on the basis of the number of conductors in parallel.

(B) Wire-Bending Space at Terminals. Wire-bending space at each terminal shall be provided in accordance with 312.6(B)(1) or (2).

(1) Conductors Not Entering or Leaving Opposite Wall. Table 312.6(A) shall apply where the conductor does not enter or leave the enclosure through the wall opposite its terminal.

Editor's Note: See 312.6(B)(2) and Table 312.6(B) of the Code for requirements where the conductor does enter or leave the enclosure through the wall opposite its terminal.

312.7 Space in Enclosures. Cabinets and cutout boxes shall have sufficient space to accommodate all conductors installed in them without crowding.

312.8 Enclosures for Switches or Overcurrent Devices. Enclosures for switches or overcurrent devices shall not be used as junction boxes, auxiliary gutters, or raceways for conductors feeding through or tapping off to other switches or overcurrent devices, unless adequate space for this purpose is provided. The conductors shall not fill the wiring space at any cross section to more than 40 percent of the cross-sectional area of the space, and the conductors, splices, and taps shall not fill the wiring space at any cross section to more than 75 percent of the cross-sectional area of that space.

Boxes and Enclosures

Table 312.6(A) Minimum Wire-Bending Space at Terminals and Minimum Width of Wiring Gutters

Wire Size (AWG or kcmil)	1 mm	1 in.	2 mm	2 in.	3 mm	3 in.	4 mm	4 in.	5 mm	5 in.
14–10	Not specified		—	—	—	—	—	—	—	—
8–6	38.1	1-1/2	—	—	—	—	—	—	—	—
4–3	50.8	2	—	—	—	—	—	—	—	—
2	63.5	2-1/2	—	—	—	—	—	—	—	—
1	76.2	3	—	—	—	—	—	—	—	—
1/0–2/0	88.9	3-1/2	—	—	—	—	—	—	—	—
3/0–4/0	102	4	127	5	—	—	—	—	—	—
250	114	4-1/2	152	6	203	8	—	—	—	—
300–350	127	5	152	6	203	8	254	10	—	—
400–500	152	6	203	8	254	10	305	12	—	—
600–700	203	8	203	8	254	10	305	12	356	14
750–900	203	8	254	10	305	12	356	14	406	16
1000–1250	254	10	305	12	356	14	406	16	457	18
1500–2000	305	12	—	—	—	—	—	—	—	—

Note: Bending space at terminals shall be measured in a straight line from the end of the lug or wire connector (in the direction that the wire leaves the terminal) to the wall, barrier, or obstruction.

CHAPTER 12

CABLES

INTRODUCTION

This chapter contains provisions from Article 330—Metal-Clad Cable (Type MC), Article 332—Mineral-Insulated, Metal Sheathed Cable: Type MI, and Article 336—Power and Control Tray Cable: Type TC. Article 330 covers the use and installation of metal-clad cable, Type MC. Part I contains the definition of metal-clad cable, and Part II covers the installation. Some of the topics covered in Part II include: uses permitted, uses not permitted, bending radius, securing and supporting, and fittings.

Articles 332 and 336 similarly cover the use and installation of mineral-insulated, metal sheathed cable, Type MI, and power and control tray cable, Type TC.

ARTICLE 330
Metal-Clad Cable: Type MC

I. GENERAL

330.2 Definition.

Metal Clad Cable, Type MC. A factory assembly of one or more insulated circuit conductors with or without optical fiber members enclosed in an armor of interlocking metal tape, or a smooth or corrugated metallic sheath.

II. INSTALLATION

330.10 Uses Permitted.

(A) General Uses. Where not subject to physical damage, Type MC cables shall be permitted as follows:

(1) For services, feeders, and branch circuits
(2) For power, lighting, control, and signal circuits
(3) Indoors or outdoors
(4) Where exposed or concealed
(5) Direct buried where identified for such use
(6) In cable tray
(7) In any raceway
(8) As open runs of cable
(9) As aerial cable on a messenger
(10) In hazardous (classified) locations as permitted in Articles 501, 502, 503, 504, and 505
(11) In dry locations and embedded in plaster finish on brick or other masonry except in damp or wet locations
(12) In wet locations where any of the following conditions are met:

    a. The metallic covering is impervious to moisture.
    b. A lead sheath or moisture-impervious jacket is provided under the metal covering.
    c. The insulated conductors under the metallic covering are listed for use in wet locations.

(13) Where single-conductor cables are used, all phase conductors and, where used, the neutral conductor

Cables

shall be grouped together to minimize induced voltage on the sheath.

330.12 Uses Not Permitted. Type MC cable shall not be used where exposed to the following destructive corrosive conditions, unless the metallic sheath is suitable for the conditions or is protected by material suitable for the conditions:

(1) Direct burial in the earth
(2) In concrete
(3) Where exposed to cinder fills, strong chlorides, caustic alkalis, or vapors of chlorine or of hydrochloric acids

330.17 Through or Parallel to Framing Members. Type MC cable shall be protected in accordance with 300.4 where installed through or parallel to framing members.

330.24 Bending Radius. Bends in Type MC cable shall be made so that the cable will not be damaged. The radius of the curve of the inner edge of any bend shall not be less than shown in 330.24(A) through (C).

(A) Smooth Sheath.

(1) Ten times the external diameter of the metallic sheath for cable not more than 19 mm (3/4 in.) in external diameter
(2) Twelve times the external diameter of the metallic sheath for cable more than 19 mm (3/4 in.) but not more than 38 mm (1-1/2 in.) in external diameter
(3) Fifteen times the external diameter of the metallic sheath for cable more than 38 mm (1-1/2 in.) in external diameter

(B) Interlocked-Type Armor or Corrugated Sheath. Seven times the external diameter of the metallic sheath.

(C) Shielded Conductors. Twelve times the overall diameter of one of the individual conductors or seven times the overall diameter of the multiconductor cable, whichever is greater.

330.30. Securing and Supporting. Type MC cable shall be supported and secured at intervals not exceeding 1.8 m (6 ft).

(A) Horizontal Runs Through Holes and Notches. In other than vertical runs, cables installed in accordance with 300.4 shall be considered supported and secured where such support does not exceed 1.8-m (6-ft) intervals.

(B) Unsupported Cables. Type MC cable shall be permitted to be unsupported where the cable:

(1) Is fished between access points, where concealed in finished buildings or structures and supporting is impracticable
(2) Is not more than 1.8 m (6 ft) from the last point of support for connections within an accessible ceiling to luminaire(s) [lighting fixture(s)] or equipment

330.40 Boxes and Fitting. Fittings used for connecting Type MC cable to boxes, cabinets, or other equipment shall be listed and identified for such use.

ARTICLE 332
Mineral-Insulated, Metal-Sheathed Cable: Type MI

I. GENERAL

332.2 Definition.

Mineral-Insulated, Metal-Sheathed Cable, Type MI. A factory assembly of one or more conductors insulated with a highly compressed refractory mineral insulation and enclosed in a liquidtight and gastight continuous copper or alloy steel sheath.

II. INSTALLATION

332.10 Uses Permitted. Type MI cable shall be permitted as follows:

(1) For services, feeders, and branch circuits
(2) For power, lighting, control, and signal circuits
(3) In dry, wet, or continuously moist locations

Cables

(4) Indoors or outdoors
(5) Where exposed or concealed
(6) Embedded in plaster, concrete, fill, or other masonry, whether above or below grade
(7) In any hazardous (classified) location
(8) Where exposed to oil and gasoline
(9) Where exposed to corrosive conditions not deteriorating to its sheath
(10) In underground runs where suitably protected against physical damage and corrosive conditions

332.12 Uses Not Permitted. Type MI cable shall not be used where exposed to conditions that are destructive and corrosive to the metallic sheath unless additionally protected by materials suitable for the conditions.

332.17 Through or Parallel to Framing Members. Type MI cable shall be protected in accordance with 300.4 where installed through or parallel to framing members.

332.24 Bending Radius. Bends in Type MI cable shall be made so that the cable will not be damaged. The radius of the inner edge of any bend shall not be less than shown as follows:

(1) Five times the external diameter of the metallic sheath for cable not more than 19 mm (3/4 in.) in external diameter
(2) Ten times the external diameter of the metallic sheath for cable greater than 19 mm (3/4 in.) but not more than 25 mm (1 in.) in external diameter

332.30 Securing and Supporting. Type MI cable shall be supported securely at intervals not exceeding 1.8 m (6 ft) by straps, staples, hangers, or similar fittings designed and installed so as not to damage the cable.

(A) Horizontal Runs Through Holes and Notches. In other than vertical runs, cables installed in accordance with 300.4 shall be considered supported and secured where such support does not exceed 1.8-m (6-ft) intervals.

(B) Unsupported Cable. Type MI cable shall be permitted to be unsupported where the cable is fished.

(C) Cable Trays. Type MI cable installed in cable trays shall comply with 392.8(B).

332.31 Single Conductors. Where single-conductor cables are used, all phase conductors and, where used, the neutral conductor shall be grouped together to minimize induced voltage on the sheath.

332.40 Boxes and Fittings.

(A) Fittings. Fittings used for connecting Type MI cable to boxes, cabinets, or other equipment shall be identified for such use.

(B) Terminal Seals. Where Type MI cable terminates, an end seal fitting shall be installed immediately after stripping to prevent the entrance of moisture into the insulation. The conductors extending beyond the sheath shall be individually provided with an insulating material.

332.80 Ampacity. The ampacity of Type MI cable shall be determined in accordance with 310.15. The conductor temperature at the end seal fitting shall not exceed the temperature rating of the listed end seal fitting, and the installation shall not exceed the temperature ratings of terminations or equipment.

(A) Type MI Cable Installed in Cable Tray. The ampacities for Type MI cable installed in cable tray shall be determined in accordance with 392.11.

(B) Single Type MI Conductors Grouped Together. Where single Type MI conductors are grouped together in a triangular or square configuration, as required by 332.31, and installed on a messenger or as open runs with a maintained free air space of not less than 2.15 times one conductor diameter (2.15 x O.D.) of the largest conductor contained within the configuration and adjacent conductor configurations or cables, the ampacity of the conductors shall not exceed the allowable ampacities of Table 310.17.

Cables

ARTICLE 336
Power and Control Tray Cable: Type TC

I. GENERAL

336.2 Definition.

Power and Control Tray Cable, Type TC. A factory assembly of two or more insulated conductors, with or without associated bare or covered grounding conductors, under a nonmetallic jacket, for installation in cable trays, in raceways, or where supported by a messenger wire.

II. INSTALLATION

336.10 Uses Permitted. Type TC tray cable shall be permitted to be used in the following:

(1) For power, lighting, control, and signal circuits.
(2) In cable trays, or in raceways, or where supported in outdoor locations by a messenger wire.
(3) In cable trays in hazardous (classified) locations as permitted in Articles 392, 501, 502, 504, and 505 in industrial establishments where the conditions of maintenance and supervision ensure that only qualified persons service the installation.
(4) For Class I circuits as permitted in Article 725.
(5) For non—power-limited fire alarm circuits if conductors comply with the requirements of 760.27.
(6) In industrial establishments where the conditions of maintenance and supervision ensure that only qualified persons service the installation, and where the cable is continuously supported and protected against physical damage using mechanical protection, such as struts, angles, or channel, Type TC tray cable that complies with the crush and impact requirements of Type MC cable and is identified for such use shall be permitted between a cable tray and the utilization equipment or device. The cable shall be secured at intervals not exceeding 1.8 m (6 ft). Equipment grounding for the utilization equipment shall be provided by an equipment grounding conductor within the cable.

(7) Where installed in wet locations, Type TC cable shall also be resistant to moisture and corrosive agents.

FPN: See 310.10 for temperature limitation of conductors.

336.12 Uses Not Permitted. Type TC tray cable shall not be used in the following:

(1) Installed where it will be exposed to physical damage
(2) Installed as open cable on brackets or cleats, except as permitted in 340.10(6)
(3) Used where exposed to direct rays of the sun, unless identified as sunlight resistant
(4) Direct buried, unless identified for such use

336.24 Bending Radius. Bends in Type TC cable shall be made so as not to damage the cable. For Type TC cable without metal shielding, the minimum bending radius shall be as follows:

(1) Four times the overall diameter for cables 25 mm (1 in.) or less in diameter
(2) Five times the overall diameter for cables larger than 25 mm (1 in.) but not more than 50 mm (2 in.) in diameter
(3) Six times the overall diameter for cables larger than 50 mm (2 in.) in diameter

Type TC cables with metallic shielding shall have a minimum bending radius of not less than 12 times the cable overall diameter.

336.80 Ampacity. The ampacity of Type TC tray cable shall be determined in accordance with 392.11 for 14 AWG and larger conductors, in accordance with 402.5 for 18 AWG through 16 AWG conductors where installed in cable tray, and in accordance with 310.15 where installed in a raceway or as messenger supported wiring.

CHAPTER 13

RACEWAYS

INTRODUCTION

This chapter contains provisions from Article 344—Rigid Metal Conduit (Type RMC), Article 348—Flexible Metal Conduit (Type FMC), Article 350—Liquidtight Flexible Metal Conduit (Type LFMC), Article 352—Rigid Nonmetallic Conduit (Type RNC), Article 356—Liquidtight Flexible Nonmetallic Conduit (Type LFNC), Article 358—Electrical Metallic Tubing (Type EMT), Article 362—Electrical Nonmetallic Tubing (Type ENT), Article 368—Busways, Article 376—Metal Wireways, Article 378—Nonmetallic Wireways, Article 386—Surface Metal Raceways, and Article 388—Surface Nonmetallic Raceways. Contained in each conduit and tubing article are provisions pertaining to the use and installation for the particular raceway listed. Part I provides a definition of the raceway, and Part II covers its installation provisions. Some of the topics covered in most of these raceway articles include: uses permitted, uses not permitted, bends (how made and number in one run), trimming (or reaming and threading), securing and supporting, and joints (or couplings and connectors). Additional topics covered in Articles 376, 378, 386, and 388 include: size of conductors, number of conductors, and splices and taps.

Although the requirements in each article may be different, the arrangement and numbering in each article follow the same format. For example, in each article 3__.10 covers uses permitted, 3__.12 covers uses not permitted, 3__.24 covers bends—how made, 3__.26 covers bends—number in one run, 3__.28 covers trimming (or reaming and threading), 3__.30 covers securing and supporting, and 3__.42 covers couplings and connectors.

ARTICLE 344
Rigid Metal Conduit: Type RMC

I. GENERAL

344.2 Definition.

Rigid Metal Conduit (RMC). A threadable raceway of circular cross section designed for the physical protection and routing of conductors and cables and for use as an equipment grounding conductor when installed with its integral or associated coupling and appropriate fittings. RMC is generally made of steel (ferrous) with protective coatings or aluminum (nonferrous). Special use types are silicon bronze and stainless steel.

II. INSTALLATION

344.10 Uses Permitted.

(A) All Atmospheric Conditions and Occupancies. Use of RMC shall be permitted under all atmospheric conditions and occupancies. Ferrous raceways and fittings protected from corrosion solely by enamel shall be permitted only indoors and in occupancies not subject to severe corrosive influences.

(B) Corrosion Environments. RMC, elbows, couplings, and fittings shall be permitted to be installed in concrete, in direct contact with the earth, or in areas subject to severe corrosive influences where protected by corrosion protection and judged suitable for the condition.

(C) Cinder Fill. RMC shall be permitted to be installed in or under cinder fill where subject to permanent moisture where protected on all sides by a layer of noncinder concrete not less than 50 mm (2 in.) thick; where the conduit is not less than 450 mm (18 in.) under the fill; or where protected by corrosion protection and judged suitable for the condition.

(D) Wet Locations. All supports, bolts, straps, screws, and so forth, shall be of corrosion-resistant materials or protected against corrosion by corrosion-resistant materials.

Raceways

Table 344.24 Radius of Conduit Bends

Conduit Size		One Shot and Full Shoe Benders		Other Bends	
Metric Designator	Trade Size	mm	in.	mm	in.
16	1/2	101.6	4	101.6	4
21	3/4	114.3	4-1/2	127	5
27	1	146.05	5-3/4	152.4	6
35	1-1/4	184.15	7-1/4	203.2	8
41	1-1/2	209.55	8-1/4	254	10
53	2	241.3	9-1/2	304.8	12
63	2-1/2	266.7	10-1/2	381	15
78	3	330.2	13	457.2	18
91	3-1/2	381	15	533.4	21
103	4	406.4	16	609.6	24
129	5	606.9	24	762	30
155	6	762	30	914.4	36

344.24 Bends—How Made. Bends of RMC shall be made so that the conduit is not damaged and the internal diameter of the conduit is not effectively reduced. The radius of the curve of any field bend to the centerline of the conduit shall not be less than indicated in Table 344.24.

344.26 Bends—Number in One Run. There shall not be more than the equivalent of four quarter bends (360 degrees total) between pull points, for example, conduit bodies and boxes.

344.28 Reaming and Threading. All cut ends shall be reamed or otherwise finished to remove rough edges. Where conduit is threaded in the field, a standard cutting die with a 1 in 16 taper (3/4-in. taper per foot) shall be used.

344.30 Securing and Supporting. RMC shall be installed as a complete system as provided in Article 300 and shall be securely fastened in place and supported in accordance with 344.30(A) and (B).

(A) Securely Fastened. RMC shall be securely fastened within 900 mm (3 ft) of each outlet box, junction box, device box, cabinet, conduit body, or other conduit termination. Fastening shall be permitted to be increased to a distance of 1.5 m (5 ft) where structural members do not readily permit fastening within 900 mm (3 ft). Where approved, conduit shall not be required to be securely fastened within 900 mm (3 ft) of the service head for above-the-roof termination of a mast.

(B) Supports. RMC shall be supported in accordance with one of the following.

(1) Conduit shall be supported at intervals not exceeding 3 m (10 ft).
(2) The distance between supports for straight runs of conduit shall be permitted in accordance with Table 346.30(B)(2), provided the conduit is made up with threaded couplings, and such supports prevent transmission of stresses to termination where conduit is deflected between supports.
(3) Exposed vertical risers from industrial machinery or fixed equipment shall be permitted to be supported at intervals not exceeding 6 m (20 ft), if the conduit is made up with threaded couplings, the conduit is firmly supported at the top and bottom of the riser, and no other means of intermediate support is readily available.
(4) Horizontal runs of RMC supported by openings through framing members at intervals not exceeding 3 m (10 ft) and securely fastened within 900 mm (3 ft) of termination points shall be permitted.

Table 344.30(B)(2) Supports for Rigid Metal Conduit

Conduit Size		Maximum Distance Between Rigid Metal Conduit Supports	
Metric Designator	Trade Size	m	ft
16–21	1/2–3/4	3.0	10
27	1	3.7	12
35–41	1-1/4–1-1/2	4.3	14
53–63	2–2-1/2	4.9	16
78 and larger	3 and larger	6.1	20

344.42 Couplings and Connectors.

(A) Threadless. Threadless couplings and connectors used with conduit shall be made tight. Where buried in masonry or concrete, they shall be the concretetight type. Where installed in wet locations, they shall be the raintight type. Threadless couplings and connectors shall not be used on threaded conduit ends unless listed for the purpose.

(B) Running Threads. Running threads shall not be used on conduit for connection at couplings.

344.46 Bushings.
Where a conduit enters a box, fitting, or other enclosure, a bushing shall be provided to protect the wire from abrasion unless the design of the box, fitting, or enclosure is such as to afford equivalent protection.

344.60 Grounding.
RMC shall be permitted as an equipment grounding conductor.

ARTICLE 348
Flexible Metal Conduit: Type FMC

I. GENERAL

348.2 Definition.

Flexible Metal Conduit (FMC). A raceway of circular cross section made of helically wound, formed, interlocked metal strip.

II. INSTALLATION

348.10 Uses Permitted. FMC shall be permitted to be used in exposed and concealed locations.

348.12 Uses Not Permitted. FMC shall not be used in the following:

(1) In wet locations unless the conductors are approved for the specific conditions and the installation is such that liquid is not likely to enter raceways or enclosures to which the conduit is connected
(2) In hoistways, other than as permitted in 620.21(A)(1)
(3) In storage battery rooms

(4) In any hazardous (classified) location other than as permitted in 501.4(B) and 504.20
(5) Where exposed to materials having a deteriorating effect on the installed conductors, such as oil or gasoline
(6) Underground or embedded in poured concrete or aggregate
(7) Where subject to physical damage

348.26 Bends—Number in One Run. There shall not be more than the equivalent of four quarter bends (360 degrees total) between pull points, for example, conduit bodies and boxes.

348.28 Trimming. All cut ends shall be trimmed or otherwise finished to remove rough edges, except where fittings that thread into the convolutions are used.

348.30 Securing and Supporting. FMC shall be securely fastened in place and supported in accordance with 348.30(A) and (B).

(A) Securely Fastened. FMC shall be securely fastened in place by an approved means within 300 mm (12 in.) of each box, cabinet, conduit body, or other conduit termination and shall be supported and secured at intervals not to exceed 1.4 m (4-1/2 ft).

Exception No. 1: Where FMC is fished.

Exception No. 2: Lengths not exceeding 900 mm (3 ft) at terminals where flexibility is required.

Exception No. 3: Lengths not exceeding 1.8 m (6 ft) from a luminaire (fixture) terminal connection for tap connections to luminaires (light fixtures) as permitted in 410.67(C).

(B) Supports. Horizontal runs of flexible metal conduit FMC supported by openings through framing members at intervals not greater than 1.4 m (4-1/2 ft) and securely fastened within 300 mm (12 in.) of termination points shall be permitted.

348.42 Couplings and Connectors. Angle connectors shall not be used for concealed raceway installations.

348.60 Grounding and Bonding. Where used to connect equipment where flexibility is required, an equipment grounding conductor shall be installed.

Where required or installed, equipment grounding conductors shall be installed in accordance with 250.134(B).

Where required or installed, equipment bonding jumpers shall be installed in accordance with 250.102.

ARTICLE 350
Liquidtight Flexible Metal Conduit: Type LFMC

I. GENERAL

350.2 Definition.

Liquidtight Flexible Metal Conduit (LFMC). A raceway of circular cross section having an outer liquidtight, nonmetallic, sunlight-resistant jacket over an inner flexible metal core with associated couplings, connectors, and fittings for the installation of electric conductors.

II. INSTALLATION

350.10 Uses Permitted. LFMC shall be permitted to be used in exposed or concealed locations as follows:

(1) Where conditions of installation, operation, or maintenance require flexibility or protection from liquids, vapors, or solids
(2) As permitted by 501.4(B), 502.4, 503.3, and 504.20 and in other hazardous (classified) locations where specifically approved, and by 553.7(B)
(3) For direct burial where listed and marked for the purpose

350.12 Uses Not Permitted. LFMC shall not be used as follows:

(1) Where subject to physical damage
(2) Where any combination of ambient and conductor temperature produces an operating temperature in excess of that for which the material is approved

350.26 Bends—Number in One Run. There shall not be more than the equivalent of four quarter bends (360 degrees total) between pull points, for example, conduit bodies and boxes.

350.30 Securing and Supporting. LFMC shall be securely fastened in place and supported in accordance with 350.30(A) and (B).

(A) Securely Fastened. LFMC shall be securely fastened in place by an approved means within 300 mm (12 in.) of each box, cabinet, conduit body, or other conduit termination and shall be supported and secured at intervals not to exceed 1.4 m (4-1/2 ft).

Exception No. 1: Where LFMC is fished.

Exception No. 2: Lengths not exceeding 900 mm (3 ft) at terminals where flexibility is necessary.

Exception No. 3: Lengths not exceeding 1.8 m (6 ft) from a luminaire (fixture) terminal connection for tap conductors to luminaires (lighting fixtures), as permitted in 410.67(C).

(B) Supports. Horizontal runs of LFMC supported by openings through framing members at intervals not greater than 1.4 m (4-1/2 ft) and securely fastened within 300 mm (12 in.) of termination points shall be permitted.

350.42 Couplings and Connectors. Angle connectors shall not be used for concealed raceway installations.

350.60 Grounding and Bonding. Where used to connect equipment where flexibility is required, an equipment grounding conductor shall be installed.

Where required or installed, equipment grounding conductors shall be installed in accordance with 250.134(B).

Where required or installed, equipment bonding jumpers shall be installed in accordance with 250.102.

ARTICLE 352
Rigid Nonmetallic Conduit: Type RNC

I. GENERAL

352.2 Definition.

Rigid Nonmetallic Conduit (RNC). A nonmetallic raceway of circular cross section, with integral or associated couplings, connectors, and fittings for the installation of electrical conductors.

II. INSTALLATION

352.10 Uses Permitted. The use of RNC shall be permitted under the following conditions.

(A) Concealed. In walls, floors, and ceilings.

(B) Corrosive Influences. In locations subject to severe corrosive influences as covered in 300.6 and where subject to chemicals for which the materials are specifically approved.

(C) Cinders. In cinder fill.

(D) Wet Locations. In portions of dairies, laundries, canneries, or other wet locations and in locations where walls are frequently washed, the entire conduit system including boxes and fittings used therewith shall be installed and equipped so as to prevent water from entering the conduit. All supports, bolts, straps, screws, and so forth, shall be of corrosion-resistant materials or be protected against corrosion by approved corrosion-resistant materials.

(E) Dry and Damp Locations. In dry and damp locations not prohibited by 352.12.

(F) Exposed. For exposed work where not subject to physical damage if identified for such use.

(G) Underground Installations. For underground installations, see 300.5 and 300.50. Conduits listed for the purpose shall be permitted to be installed underground in continuous lengths from a reel.

(H) Support of Conduit Bodies. Rigid nonmetallic conduit shall be permitted to support nonmetallic conduit bodies not larger than the largest trade size of an entering raceway. The conduit bodies shall not contain devices or support luminaires (fixtures) or other equipment.

352.12 Uses Not Permitted. RNC shall not be used in the following locations.

(B) Support of Luminaires (Fixtures). For the support of luminaires (fixtures) or other equipment not described in 352.10(H).

(C) Physical Damage. Where subject to physical damage unless identified for such use.

(D) Ambient Temperatures. Where subject to ambient temperatures in excess of 50°C (122°F) unless listed otherwise.

(E) Insulation Temperature Limitations. For conductors whose insulation temperature limitations would exceed those for which the conduit is listed.

(F) Theaters and Similar Locations. In theaters and similar locations, except as provided in Articles 518 and 520.

352.24 Bends—How Made. Bends shall be made so that the conduit will not be damaged and the internal diameter of the conduit will not be effectively reduced. Field bends shall be made only with bending equipment identified for the purpose. The radius of the curve to the centerline of such bends shall not be less than shown in Table 344.24, column "Other Bends."

352.26 Bends—Number in One Run. There shall not be more than the equivalent of four quarter bends (360 degrees total) between pull points, for example, conduit bodies and boxes.

352.28 Trimming. All cut ends shall be trimmed inside and outside to remove rough edges.

352.30 Securing and Supporting. RNC shall be installed as a complete system as provided in 300.18 and shall be fastened so that movement from thermal expansion or contraction is permitted. RNC shall be securely fastened and supported in accordance with 352.30(A) and (B).

(A) Securely Fastened. RNC shall be securely fastened within 900 mm (3 ft) of each outlet box, junction box, device box, conduit body, or other conduit termination. Conduit listed for securing at other than 900 mm (3 ft) shall be permitted to be installed in accordance with the listing.

(B) Supports. RNC shall be supported as required in Table 352.30(B). Conduit listed for support at spacings other than as shown in Table 352.30(B) shall be permitted to be installed in accordance with the listing. Horizontal runs of RNC supported by openings through framing members at intervals not exceeding those in Table 352.30(B) and securely fastened within 900 mm (3 ft) of termination points shall be permitted.

Raceways

Table 352.30(B) Support of Rigid Nonmetallic Conduit (RNC)

Conduit Size		Maximum Spacing Between Supports	
Metric Designator	Trade Size	mm or m	ft
16–27	1/2–1	900 mm	3
35–53	1-1/4–2	1.5 m	5
63–78	2-1/2–3	1.8 m	6
91–129	3-1/2–5	2.1 m	7
155	6	2.5 m	8

352.44 Expansion Fittings. Expansion fittings for RNC shall be provided to compensate for thermal expansion and contraction where the length change, in accordance with Table 352.44(A) or (B), is expected to be 6 mm (1/4 in.) or greater in a straight run between securely mounted items such as boxes, cabinets, elbows, or other conduit terminations.

352.46 Bushings. Where a conduit enters a box, fitting, or other enclosure, a bushing or adapter shall be provided to protect the wire from abrasion unless the box, fitting, or enclosure design provides equivalent protection.

352.48 Joints. All joints between lengths of conduit, and between conduit and couplings, fittings, and boxes, shall be made by an approved method.

ARTICLE 356
Liquidtight Flexible Nonmetallic Conduit: Type LFNC

I. GENERAL

356.2 Definition.

Liquidtight Flexible Nonmetallic Conduit (LFNC). A raceway of circular cross section of various types as follows:

(1) A smooth seamless inner core and cover bonded together and having one or more reinforcement layers

Table 352.44(A) Expansion Characteristics of PVC Rigid Nonmetallic Conduit Coefficient of Thermal Expansion = 6.084 × 10⁻⁵ mm/mm/°C (3.38 × 10⁻⁵ in./in./°F)

Temperature Change (°C)	Length Change of PVC Conduit (mm/m)	Temperature Change (°F)	Length Change of PVC Conduit (in./100 ft)	Temperature Change (°F)	Length Change of PVC Conduit (in./100 ft)
5	0.30	5	0.20	105	4.26
10	0.61	10	0.41	110	4.46
15	0.91	15	0.61	115	4.66
20	1.22	20	0.81	120	4.87
25	1.52	25	1.01	125	5.07
30	1.83	30	1.22	130	5.27
35	2.13	35	1.42	135	5.48
40	2.43	40	1.62	140	5.68
45	2.74	45	1.83	145	5.88
50	3.04	50	2.03	150	6.08
55	3.35	55	2.23	155	6.29
60	3.65	60	2.43	160	6.49
65	3.95	65	2.64	165	6.69

70	4.26	70	2.84	170	6.90
75	4.56	75	3.04	175	7.10
80	4.87	80	3.24	180	7.30
85	5.17	85	3.45	185	7.50
90	5.48	90	3.65	190	7.71
95	5.78	95	3.85	195	7.91
100	6.08	100	4.06	200	8.11

Table 352.44(B) Expansion Characteristics of Reinforced Thermosetting Resin Conduit (RTRC) Coefficient of Thermal Expansion = 2.7×10^{-5} mm/mm/°C (1.5×10^{-5} in./in./°F)

Temperature Change (°C)	Length Change of RTRC Conduit (mm/m)	Temperature Change (°F)	Length Change of RTRC Conduit (in./100 ft)	Temperature Change (°F)	Length Change of RTRC Conduit (in./100 ft)
5	0.14	5	0.09	105	1.89
10	0.27	10	0.18	110	1.98
15	0.41	15	0.27	115	2.07
20	0.54	20	0.36	120	2.16
25	0.68	25	0.45	125	2.25
30	0.81	30	0.54	130	2.34
35	0.95	35	0.63	135	2.43
40	1.08	40	0.72	140	2.52
45	1.22	45	0.81	145	2.61
50	1.35	50	0.90	150	2.70
55	1.49	55	0.99	155	2.79
60	1.62	60	1.08	160	2.88

Raceways

65	1.76	65	1.17	165	2.97
70	1.89	70	1.26	170	3.06
75	2.03	75	1.35	175	3.15
80	2.16	80	1.44	180	3.24
85	2.30	85	1.53	185	3.33
90	2.43	90	1.62	190	3.42
95	2.57	95	1.71	195	3.51
100	2.70	100	1.80	200	3.60

between the core and covers, designated as Type LFNC-A

(2) A smooth inner surface with integral reinforcement within the conduit wall, designated as Type LFNC-B
(3) A corrugated internal and external surface without integral reinforcement within the conduit wall, designated as LFNC-C.

LFNC is flame resistant and with fittings and is approved for the installation of electrical conductors.

FPN: FNMC is an alternative designation for LFNC.

II. INSTALLATION

356.10 Uses Permitted. LFNC shall be permitted to be used in exposed or concealed locations for the following purposes:

(1) Where flexibility is required for installation, operation, or maintenance
(2) Where protection of the contained conductors is required from vapors, liquids, or solids
(3) For outdoor locations where listed and marked as suitable for the purpose
(4) For direct burial where listed and marked for the purpose
(5) Type LFNC-B shall be permitted to be installed in lengths longer than 1.8 m (6 ft) where secured in accordance with 356.30
(6) Type LFNC-B as a listed manufactured prewired assembly, metric designator 16 through 27 (trade size 1/2 through 1) conduit

356.12 Uses Not Permitted. LFNC shall not be used as follows:

(1) Where subject to physical damage
(2) Where any combination of ambient and conductor temperatures is in excess of that for which the LFNC is approved
(3) In lengths longer than 1.8 m (6 ft), except as permitted by 356.10(5) or where a longer length is approved as essential for a required degree of flexibility

(4) Where voltage of the contained conductors is in excess of 600 volts, nominal

356.26 Bends—Number in One Run. There shall not be more than the equivalent of four quarter bends (360 degrees total) between pull points, for example, conduit bodies and boxes.

356.28 Trimming. All cut ends of conduit shall be trimmed inside and outside to remove rough edges.

356.30 Securing and Supporting. Type LFNC-B shall be securely fastened and supported in accordance with one of the following:

(1) The conduit shall be securely fastened at intervals not exceeding 900 mm (3 ft) and within 300 mm (12 in.) on each side of every outlet box, junction box, cabinet, or fitting.
(2) Securing or supporting of the conduit shall not be required where it is fished, installed in lengths not exceeding 900 mm (3 ft) at terminals where flexibility is required, or installed in lengths not exceeding 1.8 m (6 ft) from a luminaire (fixture) terminal connection for tap conductors to luminaires (lighting fixtures) permitted in 410.67(C).
(3) Horizontal runs of LFNC supported by openings through framing members at intervals not exceeding 900 mm (3 ft) and securely fastened within 300 mm (12 in.) of termination points shall be permitted.

356.42 Couplings and Connectors. Angle connectors shall not be used for concealed raceway installations.

ARTICLE 358
Electrical Metallic Tubing: Type EMT

I. GENERAL

358.2 Definition.

Electrical Metallic Tubing (EMT). An unthreaded thinwall raceway of circular cross section designed for the physical protection and routing of conductors and cables and for use as an equipment grounding conductor when installed utilizing appropriate fittings. EMT is generally made of steel (ferrous) with protective coatings or aluminum (nonferrous).

II. INSTALLATION

358.10 Uses Permitted.

(A) Exposed and Concealed. The use of EMT shall be permitted for both exposed and concealed work.

(B) Corrosion Protection. Ferrous or nonferrous EMT, elbows, couplings, and fittings shall be permitted to be installed in concrete, in direct contact with the earth, or in areas subject to severe corrosive influences where protected by corrosion protection and judged suitable for the condition.

(C) Wet Locations. All supports, bolts, straps, screws, and so forth shall be of corrosion-resistant materials or protected against corrosion by corrosion-resistant materials.

358.12 Uses Not Permitted. EMT shall not be used under the following conditions:

(1) Where, during installation or afterward, it will be subject to severe physical damage
(2) Where protected from corrosion solely by enamel
(3) In cinder concrete or cinder fill where subject to permanent moisture unless protected on all sides by a layer of noncinder concrete at least 50 mm (2 in.) thick or unless the tubing is at least 450 mm (18 in.) under the fill
(4) In any hazardous (classified) location except as permitted by 502.4, 503.3, and 504.20
(5) For the support of luminaires (fixtures) or other equipment except conduit bodies no larger than the largest trade size of the tubing

358.24 Bends—How Made. Bends shall be made so that the tubing is not damaged and the internal diameter of the tubing is not effectively reduced. The radius of the curve of any field bend to the centerline of the conduit shall not be

less than shown in Table 344.24 for one-shot and full shoe benders.

358.26 Bends—Number in One Run. There shall not be more than the equivalent of four quarter bends (360 degrees total) between pull points, for example, conduit bodies and boxes.

358.28 Reaming and Threading.

(A) Reaming. All cut ends of EMT shall be reamed or otherwise finished to remove rough edges.

358.30 Securing and Supporting. EMT shall be installed as a complete system as provided in Article 300 and shall be securely fastened in place and supported in accordance with 358.30(A) and (B).

(A) Securely Fastened. EMT shall be securely fastened in place at least every 3 m (10 ft). In addition, each EMT run between termination points shall be securely fastened within 900 mm (3 ft) of each outlet box, junction box, device box, cabinet, conduit body, or other tubing termination.

Exception No. 1: Fastening of unbroken lengths shall be permitted to be increased to a distance of 1.5 m (5 ft) where structural members do not readily permit fastening within 900 mm (3 ft).

Exception No. 2: For concealed work in finished buildings or prefinished wall panels where such securing is impracticable, unbroken lengths (without coupling) of EMT shall be permitted to be fished.

(B) Supports. Horizontal runs of EMT supported by openings through framing members at intervals not greater than 3 m (10 ft) and securely fastened within 900 mm (3 ft) of termination points shall be permitted.

358.42 Couplings and Connectors. Couplings and connectors used with EMT shall be made up tight. Where buried in masonry or concrete, they shall be concretetight type. Where installed in wet locations, they shall be of the raintight type.

358.60 Grounding. EMT shall be permitted as an equipment grounding conductor.

ARTICLE 362
Electrical Nonmetallic Tubing: Type ENT

I. GENERAL

362.2 Definition.

Electrical Nonmetallic Tubing (ENT). A nonmetallic pliable corrugated raceway of circular cross section with integral or associated couplings, connectors, and fittings for the installation of electric conductors. ENT is composed of a material that is resistant to moisture and chemical atmospheres and is flame retardant.

A pliable raceway is a raceway that can be bent by hand with a reasonable force, but without other assistance.

II. INSTALLATION

362.10 Uses Permitted. For the purpose of this article, the first floor of a building shall be that floor that has 50 percent or more of the exterior wall surface area level with or above finished grade. One additional level that is the first level and not designed for human habitation and used only for vehicle parking, storage, or similar use shall be permitted. The use of ENT and fittings shall be permitted in the following:

(1) In any building not exceeding three floors above grade

　a. For exposed work, where not prohibited by 362.12
　b. Concealed within walls, floors, and ceilings

(2) In any building exceeding three floors above grade, ENT shall be concealed within walls, floors, and ceilings where the walls, floors, and ceilings provide a thermal barrier of material that has at least a 15-minute finish rating as identified in listings of fire-rated assemblies. The 15-minute-finish-rated thermal barrier shall be permitted to be used for combustible or noncombustible walls, floors, and ceilings.

Raceways

Exception: Where a fire sprinkler system(s) is installed in accordance with NFPA 13-1999, Standard for the Installation of Sprinkler Systems, on all floors, ENT is permitted to be used within walls, floors, and ceilings, exposed or concealed, in buildings exceeding three floors above grade.

(3) In locations subject to severe corrosive influences as covered in 300.6 and where subject to chemicals for which the materials are specifically approved.
(4) In concealed, dry, and damp locations not prohibited by 362.12.
(5) Above suspended ceilings where the suspended ceilings provide a thermal barrier of material that has at least a 15-minute finish rating as identified in listings of fire-rated assemblies, except as permitted in 362.10(1)(a).

Exception: Where a fire sprinkler system(s) is installed in accordance with NFPA 13-1999, Standard for the Installation of Sprinkler Systems, on all floors, ENT is permitted to be used within walls, floors, and ceilings, exposed or concealed, in buildings exceeding three floors above grade.

(6) Encased in poured concrete, or embedded in a concrete slab on grade where ENT is placed on sand or approved screenings, provided fittings identified for this purpose are used for connections.
(7) For wet locations indoors as permitted in this section or in a concrete slab on or below grade, with fittings listed for the purpose.
(8) Metric designator 16 through 27 (trade size 1/2 through 1) as listed manufactured prewired assembly.

362.12 Uses Not Permitted. ENT shall not be used in the following:

(1) In hazardous (classified) locations, except as permitted by 504.20 and 505.15(A)(1)
(2) For the support of luminaires (fixtures) and other equipment
(3) Where subject to ambient temperatures in excess of 50°C (122°F) unless listed otherwise

(4) For conductors whose insulation temperature limitations would exceed those for which the tubing is listed
(5) For direct earth burial
(6) Where the voltage is over 600 volts
(7) In exposed locations, except as permitted by 362.10(1), 362.10(5), and 362.10(7)
(8) In theaters and similar locations, except as provided in Articles 518 and 520
(9) Where exposed to the direct rays of the sun, unless identified as sunlight resistant
(10) Where subject to physical damage

362.24 Bends—How Made. Bends shall be made so that the tubing will not be damaged and that the internal diameter of the tubing will not be effectively reduced. Bends shall be permitted to be made manually without auxiliary equipment, and the radius of the curve to the centerline of such bends shall not be less than shown in Table 344.24 using the column "Other Bends."

362.26 Bends—Number in One Run. There shall not be more than the equivalent of four quarter bends (360 degrees total) between pull points, for example, conduit bodies and boxes.

362.28 Trimming. All cut ends shall be trimmed inside and outside to remove rough edges.

362.30 Securing and Supporting. ENT shall be installed as a complete system as provided in Article 300 and shall be securely fastened in place and supported in accordance with 362.30(A) and (B).

(A) Securely Fastened. ENT shall be securely fastened at intervals not exceeding 900 mm (3 ft). In addition, ENT shall be securely fastened in place within 900 mm (3 ft) of each outlet box, device box, junction box, cabinet, or fitting where it terminates.

Exception: Lengths not exceeding a distance of 1.8 m (6 ft) from a luminaire (fixture) terminal connection for tap connections to lighting luminaires (fixtures) shall be permitted without being secured.

(B) Supports. Horizontal runs of ENT supported by openings in framing members at intervals not exceeding 900 mm (3 ft) and securely fastened within 900 mm (3 ft) of termination points shall be permitted.

362.46 Bushings. Where a tubing enters a box, fitting, or other enclosure, a bushing or adapter shall be provided to protect the wire from abrasion unless the box, fitting, or enclosure design provides equivalent protection.

362.48 Joints. All joints between lengths of tubing and between tubing and couplings, fittings, and boxes shall be by an approved method.

ARTICLE 368
Busways

I. GENERAL REQUIREMENTS

368.2 Definition.

Busway. A grounded metal enclosure containing factory-mounted, bare or insulated conductors, which are usually copper or aluminum bars, rods, or tubes.

FPN: For cablebus, refer to Article 370.

368.4 Use.

(A) Uses Permitted. Busways shall be permitted to be installed where they are located as follows:

(1) Located in the open and are visible, except as permitted in 368.6, or
(2) Installed behind access panels, provided the busways are totally enclosed, of nonventilating-type construction, and installed so that the joints between sections and at fittings are accessible for maintenance purposes. Where installed behind access panels, means of access shall be provided, and the following conditions shall be met:

  a. The space behind the access panels shall not be used for air-handling purposes, or

b. Where the space behind the access panels is used for environmental air, other than ducts and plenums, there shall be no provisions for plug-in connections, and the conductors shall be insulated.

(B) Uses Not Permitted. Busways shall not be installed as follows:

(1) Where subject to severe physical damage or corrosive vapors
(2) In hoistways
(3) In any hazardous (classified) location, unless specifically approved for such use
(4) Outdoors or in wet or damp locations unless identified for such use

Lighting busway and trolley busway shall not be installed less than 2.5 m (8 ft) above the floor or working platform unless provided with a cover identified for the purpose.

368.5 Support. Busways shall be securely supported at intervals not exceeding 1.5 m (5 ft) unless otherwise designed and marked.

368.6 Through Walls and Floors.

(A) Walls. Unbroken lengths of busway shall be permitted to be extended through dry walls.

(B) Floors. Floor penetrations shall comply with (1) and (2):

(1) Busways shall be permitted to be extended vertically through dry floors if totally enclosed (unventilated) where passing through and for a minimum distance of 1.8 m (6 ft) above the floor to provide adequate protection from physical damage.
(2) In other than industrial establishments, where a vertical riser penetrates two or more dry floors, a minimum 100 mm (4 in.) high curb shall be installed around all floor openings for riser busways to prevent liquids from entering the opening. The curb shall be installed within 300 mm (12 in.) of the floor opening. Electrical equipment shall be located so that it will not be damaged by liquids that are retained by the curb.

368.7 Dead Ends. A dead end of a busway shall be closed.

368.8 Branches from Busways. Branches from busways shall be permitted to be made in accordance with 368.8(A), (B), and (C).

(A) General. Branches from busways shall be made in accordance with Articles 320, 330, 332, 342, 344, 348, 350, 352, 356, 358, 362, 368, 384, 386, and 388. Where a separate equipment grounding conductor is used, connection of the equipment grounding conductor to the busway shall comply with 250.8 and 250.12.

(B) Cord and Cable Assemblies. Suitable cord and cable assemblies approved for extra-hard usage or hard usage and listed bus drop cable shall be permitted as branches from busways for the connection of portable equipment or the connection of stationary equipment to facilitate their interchange in accordance with 400.7 and 400.8 and the following conditions:

(1) The cord or cable shall be attached to the building by an approved means.
(2) The length of the cord or cable from a busway plug-in device to a suitable tension take-up support device shall not exceed 1.8 m (6 ft).

Exception: In industrial establishments only, where the conditions of maintenance and supervision ensure that only qualified persons service the installation, lengths exceeding 1.8 m (6 ft) shall be permitted between the busway plug-in device and the tension take-up support device where the cord or cable is supported at intervals not exceeding 2.5 m (8 ft).

(3) The cord or cable shall be installed as a vertical riser from the tension take-up support device to the equipment served.
(4) Strain relief cable grips shall be provided for the cord or cable at the busway plug-in device and equipment terminations.

(C) Branches from Trolley-Type Busways. Suitable cord and cable assemblies approved for extra-hard usage or hard usage and listed bus drop cable shall be permitted as branches from trolley-type busways for the connection of movable equipment in accordance with 400.7 and 400.8.

368.10 Rating of Overcurrent Protection—Feeders. A busway shall be protected against overcurrent in accordance with the allowable current rating of the busway.

Exception No. 1: The applicable provisions of 240.4 shall be permitted.

Exception No. 2: Where used as transformer secondary ties, the provisions of 450.6(A)(3) shall be permitted.

368.11 Reduction in Ampacity Size of Busway. Overcurrent protection shall be required where busways are reduced in ampacity.

Exception: For industrial establishments only, omission of overcurrent protection shall be permitted at points where busways are reduced in ampacity, provided that the length of the busway having the smaller ampacity does not exceed 15 m (50 ft) and has an ampacity at least equal to one-third the rating or setting of the overcurrent device next back on the line, and provided that such busway is free from contact with combustible material.

368.12 Feeder or Branch Circuits. Where a busway is used as a feeder, devices or plug-in connections for tapping off feeder or branch circuits from the busway shall contain the overcurrent devices required for the protection of the feeder or branch circuits. The plug-in device shall consist of an externally operable circuit breaker or an externally operable fusible switch. Where such devices are mounted out of reach and contain disconnecting means, suitable means such as ropes, chains, or sticks shall be provided for operating the disconnecting means from the floor.

Exception No. 1: As permitted in 240.21.

Exception No. 2: For fixed or semifixed luminaires (lighting fixtures), where the branch-circuit overcurrent device is part of the luminaire (fixture) cord plug on cord-connected luminaires (fixtures).

Exception No. 3: Where luminaires (fixtures) without cords are plugged directly into the busway and the overcurrent device is mounted on the luminaire (fixture).

368.13 Rating of Overcurrent Protection—Branch Circuits. A busway used as a branch circuit shall be protected against overcurrent in accordance with 210.20. Where so used, the circuit shall comply with the applicable requirements of Articles 210, 430, and 440.

ARTICLE 376
Metal Wireways

I. GENERAL

376.2 Definition.

Metal Wireways. Sheet metal troughs with hinged or removable covers for housing and protecting electric wires and cable and in which conductors are laid in place after the wireway has been installed as a complete system.

II. INSTALLATION

376.10 Uses Permitted. The use of metal wireways shall be permitted in the following:

(1) For exposed work
(2) In concealed spaces as permitted in 376.10(4)
(3) In hazardous (classified) locations as permitted by 501.4(B) for Class I, Division 2 locations; 502.4(B) for Class II, Division 2 locations; and 504.20 for intrinsically safe wiring. Where installed in wet locations, wireways shall be listed for the purpose.
(4) As extensions to pass transversely through walls if the length passing through the wall is unbroken. Access to the conductors shall be maintained on both sides of the wall.

376.12 Uses Not Permitted. Metal wireways shall not be used in the following:

(1) Where subject to severe physical damage
(2) Where subject to severe corrosive environments

376.22 Number of Conductors. The sum of the cross-sectional areas of all contained conductors at any cross section of a wireway shall not exceed 20 percent of the interior cross-sectional area of the wireway. The derating factors in 310.15(B)(2)(a) shall be applied only where the number of current-carrying conductors, including neutral conductors classified as current-carrying under the provisions of 310.15(B)(4), exceeds 30. Conductors for signaling circuits or controller conductors between a motor and its starter and used only for starting duty shall not be considered as current-carrying conductors.

376.23 Insulated Conductors. Insulated conductors installed in a metallic wireway shall comply with 376.23(A) and (B).

(A) Deflected Insulated Conductors. Where insulated conductors are deflected within a metallic wireway, either at the ends or where conduits, fittings, or other raceways or cables enter or leave the metallic wireway, or where the direction of the metallic wireway is deflected greater than 30 degrees, dimensions corresponding to 312.6(A) shall apply.

(B) Metallic Wireways Used as Pullboxes. Where insulated conductors 4 AWG or larger are pulled through a wireway, the distance between raceway and cable entries enclosing the same conductor shall not be less than that required in 314.28(A)(1) for straight pulls and 314.28(A)(2) for angle pulls.

376.30 Securing and Supporting. Metal wireways shall be supported in accordance with 376.30(A) and (B).

(A) Horizontal Support. Wireways shall be supported where run horizontally at each end and at intervals not to exceed 1.5 m (5 ft) or for individual lengths longer than 1.5 m (5 ft) at each end or joint, unless listed for other support intervals. The distance between supports shall not exceed 3 m (10 ft).

(B) Vertical Support. Vertical runs of wireways shall be securely supported at intervals not exceeding 4.5 m (15 ft) and shall not have more than one joint between supports. Adjoining wireway sections shall be securely fastened together to provide a rigid joint.

376.56 Splices and Taps. Splices and taps shall be permitted within a wireway provided they are accessible. The conductors, including splices and taps, shall not fill the wireway to more than 75 percent of its area at that point.

376.58 Dead Ends. Dead ends of metal wireways shall be closed.

376.70 Extensions from Metal Wireways. Extensions from wireways shall be made with cord pendants installed in accordance with 400.10 or any wiring method in Chapter 3 that includes a means for equipment grounding. Where a separate equipment grounding conductor is employed, connection of the equipment grounding conductors in the wiring method to the wireway shall comply with 250.8 and 250.12.

ARTICLE 378
Nonmetallic Wireways

I. GENERAL

378.2 Definition.

Nonmetallic Wireways. Flame retardant, nonmetallic troughs with removable covers for housing and protecting electric wires and cables in which conductors are laid in place after the wireway has been installed as a complete system.

II. INSTALLATION

378.10 Uses Permitted. The use of nonmetallic wireways shall be permitted in the following:

(1) Only for exposed work, except as permitted in 378.10(4).
(2) Where subject to corrosive environments where identified for the use.
(3) In wet locations where listed for the purpose.
(4) As extensions to pass transversely through walls if the length passing through the wall is unbroken. Access to the conductors shall be maintained on both sides of the wall.

378.12 Uses Not Permitted. Nonmetallic wireways shall not be used in the following:

(1) Where subject to physical damage
(2) In any hazardous (classified) location, except as permitted in 504.20
(3) Where exposed to sunlight unless listed and marked as suitable for the purpose

(4) Where subject to ambient temperatures other than those for which nonmetallic wireway is listed
(5) For conductors whose insulation temperature limitations would exceed those for which the nonmetallic wireway is listed

378.22 Number of Conductors. The sum of cross-sectional areas of all contained conductors at any cross section of the nonmetallic wireway shall not exceed 20 percent of the interior cross-sectional area of the nonmetallic wireway. Conductors for signaling circuits or controller conductors between a motor and its starter and used only for starting duty shall not be considered as current-carrying conductors.

The derating factors specified in 310.15(B)(2)(a) shall be applicable to the current-carrying conductors up to and including the 20 percent fill specified above.

378.23 Insulated Conductors. Insulated conductors installed in a nonmetallic wireway shall comply with 378.23(A) and (B).

(A) Deflected Insulated Conductors. Where insulated conductors are deflected within a nonmetallic wireway, either at the ends or where conduits, fittings, or other raceways or cables enter or leave the nonmetallic wireway, or where the direction of the nonmetallic wireway is deflected greater than 30 degrees, dimensions corresponding to 312.6(A) shall apply.

(B) Nonmetallic Wireways Used as Pull Boxes. Where insulated conductors 4 AWG or larger are pulled through a wireway, the distance between raceway and cable entries enclosing the same conductor shall not be less than that required in 314.28(A)(1) for straight pulls and in 314.28(A)(2) for angle pulls.

378.30 Securing and Supporting. Nonmetallic wireway shall be supported in accordance with 378.30(A) and (B).

(A) Horizontal Support. Nonmetallic wireways shall be supported where run horizontally at intervals not to exceed 900 mm (3 ft), and at each end or joint, unless listed for other support intervals. In no case shall the distance between supports exceed 3 m (10 ft).

(B) Vertical Support. Vertical runs of nonmetallic wireway shall be securely supported at intervals not exceeding 1.2 m (4 ft), unless listed for other support intervals, and shall not have more than one joint between supports. Adjoining nonmetallic wireway sections shall be securely fastened together to provide a rigid joint.

378.44 Expansion Fittings. Expansion fittings for nonmetallic wireway shall be provided to compensate for thermal expansion and contraction where the length change is expected to be 6 mm (0.25 in.) or greater in a straight run.

378.56 Splices and Taps. Splices and taps shall be permitted within a nonmetallic wireway, provided they are accessible. The conductors, including splices and taps, shall not fill the nonmetallic wireway to more than 75 percent of its area at that point.

378.58 Dead Ends. Dead ends of nonmetallic wireway shall be closed using listed fittings.

378.70 Extensions from Nonmetallic Wireways. Extensions from nonmetallic wireway shall be made with cord pendants or any wiring method of Chapter 3. A separate equipment grounding conductor shall be installed in, or an equipment grounding connection shall be made to, any of the wiring methods used for the extension.

ARTICLE 386
Surface Metal Raceways

I. GENERAL

386.2 Definition.

Surface Metal Raceway. A metallic raceway that is intended to be mounted to the surface of a structure, with associated couplings, connectors, boxes, and fittings for the installation of electrical conductors.

II. INSTALLATION

386.10 Uses Permitted. The use of surface metal raceways shall be permitted in the following:

(1) In dry locations.
(2) In Class I, Division 2 hazardous (classified) locations as permitted in 501.4(B)(3).
(3) Under raised floors, as permitted in 645.5(D)(2).
(4) Extension through walls and floors. Surface metal raceway shall be permitted to pass transversely through dry walls, dry partitions, and dry floors if the length passing through is unbroken. Access to the conductors shall be maintained on both sides of the wall, partition, or floor.

386.12 Uses Not Permitted. Surface metal raceways shall not be used in the following:

(1) Where subject to severe physical damage, unless otherwise approved
(2) Where the voltage is 300 volts or more between conductors, unless the metal has a thickness of not less than 1.02 mm (0.040 in.) nominal
(3) Where subject to corrosive vapors
(4) In hoistways
(5) Where concealed, except as permitted in 386.10(4)

386.21 Size of Conductors. No conductor larger than that for which the raceway is designed shall be installed in surface metal raceway.

386.22 Number of Conductors or Cables. The number of conductors or cables installed in surface metal raceway shall not be greater than the number for which the raceway is designed. Cables shall be permitted to be installed where such use is permitted by the respective cable articles.

The derating factors of 310.15(B)(2)(a) shall not apply to conductors installed in surface metal raceways where all of the following conditions are met:

(1) The cross-sectional area of the raceway exceeds 2500 mm^2 (4 in.^{2})
(2) The current-carrying conductors do not exceed 30 in number
(3) The sum of the cross-sectional areas of all contained conductors does not exceed 20 percent of the interior cross-sectional area of the surface metal raceway

386.56 Splices and Taps. Splices and taps shall be permitted in surface metal raceways having a removable cover that is accessible after installation. The conductors, including splices and taps, shall not fill the raceway to more than 75 percent of its area at that point. Splices and taps in surface metal raceways without removable covers shall be made only in junction boxes. All splices and taps shall be made by approved methods.

Taps of Type FC cable installed in surface metal raceway shall be made in accordance with 322.56(B).

386.60 Grounding. Surface metal raceway enclosures providing a transition from other wiring methods shall have a means for connecting an equipment grounding conductor.

386.70 Combination Raceways. When combination surface metal raceways are used both for signaling and for lighting and power circuits, the different systems shall be run in separate compartments identified by sharply contrasting colors of the interior finish, and the same relative position of compartments shall be maintained throughout the premises.

ARTICLE 388
Surface Nonmetallic Raceways

I. GENERAL

388.2 Definition.

Surface Nonmetallic Raceway. A nonmetallic raceway that is intended to be mounted to the surface of a structure, with associated couplings, connectors, boxes, and fittings for the installation of electrical conductors.

II. INSTALLATION

388.10 Uses Permitted. Surface nonmetallic raceway shall be permitted as follows:

(1) The use of surface nonmetallic raceways shall be permitted in dry locations.
(2) Extension through walls and floors shall be permitted. Surface nonmetallic raceway shall be permitted to pass transversely through dry walls, dry partitions, and dry floors if the length passing through is unbroken.

Access to the conductors shall be maintained on both sides of the wall, partition, or floor.

388.12 Uses Not Permitted. Surface nonmetallic raceways shall not be used in the following:

(1) Where concealed, except as permitted in 388.10(2)
(2) Where subject to severe physical damage
(3) Where the voltage is 300 volts or more between conductors, unless listed for higher voltage
(4) In hoistways
(5) In any hazardous (classified) location except Class I, Division 2 locations as permitted in 501.4(B)(3)
(6) Where subject to ambient temperatures exceeding those for which the nonmetallic raceway is listed
(7) For conductors whose insulation temperature limitations would exceed those for which the nonmetallic raceway is listed

388.21 Size of Conductors. No conductor larger than that for which the raceway is designed shall be installed in surface nonmetallic raceway.

388.22 Number of Conductors or Cables. The number of conductors or cables installed in surface nonmetallic raceway shall not be greater than the number for which the raceway is designed. Cables shall be permitted to be installed where such use is permitted by the respective cable articles.

388.56 Splices and Taps. Splices and taps shall be permitted in surface nonmetallic raceways having a removable cover that is accessible after installation. The conductors, including splices and taps, shall not fill the raceway to more than 75 percent of its area at that point. Splices and taps in surface nonmetallic raceways without removable covers shall be made only in junction boxes. All splices and taps shall be made by approved methods.

388.60 Grounding. Where equipment grounding is required by Article 250, a separate equipment grounding conductor shall be installed in the raceway.

388.70 Combination Raceways. When combination surface nonmetallic raceways are used both for signaling and for lighting and power circuits, the different systems shall be run in separate compartments identified by sharply contrasting colors of the interior finish.

CHAPTER 14

WIRING METHODS AND CONDUCTORS

INTRODUCTION

This chapter contains provisions from Articles 300 and 310. Article 300 covers wiring methods for all wiring installations unless modified by other articles and provides general installation requirements for both cables and raceways. Some of the topics include: conductors; protection against physical damage; underground installations; raceways exposed to different temperatures; securing and supporting; boxes, conduit bodies, or fittings; supporting conductors in vertical raceways; and induced currents in metal enclosures or metal raceways.

Article 310 covers general requirements for conductors, including their ampacity ratings. For many commercial and industrial installations, the allowable ampacities for insulated conductors specified in Table 310.16 are used. Under certain conditions of use, the allowable ampacity of each conductor must be reduced.

ARTICLE 300
Wiring Methods

I. GENERAL REQUIREMENTS

300.1 Scope.

(A) All Wiring Installations. This article covers wiring methods for all wiring installations unless modified by other articles.

(C) Metric Designators and Trade Sizes. Metric designators and trade sizes for conduit, tubing, and associated fitting and accessories shall be as designated in Table 300.1(C).

300.3 Conductors.

(B) Conductors of the Same Circuit. All conductors of the same circuit and, where used, the grounded conductor and all equipment grounding conductors and bonding conductors shall be contained within the same raceway, auxiliary gutter, cable tray, cablebus assembly, trench, cable, or cord, unless otherwise permitted in accordance with 300.3(B)(1) through (4).

(1) Paralleled Installations. Conductors shall be permitted to be run in parallel in accordance with the provisions of

Table 300.1(C) Metric Designator and Trade Sizes

Metric Designator	Trade Size	Metric Designator	Trade Size
12	3/8	63	2-1/2
16	1/2	78	3
21	3/4	91	3-1/2
27	1	103	4
35	1-1/4	129	5
41	1-1/2	155	6
53	2		

Note: The metric designators and trade sizes are for identification purposes only and are not actual dimensions.

Wiring Methods and Conductors

310.4. The requirement to run all circuit conductors within the same raceway, auxiliary gutter, cable tray, trench, cable, or cord shall apply separately to each portion of the paralleled installation, and the equipment grounding conductors shall comply with the provisions of 250.122. Parallel runs in cable tray shall comply with the provisions of 392.8(D).

(C) Conductors of Different Systems.

(1) 600 Volts, Nominal, or Less. Conductors of circuits rated 600 volts, nominal, or less, ac circuits, and dc circuits shall be permitted to occupy the same equipment wiring enclosure, cable, or raceway. All conductors shall have an insulation rating equal to at least the maximum circuit voltage applied to any conductor within the enclosure, cable, or raceway.

300.4 Protection Against Physical Damage. Where subject to physical damage, conductors shall be adequately protected.

(A) Cables and Raceways Through Wood Members.

(1) Bored Holes. In both exposed and concealed locations, where a cable- or raceway-type wiring method is installed through bored holes in joists, rafters, or wood members, holes shall be bored so that the edge of the hole is not less than 32 mm (1-1/4 in.) from the nearest edge of the wood member. Where this distance cannot be maintained, the cable or raceway shall be protected from penetration by screws or nails by a steel plate or bushing, at least 1.6 mm (1/16 in.) thick, and of appropriate length and width installed to cover the area of the wiring.

Exception: Steel plates shall not be required to protect rigid metal conduit, intermediate metal conduit, rigid nonmetallic conduit, or electrical metallic tubing.

(2) Notches in Wood. Where there is no objection because of weakening the building structure, in both exposed and concealed locations, cables or raceways shall be permitted to be laid in notches in wood studs, joists, rafters, or other wood members where the cable or raceway at those points is protected against nails or screws by a steel plate at least 1.6 mm (1/16 in.) thick installed before the building finish is applied.

Exception: Steel plates shall not be required to protect rigid metal conduit, intermediate metal conduit, rigid nonmetallic conduit, or electrical metallic tubing.

(B) Nonmetallic-Sheathed Cables and Electrical Nonmetallic Tubing Through Metal Framing Members.

(1) Nonmetallic-Sheathed Cable. In both exposed and concealed locations where nonmetallic-sheathed cables pass through either factory or field punched, cut, or drilled slots or holes in metal members, the cable shall be protected by listed bushings or listed grommets covering all metal edges that are securely fastened in the opening prior to installation of the cable.

(2) Nonmetallic-Sheathed Cable and Electrical Nonmetallic Tubing. Where nails or screws are likely to penetrate nonmetallic-sheathed cable or electrical nonmetallic tubing, a steel sleeve, steel plate, or steel clip not less than 1.6 mm (1/16 in.) in thickness shall be used to protect the cable or tubing.

(C) Cables Through Spaces Behind Panels Designed to Allow Access. Cables or raceway-type wiring methods, installed behind panels designed to allow access, shall be supported according to their applicable articles.

(D) Cables and Raceways Parallel to Framing Members. In both exposed and concealed locations, where a cable- or raceway-type wiring method is installed parallel to framing members, such as joists, rafters, or studs, the cable or raceway shall be installed and supported so that the nearest outside surface of the cable or raceway is not less than 32 mm (1-1/4 in.) from the nearest edge of the framing member where nails or screws are likely to penetrate. Where this distance cannot be maintained, the cable or raceway shall be protected from penetration by nails or screws by a steel plate, sleeve, or equivalent at least 1.6 mm (1/16 in.) thick.

Exception No. 1: Steel plates, sleeves, or the equivalent shall not be required to protect rigid metal conduit, inter-

mediate metal conduit, rigid nonmetallic conduit, or electrical metallic tubing.

Exception No. 2: For concealed work in finished buildings, or finished panels for prefabricated buildings where such supporting is impracticable, it shall be permissible to fish the cables between access points.

(E) Cables and Raceways Installed in Shallow Grooves. Cable- or raceway-type wiring methods installed in a groove, to be covered by wallboard, siding, paneling, carpeting, or similar finish, shall be protected by 1.6 mm (1/16 in.) thick steel plate, sleeve, or equivalent or by not less than 32 mm (1-1/4 in.) free space for the full length of the groove in which the cable or raceway is installed.

Exception: Steel plates, sleeves, or the equivalent shall not be required to protect rigid metal conduit, intermediate metal conduit, rigid nonmetallic conduit, or electrical metallic tubing.

(F) Insulated Fittings. Where raceways containing ungrounded conductors 4 AWG or larger enter a cabinet, box enclosure, or raceway, the conductors shall be protected by a substantial fitting providing a smoothly rounded insulating surface, unless the conductors are separated from the fitting or raceway by substantial insulating material that is securely fastened in place.

Exception: Where threaded hubs or bosses that are an integral part of a cabinet, box enclosure, or raceway provide a smoothly rounded or flared entry for conductors.

Conduit bushings constructed wholly of insulating material shall not be used to secure a fitting or raceway. The insulating fitting or insulating material shall have a temperature rating not less than the insulation temperature rating of the installed conductors.

300.5 Underground Installations.

(A) Minimum Cover Requirements. Direct-buried cable or conduit or other raceways shall be installed to meet the minimum cover requirements of Table 300.5.

Table 300.5 Minimum Cover Requirements, 0 to 600 Volts, Nominal, Burial in Millimeters (Inches)

	Type of Wiring Method or Circuit									
	Column 1 Direct Burial Cables or Conductors		Column 2 Rigid Metal Conduit or Intermediate Metal Conduit		Column 3 Nonmetallic Raceways Listed for Direct Burial Without Concrete Encasement or Other Approved Raceways		Column 4 Residential Branch Circuits Rated 120 Volts or Less with GFCI Protection and Maximum Overcurrent Protection of 20 Amperes		Column 5 Circuits for Control of Irrigation and Landscape Lighting Limited to Not More than 30 Volts and Installed with Type UF or in Other Identified Cable or Raceway	
Location of Wiring Method or Circuit	mm	in.	mm	in.	mm	in.	mm	in.	mm	in.
All locations not specified below	600	24	150	6	450	18	300	12	150	6
In trench below 50-mm (2-in.) thick concrete or equivalent	450	18	150	6	300	12	50	6	50	6
Under a building	0 (in raceway only)	0	0	0	0	0	0 (in raceway only)	0	0 (in raceway only)	0
Under minimum of 102-mm (4-in.) thick concrete exterior slab with no vehicular traffic and the slab extending not less than 152 mm (6 in.) beyond the underground installation	450	18	100	4	100	4	150 (direct burial) 100 (in raceway)	6 4	150 (direct burial)	6

Under streets, highways, roads, alleys, driveways, and parking lots	600	24	600	24	600	24	600	24	600	24
One- and two-family dwelling driveways and outdoor parking areas, and used only for dwelling-related purposes	450	18	450	18	450	18	300	12	450	18
In or under airport runways, including adjacent areas where trespassing prohibited	450	18	450	18	450	18	450	18	450	18

Notes:
1. Cover is defined as the shortest distance in millimeters (inches) measured between a point on the top surface of any direct-buried conductor, cable, conduit, or other raceway and the top surface of finished grade, concrete, or similar cover.
2. Raceways approved for burial only where concrete encased shall require concrete envelope not less than 50 mm (2 in.) thick.
3. Lesser depths shall be permitted where cables and conductors rise for terminations or splices or where access is otherwise required.
4. Where one of the wiring method types listed in Columns 1–3 is used for one of the circuit types in Columns 4 and 5, the shallower depth of burial shall be permitted.
5. Where solid rock prevents compliance with the cover depths specified in this table, the wiring shall be installed in metal or nonmetallic raceway permitted for direct burial. The raceways shall be covered by a minimum of 50 mm (2 in.) of concrete extending down to rock.

(B) Grounding. All underground installations shall be grounded and bonded in accordance with Article 250.

(C) Underground Cables Under Buildings. Underground cable installed under a building shall be in a raceway that is extended beyond the outside walls of the building.

(D) Protection from Damage. Direct-buried conductors and cables shall be protected from damage in accordance with (1) through (5).

(1) Emerging from Grade. Direct-buried conductors and enclosures emerging from grade shall be protected by enclosures or raceways extending from the minimum cover distance required by 300.5(A) below grade to a point at least 2.5 m (8 ft) above finished grade. In no case shall the protection be required to exceed 450 mm (18 in.) below finished grade.

(2) Conductors Entering Buildings. Conductors entering a building shall be protected to the point of entrance.

(3) Service Conductors. Underground service conductors that are not encased in concrete and that are buried 450 mm (18 in.) or more below grade shall have their location identified by a warning ribbon that is placed in the trench at least 300 mm (12 in.) above the underground installation.

(4) Enclosure or Raceway Damage. Where the enclosure or raceway is subject to physical damage, the conductors shall be installed in rigid metal conduit, intermediate metal conduit, Schedule 80 rigid nonmetallic conduit, or equivalent.

(5) Listing. Cables and insulated conductors installed in enclosures or raceways in underground installations shall be listed for use in wet locations.

(E) Splices and Taps. Direct-buried conductors or cables shall be permitted to be spliced or tapped without the use of splice boxes. The splices or taps shall be made in accordance with 110.14(B).

(F) Backfill. Backfill that contains large rocks, paving materials, cinders, large or sharply angular substances, or corrosive material shall not be placed in an excavation where materials may damage raceways, cables, or other substructures or prevent adequate compaction of fill or

Wiring Methods and Conductors

contribute to corrosion of raceways, cables, or other substructures.

Where necessary to prevent physical damage to the raceway or cable, protection shall be provided in the form of granular or selected material, suitable running boards, suitable sleeves, or other approved means.

(G) Raceway Seals. Conduits or raceways through which moisture may contact energized live parts shall be sealed or plugged at either or both ends.

(H) Bushing. A bushing, or terminal fitting, with an integral bushed opening shall be used at the end of a conduit or other raceway that terminates underground where the conductors or cables emerge as a direct burial wiring method. A seal incorporating the physical protection characteristics of a bushing shall be permitted to be used in lieu of a bushing.

(I) Conductors of the Same Circuit. All conductors of the same circuit and, where used, the grounded conductor and all equipment grounding conductors shall be installed in the same raceway or cable or shall be installed in close proximity in the same trench.

Exception No. 1: Conductors in parallel in raceways or cables shall be permitted, but each raceway or cable shall contain all conductors of the same circuit including grounding conductors.

Exception No. 2: Isolated phase, polarity, grounded conductor, and equipment grounding and bonding conductor installations shall be permitted in nonmetallic raceways or cables with a nonmetallic covering or nonmagnetic sheath in close proximity where conductors are paralleled as permitted in 310.4, and where the conditions of 300.20(B) are met.

(J) Ground Movement. Where direct-buried conductors, raceways, or cables are subject to movement by settlement or frost, direct-buried conductors, raceways, or cables shall be arranged to prevent damage to the enclosed conductors or to equipment connected to the raceways.

(K) Directional Boring. Cables or raceways installed using directional boring equipment shall be approved for the purpose.

300.6 Protection Against Corrosion. Metal raceways, cable trays, cablebus, auxiliary gutters, cable armor, boxes, cable sheathing, cabinets, elbows, couplings, fittings, supports, and support hardware shall be of materials suitable for the environment in which they are to be installed.

(A) General. Ferrous raceways, cable trays, cablebus, auxiliary gutters, cable armor, boxes, cable sheathing, cabinets, metal elbows, couplings, fittings, supports, and support hardware shall be suitably protected against corrosion inside and outside (except threads at joints) by a coating of approved corrosion-resistant material such as zinc, cadmium, or enamel. Where protected from corrosion solely by enamel, they shall not be used outdoors or in wet locations as described in 300.6(C). Where boxes or cabinets have an approved system of organic coatings and are marked "Raintight," "Rainproof," or "Outdoor Type," they shall be permitted outdoors. Where corrosion protection is necessary and the conduit is threaded in the field, the threads shall be coated with an approved electrically conductive, corrosion-resistant compound.

(B) In Concrete or in Direct Contact with the Earth. Ferrous or nonferrous metal raceways, cable armor, boxes, cable sheathing, cabinets, elbows, couplings, fittings, supports, and support hardware shall be permitted to be installed in concrete or in direct contact with the earth, or in areas subject to severe corrosive influences where made of material judged suitable for the condition, or where provided with corrosion protection approved for the condition.

(C) Indoor Wet Locations. In portions of dairy processing facilities, laundries, canneries, and other indoor wet locations, and in locations where walls are frequently washed or where there are surfaces of absorbent materials, such as damp paper or wood, the entire wiring system, where installed exposed, including all boxes, fittings, conduits, and cable used therewith, shall be mounted so that there is at least a 6-mm (1/4-in.) airspace between it and the wall or supporting surface.

Exception: Nonmetallic raceways, boxes, and fittings shall be permitted to be installed without the airspace on a concrete, masonry, tile, or similar surface.

Wiring Methods and Conductors

300.7 Raceways Exposed to Different Temperatures.

(A) Sealing. Where portions of a cable, raceway, or sleeve are known to be subjected to different temperatures and where condensation is known to be a problem, as in cold storage areas of buildings or where passing from the interior to the exterior of a building, the raceway or sleeve shall be filled with an approved material to prevent the circulation of warm air to a colder section of the raceway or sleeve. An explosionproof seal shall not be required for this purpose.

(B) Expansion Fittings. Raceways shall be provided with expansion fittings where necessary to compensate for thermal expansion and contraction.

300.8 Installation of Conductors with Other Systems.

Raceways or cable trays containing electric conductors shall not contain any pipe, tube, or equal for steam, water, air, gas, drainage, or any service other than electrical.

300.10 Electrical Continuity of Metal Raceways and Enclosures.

Metal raceways, cable armor, and other metal enclosures for conductors shall be metallically joined together into a continuous electric conductor and shall be connected to all boxes, fittings, and cabinets so as to provide effective electrical continuity. Unless specifically permitted elsewhere in this *Code*, raceways and cable assemblies shall be mechanically secured to boxes, fittings, cabinets, and other enclosures.

Exception No. 1: Short sections of raceways used to provide support or protection of cable assemblies from physical damage shall not be required to be made electrically continuous.

Exception No. 2: Equipment enclosures to be isolated, as permitted by 250.96(B), shall not be required to be metallically joined to the metal raceway.

300.11 Securing and Supporting.

(A) Secured in Place. Raceways, cable assemblies, boxes, cabinets, and fittings shall be securely fastened in place. Support wires that do not provide secure support shall not be permitted as the sole support. Support wires and associated fittings that provide secure support and that are installed in addition to the ceiling grid support wires shall be permitted as the sole support. Where independent support wires are used, they shall be secured at both ends. Cables and raceways shall not be supported by ceiling grids.

(1) Fire-Rated Assemblies. Wiring located within the cavity of a fire-rated floor–ceiling or roof–ceiling assembly shall not be secured to, or supported by, the ceiling assembly, including the ceiling support wires. An independent means of secure support shall be provided. Where independent support wires are used, they shall be distinguishable by color, tagging, or other effective means from those that are part of the fire-rated design.

Exception: The ceiling support system shall be permitted to support wiring and equipment that have been tested as part of the fire-rated assembly.

(2) Non–Fire-Rated Assemblies. Wiring located within the cavity of a non–fire-rated floor–ceiling or roof–ceiling assembly shall not be secured to, or supported by, the ceiling assembly, including the ceiling support wires. An independent means of secure support shall be provided.

Exception: The ceiling support system shall be permitted to support branch-circuit wiring and associated equipment where installed in accordance with the ceiling system manufacturer's instructions.

(B) Raceways Used as Means of Support. Raceways shall only be used as a means of support for other raceways, cables, or nonelectric equipment under the following conditions:

(1) Where the raceway or means of support is identified for the purpose; or
(2) Where the raceway contains power supply conductors for electrically controlled equipment and is used to support Class 2 circuit conductors or cables that are solely for the purpose of connection to the equipment control circuits; or
(3) Where the raceway is used to support boxes or conduit bodies in accordance with 314.23 or to support luminaires (fixtures) in accordance with 410.16(F)

Wiring Methods and Conductors

(C) Cables Not Used as Means of Support. Cable wiring methods shall not be used as a means of support for other cables, raceways, or nonelectrical equipment.

300.12 Mechanical Continuity—Raceways and Cables. Metal or nonmetallic raceways, cable armors, and cable sheaths shall be continuous between cabinets, boxes, fittings, or other enclosures or outlets.

Exception: Short sections of raceways used to provide support or protection of cable assemblies from physical damage shall not be required to be mechanically continuous.

300.13 Mechanical and Electrical Continuity—Conductors.

(A) General. Conductors in raceways shall be continuous between outlets, boxes, devices, and so forth. There shall be no splice or tap within a raceway unless permitted by 300.15; 368.8(A); 376.56; 378.56; 384.56; 386.56; 388.56; or 390.6.

(B) Device Removal. In multiwire branch circuits, the continuity of a grounded conductor shall not depend on device connections such as lampholders, receptacles, and so forth, where the removal of such devices would interrupt the continuity.

300.14 Length of Free Conductors at Outlets, Junctions, and Switch Points. At least 150 mm (6 in.) of free conductor, measured from the point in the box where it emerges from its raceway or cable sheath, shall be left at each outlet, junction, and switch point for splices or the connection of luminaires (fixtures) or devices. Where the opening to an outlet, junction, or switch point is less than 200 mm (8 in.) in any dimension, each conductor shall be long enough to extend at least 75 mm (3 in.) outside the opening.

Exception: Conductors that are not spliced or terminated at the outlet, junction, or switch point shall not be required to comply with 300.14.

300.15 Boxes, Conduit Bodies, or Fittings—Where Required. A box shall be installed at each outlet and switch point for concealed knob-and-tube wiring.

Fittings and connectors shall be used only with the specific wiring methods for which they are designed and listed.

Where the wiring method is conduit, tubing, Type AC cable, Type MC cable, Type MI cable, nonmetallic-sheathed cable, or other cables, a box or conduit body complying with Article 314 shall be installed at each conductor splice point, outlet point, switch point, junction point, termination point, or pull point, unless otherwise permitted in 300.15(A) through (M).

(A) Wiring Methods with Interior Access. A box or conduit body shall not be required for each splice, junction, switch, pull, termination, or outlet points in wiring methods with removable covers, such as wireways, multioutlet assemblies, auxiliary gutters, and surface raceways. The covers shall be accessible after installation.

(B) Equipment. An integral junction box or wiring compartment as part of approved equipment shall be permitted in lieu of a box.

(C) Protection. A box or conduit body shall not be required where cables enter or exit from conduit or tubing that is used to provide cable support or protection against physical damage. A fitting shall be provided on the end(s) of the conduit or tubing to protect the cable from abrasion.

(F) Fitting. A fitting identified for the use shall be permitted in lieu of a box or conduit body where conductors are not spliced or terminated within the fitting. The fitting shall be accessible after installation.

(G) Direct-Buried Conductors. As permitted in 300.5(E), a box or conduit body shall not be required for splices and taps in direct-buried conductors and cables.

(H) Insulated Devices. As permitted in 334.40(B), a box or conduit body shall not be required for insulated devices supplied by nonmetallic-sheathed cable.

(I) Enclosures. A box or conduit body shall not be required where a splice, switch, terminal, or pull point is in a cabinet or cutout box, in an enclosure for a switch or over-current device as permitted in 312.8, in a motor controller as permitted in 430.10(A), or in a motor control center.

Wiring Methods and Conductors

(J) Luminaires (Fixtures). A box or conduit body shall not be required where a luminaire (fixture) is used as a raceway as permitted in 410.31 and 410.32.

(K) Embedded. A box or conduit body shall not be required for splices where conductors are embedded as permitted in 424.40, 424.41(D), 426.22(B), 426.24(A), and 427.19(A).

300.16 Raceway or Cable to Open or Concealed Wiring.

(A) Box or Fitting. A box or terminal fitting having a separately bushed hole for each conductor shall be used wherever a change is made from conduit, electrical metallic tubing, electrical nonmetallic tubing, nonmetallic-sheathed cable, Type AC cable, Type MC cable, or mineral-insulated, metal-sheathed cable and surface raceway wiring to open wiring or to concealed knob-and-tube wiring. A fitting used for this purpose shall contain no taps or splices and shall not be used at luminaire (fixture) outlets.

(B) Bushing. A bushing shall be permitted in lieu of a box or terminal where the conductors emerge from a raceway and enter or terminate at equipment, such as open switchboards, unenclosed control equipment, or similar equipment. The bushing shall be of the insulating type for other than lead-sheathed conductors.

300.18 Raceway Installations.

(A) Complete Runs. Raceways, other than busways or exposed raceways having hinged or removable covers, shall be installed complete between outlet, junction, or splicing points prior to the installation of conductors. Where required to facilitate the installation of utilization equipment, the raceway shall be permitted to be initially installed without a terminating connection at the equipment. Prewired raceway assemblies shall be permitted only where specifically permitted in this *Code* for the applicable wiring method.

300.19 Supporting Conductors in Vertical Raceways.

(A) Spacing Intervals—Maximum. Conductors in vertical raceways shall be supported if the vertical rise exceeds the values in Table 300.19(A). One cable support shall be provided at the top of the vertical raceway or as close to the top as practical. Intermediate supports shall be provided as necessary to limit supported conductor lengths to not greater than those values specified in Table 300.19(A).

Exception: Steel wire armor cable shall be supported at the top of the riser with a cable support that clamps the steel wire armor. A safety device shall be permitted at the lower end of the riser to hold the cable in the event there is slippage of the cable in the wire-armored cable support. Additional wedge-type supports shall be permitted to relieve the strain on the equipment terminals caused by expansion of the cable under load.

(B) Support Methods. One of the following methods of support shall be used.

(1) By clamping devices constructed of or employing insulating wedges inserted in the ends of the raceways. Where clamping of insulation does not adequately support the cable, the conductor also shall be clamped.

(2) By inserting boxes at the required intervals in which insulating supports are installed and secured in a satisfactory manner to withstand the weight of the conductors attached thereto, the boxes being provided with covers.

(3) In junction boxes, by deflecting the cables not less than 90 degrees and carrying them horizontally to a distance not less than twice the diameter of the cable, the cables being carried on two or more insulating supports and additionally secured thereto by tie wires if desired. Where this method is used, cables shall be supported at intervals not greater than 20 percent of those mentioned in the preceding tabulation.

300.20 Induced Currents in Metal Enclosures or Metal Raceways.

(A) Conductors Grouped Together. Where conductors carrying alternating current are installed in metal enclosures or metal raceways, they shall be arranged so as to avoid heating the surrounding metal by induction. To accomplish this, all phase conductors and, where used, the grounded conductor and all equipment grounding conductors shall be grouped together.

Wiring Methods and Conductors

Table 300.19(A) Spacings for Conductor Supports

Size of Wire	Support of Conductors in Vertical Raceways	Aluminum or Copper-Clad Aluminum		Copper	
		m	ft	m	ft
18 AWG through 8 AWG	Not greater than	30	100	30	100
6 AWG through 1/0 AWG	Not greater than	60	200	30	100
2/0 AWG through 4/0 AWG	Not greater than	55	180	25	80
Over 4/0 AWG through 350 kcmil	Not greater than	41	135	18	60
Over 350 kcmil through 500 kcmil	Not greater than	36	120	15	50
Over 500 kcmil through 750 kcmil	Not greater than	28	95	12	40
Over 750 kcmil	Not greater than	26	85	11	35

Exception No. 1: Equipment grounding conductors for certain existing installations shall be permitted to be installed separate from their associated circuit conductors where run in accordance with the provisions of 250.130(C).

Exception No. 2: A single conductor shall be permitted to be installed in a ferromagnetic enclosure and used for skin-effect heating in accordance with the provisions of 426.42 and 427.47.

(B) Individual Conductors. Where a single conductor carrying alternating current passes through metal with magnetic properties, the inductive effect shall be minimized by (1) cutting slots in the metal between the individual holes through which the individual conductors pass or (2) passing all the conductors in the circuit through an insulating wall sufficiently large for all of the conductors of the circuit.

Exception: In the case of circuits supplying vacuum or electric-discharge lighting systems or signs or X-ray apparatus, the currents carried by the conductors are so small that the inductive heating effect can be ignored where these conductors are placed in metal enclosures or pass through metal.

> FPN: Because aluminum is not a magnetic metal, there will be no heating due to hysteresis; however, induced currents will be present. They will not be of sufficient magnitude to require grouping of conductors or special treatment in passing conductors through aluminum wall sections.

300.21 Spread of Fire or Products of Combustion. Electrical installations in hollow spaces, vertical shafts, and ventilation or air-handling ducts shall be made so that the possible spread of fire or products of combustion will not be substantially increased. Openings around electrical penetrations through fire-resistant-rated walls, partitions, floors, or ceilings shall be firestopped using approved methods to maintain the fire resistance rating.

300.22 Wiring in Ducts, Plenums, and Other Air-Handling Spaces. The provisions of this section apply to the

Wiring Methods and Conductors

installation and uses of electric wiring and equipment in ducts, plenums, and other air-handling spaces.

FPN: See Article 424, Part VI, for duct heaters.

(A) Ducts for Dust, Loose Stock, or Vapor Removal. No wiring systems of any type shall be installed in ducts used to transport dust, loose stock, or flammable vapors. No wiring system of any type shall be installed in any duct, or shaft containing only such ducts, used for vapor removal or for ventilation of commercial-type cooking equipment.

(B) Ducts or Plenums Used for Environmental Air. Only wiring methods consisting of Type MI cable, Type MC cable employing a smooth or corrugated impervious metal sheath without an overall nonmetallic covering, electrical metallic tubing, flexible metallic tubing, intermediate metal conduit, or rigid metal conduit without an overall nonmetallic covering shall be installed in ducts or plenums specifically fabricated to transport environmental air. Flexible metal conduit and liquidtight flexible metal conduit shall be permitted, in lengths not to exceed 1.2 m (4 ft), to connect physically adjustable equipment and devices permitted to be in these ducts and plenum chambers. The connectors used with flexible metal conduit shall effectively close any openings in the connection. Equipment and devices shall be permitted within such ducts or plenum chambers only if necessary for their direct action upon, or sensing of, the contained air. Where equipment or devices are installed and illumination is necessary to facilitate maintenance and repair, enclosed gasketed-type luminaires (fixtures) shall be permitted.

(C) Other Space Used for Environmental Air. This section applies to space used for environmental air-handling purposes other than ducts and plenums as specified in 300.22(A) and (B). It does not include habitable rooms or areas of buildings, the prime purpose of which is not air handling.

ARTICLE 310
Conductors for General Wiring

310.1 Scope. This article covers general requirements for conductors and their type designations, insulations, markings, mechanical strengths, ampacity ratings, and uses. These requirements do not apply to conductors that form an integral part of equipment, such as motors, motor controllers, and similar equipment, or to conductors specifically provided for elsewhere in this *Code*.

310.2 Conductors.

(A) Insulated. Conductors shall be insulated.

Exception: Where covered or bare conductors are specifically permitted elsewhere in this Code.

(B) Conductor Material. Conductors in this article shall be of aluminum, copper-clad aluminum, or copper unless otherwise specified.

310.3 Stranded Conductors. Where installed in raceways, conductors of size 8 AWG and larger shall be stranded.

310.4 Conductors in Parallel. Aluminum, copper-clad aluminum, or copper conductors of size 1/0 AWG and larger, comprising each phase, neutral, or grounded circuit conductor, shall be permitted to be connected in parallel (electrically joined at both ends to form a single conductor).

The paralleled conductors in each phase, neutral, or grounded circuit conductor shall

(1) Be the same length
(2) Have the same conductor material
(3) Be the same size in circular mil area
(4) Have the same insulation type
(5) Be terminated in the same manner

Where run in separate raceways or cables, the raceways or cables shall have the same physical characteristics. Conductors of one phase, neutral, or grounded circuit conductor shall not be required to have the same physical characteristics as those of another phase, neutral, or grounded circuit conductor to achieve balance.

Wiring Methods and Conductors

Where equipment grounding conductors are used with conductors in parallel, they shall comply with the requirements of this section except that they shall be sized in accordance with 250.122.

Conductors installed in parallel shall comply with the provisions of 310.15(B)(2)(a).

310.8 Locations.

(A) Dry Locations. Insulated conductors and cables used in dry locations shall be any of the types identified in this *Code*.

(B) Dry and Damp Locations. Insulated conductors and cables used in dry and damp locations shall be Types FEP, FEPB, MTW, PFA, RHH, RHW, RHW-2, SA, THHN, THW, THW-2, THHW, THHW-2, THWN, THWN-2, TW, XHH, XHHW, XHHW-2, Z, or ZW.

(C) Wet Locations. Insulated conductors and cables used in wet locations shall be

(1) Moisture-impervious metal-sheathed;
(2) Types MTW, RHW, RHW-2, TW, THW, THW-2, THHW, THHW-2, THWN, THWN-2, XHHW, XHHW-2, ZW; or
(3) Of a type listed for use in wet locations.

(D) Locations Exposed to Direct Sunlight. Insulated conductors and cables used where exposed to direct rays of the sun shall be of a type listed for sunlight resistance or listed and marked "sunlight resistant."

310.9 Corrosive Conditions.
Conductors exposed to oils, greases, vapors, gases, fumes, liquids, or other substances having a deleterious effect on the conductor or insulation shall be of a type suitable for the application.

310.15 Ampacities for Conductors Rated 0–2000 Volts.

(B) Tables. Ampacities for conductors rated 0 to 2000 volts shall be as specified in the Allowable Ampacity Tables 310.16 through 310.19 and Ampacity Tables 310.20 through 310.23 as modified by (1) through (6).

FPN: Tables 310.16 through 310.19 are application tables for use in determining conductor sizes on loads calculated in accordance with Article 220. Allowable ampacities result from consideration of one or more of the following:

(1) Temperature compatibility with connected equipment, especially the connection points.
(2) Coordination with circuit and system overcurrent protection.
(3) Compliance with the requirements of product listings or certifications. See 110.3(B).
(4) Preservation of the safety benefits of established industry practices and standardized procedures.

(2) Adjustment Factors.

(a) More Than Three Current-Carrying Conductors in a Raceway or Cable. Where the number of current-carrying conductors in a raceway or cable exceeds three, or where single conductors or multiconductor cables are stacked or bundled longer than 600 mm (24 in.) without maintaining spacing and are not installed in raceways, the allowable ampacity of each conductor shall be reduced as shown in Table 310.15(B)(2)(a).

Exception No. 1: Where conductors of different systems, as provided in 300.3, are installed in a common raceway or cable, the derating factors shown in Table 310.15(B)(2)(a)

Table 310.15(B)(2)(a) Adjustment Factors for More Than Three Current-Carrying Conductors in a Raceway or Cable

Number of Current-Carrying Conductors	Percent of Values in Tables 310.16 through 310.19 as Adjusted for Ambient Temperature if Necessary
4–6	80
7–9	70
10–20	50
21–30	45
31–40	40
41 and above	35

shall apply to the number of power and lighting conductors only (Articles 210, 215, 220, and 230).

Exception No. 2: For conductors installed in cable trays, the provisions of 392.11 shall apply.

Exception No. 3: Derating factors shall not apply to conductors in nipples having a length not exceeding 600 mm (24 in.).

Exception No. 4: Derating factors shall not apply to underground conductors entering or leaving an outdoor trench if those conductors have physical protection in the form of rigid metal conduit, intermediate metal conduit, or rigid nonmetallic conduit having a length not exceeding 3.05 m (10 ft) and if the number of conductors does not exceed four.

(3) Bare or Covered Conductors. Where bare or covered conductors are used with insulated conductors, their allowable ampacities shall be limited to those permitted for the adjacent insulated conductors.

(4) Neutral Conductor.

(a) A neutral conductor that carries only the unbalanced current from other conductors of the same circuit shall not be required to be counted when applying the provisions of 310.15(B)(2)(a).

(b) In a 3-wire circuit consisting of two phase wires and the neutral of a 4-wire, 3-phase, wye-connected system, a common conductor carries approximately the same current as the line-to-neutral load currents of the other conductors and shall be counted when applying the provisions of 310.15(B)(2)(a).

(c) On a 4-wire, 3-phase wye circuit where the major portion of the load consists of nonlinear loads, harmonic currents are present in the neutral conductor; the neutral shall therefore be considered a current-carrying conductor.

(5) Grounding or Bonding Conductor. A grounding or bonding conductor shall not be counted when applying the provisions of 310.15(B)(2)(a).

Table 310.16 Allowable Ampacities of Insulated Conductors Rated 0 Through 2000 Volts, 60°C Through 90°C (140°F Through 194°F), Not More than Three Current-Carrying Conductors in Raceway, Cable, or Earth (Directly Buried), Based on Ambient Temperature of 30°C (86°F)

Size AWG or kcmil	Temperature Rating of Conductor (See Table 310.13)						Size AWG or kcmil
	60°C (140°F)	75°C (167°F)	90°C (194°F)	60°C (140°F)	75°C (167°F)	90°C (194°F)	
	Types TW, UF	Types RHW, THHW, THW, THWN, XHHW, USE, ZW	Types TBS, SA, SIS, FEP, FEPB, MI, RHH, RHW-2, THHN, THHW, THW-2, THWN-2, USE-2, XHH, XHHW, XHHW-2, ZW-2	Types TW, UF	Types RHW, THHW, THW, THWN, XHHW, USE	Types TBS, SA, SIS, THHN, THHW, THW-2, THWN-2, RHH, RHW-2, USE-2, XHHW, XHHW-2, ZW-2	
	Copper			Aluminum or Copper-Clad Aluminum			
18	—	—	14	—	—	—	—
16	—	—	18	—	—	—	—
14*	20	20	25	—	—	—	—
12*	25	25	30	20	20	25	12*
10*	30	35	40	25	30	35	10*
8	40	50	55	30	40	45	8
6	55	65	75	40	50	60	6
4	70	85	95	55	65	75	4
3	85	100	110	65	75	85	3
2	95	115	130	75	90	100	2

1	110	130	150	85	100	115	1
1/0	125	150	170	100	120	135	1/0
2/0	145	175	195	115	135	150	2/0
3/0	165	200	225	130	155	175	3/0
4/0	195	230	260	150	180	205	4/0
250	215	255	290	170	205	230	250
300	240	285	320	190	230	255	300
350	260	310	350	210	250	280	350
400	280	335	380	225	270	305	400
500	320	380	430	260	310	350	500
600	355	420	475	285	340	385	600
700	385	460	520	310	375	420	700
750	400	475	535	320	385	435	750
800	410	490	555	330	395	450	500
900	435	520	585	355	425	480	900
1000	455	545	615	375	445	500	1000
1250	495	590	665	405	485	545	1250
1500	520	625	705	435	520	585	1500
1750	545	650	735	455	545	615	1750
2000	560	665	750	470	560	630	2000

Continued

Table 310.16 Allowable Ampacities of Insulated Conductors Rated 0 Through 2000 Volts, 60°C Through 90°C (140°F Through 194°F), Not More than Three Current-Carrying Conductors in Raceway, Cable, or Earth (Directly Buried), Based on Ambient Temperature of 30°C (86°F) *Continued*

Size	Temperature Rating of Conductor (See Table 310.13)						Size
	60°C (140°F)	75°C (167°F)	90°C (194°F)	60°C (140°F)	75°C (167°F)	90°C (194°F)	
AWG or kcmil	Types TW, UF	Types RHW, THW, THWN, XHHW, USE, ZW	Types TBS, SA, SIS, FEP, FEPB, MI, RHH, RHW-2, THHN, THHW, THW-2, THWN-2, USE-2, XHH, XHHW, XHHW-2, ZW-2	Types TW, UF	Types RHW, THW, THWN, XHHW, USE	Types TBS, SA, SIS, THHN, THHW, THW-2, THWN-2, RHH, RHW-2, USE-2, XHH, XHHW, XHHW-2	AWG or kcmil
CORRECTION FACTORS							
Ambient Temp. (°C)	For ambient temperatures other than 30°C (86°F), multiply the allowable ampacities shown above by the appropriate factor shown below.						Ambient Temp. (°F)
21–25	1.08	1.05	1.04	1.08	1.05	1.04	70–77
26–30	1.00	1.00	1.00	1.00	1.00	1.00	78–86
31–35	0.91	0.94	0.96	0.91	0.94	0.96	87–95
36–40	0.82	0.88	0.91	0.82	0.88	0.91	96–104
41–45	0.71	0.82	0.87	0.71	0.82	0.87	105–113

46–50	0.58		0.75		0.82		0.58		0.75		0.82	114–122
51–55	0.41	0.67	0.76	0.41	0.67	0.76	123–131
56–60	—	0.58	0.71	—	0.58	0.71	132–140
61–70	—	0.33	0.58	—	0.33	0.58	141–158
71–80	—	—	0.41	—	—	0.41	159–176

*See 240.4(D)

CHAPTER 15

BRANCH CIRCUITS

INTRODUCTION

This chapter contains provisions from Article 210—Branch Circuits. The provisions within Article 210 have been divided into three chapters of this *Pocket Guide*. This chapter covers Part I (general provisions) and Part II (branch-circuit ratings). Some of the general provisions pertain to multiwire branch circuits, ground-fault circuit-interrupter protection for personnel, and required branch circuits. Part II covers the branch-circuit ratings for conductors and receptacles.

Chapter 16 contains the receptacle outlet requirements within Article 210, and Chapter 18 contains the lighting outlet requirements.

ARTICLE 210
Branch Circuits

I. GENERAL PROVISIONS

210.1 Scope. This article covers branch circuits except for branch circuits that supply only motor loads, which are covered in Article 430. Provisions of this article and Article 430 apply to branch circuits with combination loads.

210.3 Rating. Branch circuits recognized by this article shall be rated in accordance with the maximum permitted ampere rating or setting of the overcurrent device. The rating for other than individual branch circuits shall be 15, 20, 30, 40, and 50 amperes. Where conductors of higher ampacity are used for any reason, the ampere rating or setting of the specified overcurrent device shall determine the circuit rating.

210.4 Multiwire Branch Circuits.

(A) General. Branch circuits recognized by this article shall be permitted as multiwire circuits. A multiwire branch circuit shall be permitted to be considered as multiple circuits. All conductors shall originate from the same panelboard.

> FPN: A 3-phase, 4-wire, wye-connected power system used to supply power to nonlinear loads may necessitate that the power system design allow for the possibility of high harmonic neutral currents.

(C) Line-to-Neutral Loads. Multiwire branch circuits shall supply only line-to-neutral loads.

Exception No. 1: A multiwire branch circuit that supplies only one utilization equipment.

Exception No. 2: Where all ungrounded conductors of the multiwire branch circuit are opened simultaneously by the branch-circuit overcurrent device.

(D) Identification of Ungrounded Conductors. Where more than one nominal voltage system exists in a building, each ungrounded conductor of a multiwire branch circuit, where accessible, shall be identified by phase and system.

Branch Circuits

This means of identification shall be permitted to be by separate color coding, marking tape, tagging, or other approved means and shall be permanently posted at each branch-circuit panelboard.

210.7 Branch Circuit Receptacle Requirements.

(A) Receptacle Outlet Location. Receptacle outlets shall be located in branch circuits in accordance with Part III of Article 210.

(B) Receptacle Requirements. Specific requirements for receptacles are covered in Article 406.

(C) Multiple Branch Circuits. Where more than one branch circuit supplies more than one receptacle on the same yoke, a means to simultaneously disconnect the ungrounded conductors supplying those receptacles shall be provided at the panelboard where the branch circuits originated.

210.8 Ground-Fault Circuit-Interrupter Protection for Personnel.

(B) Other Than Dwelling Units. All 125-volt, single-phase, 15- and 20-ampere receptacles installed in the locations specified in (1), (2), and (3) shall have ground-fault circuit-interrupter protection for personnel:

(1) Bathrooms
(2) Rooftops
(3) Kitchens

Exception to (2): Receptacles that are not readily accessible and are supplied from a dedicated branch circuit for electric snow-melting or deicing equipment shall be permitted to be installed in accordance with the applicable provisions of Article 426.

210.11 Branch Circuits Required.
Branch circuits for lighting and for appliances, including motor-operated appliances, shall be provided to supply the loads computed in accordance with 220.3. In addition, branch circuits shall be provided for specific loads not covered by 220.3 where required elsewhere in this *Code* and for dwelling unit loads as specified in 210.11(C).

(A) Number of Branch Circuits. The minimum number of branch circuits shall be determined from the total computed load and the size or rating of the circuits used. In all installations, the number of circuits shall be sufficient to supply the load served. In no case shall the load on any circuit exceed the maximum specified by 220.4.

(B) Load Evenly Proportioned Among Branch Circuits. Where the load is computed on a volt-amperes/square meter or square foot basis, the wiring system up to and including the branch-circuit panelboard(s) shall be provided to serve not less than the calculated load. This load shall be evenly proportioned among multioutlet branch circuits within the panelboard(s). Branch-circuit overcurrent devices and circuits shall only be required to be installed to serve the connected load.

II. BRANCH-CIRCUIT RATINGS

210.19 Conductors—Minimum Ampacity and Size.

(A) Branch Circuits Not More Than 600 Volts.

(1) General. Branch-circuit conductors shall have an ampacity not less than the maximum load to be served. Where a branch circuit supplies continuous loads or any combination of continuous and noncontinuous loads, the minimum branch-circuit conductor size, before the application of any adjustment or correction factors, shall have an allowable ampacity not less than the noncontinuous load plus 125 percent of the continuous load.

(2) Multioutlet Branch Circuits. Conductors of branch circuits supplying more than one receptacle for cord-and-plug connected portable loads shall have an ampacity of not less than the rating of the branch circuit.

(4) Other Loads. Branch-circuit conductors that supply loads other than those specified in 210.2 and other than cooking appliances as covered in 210.19(A)(3) shall have an ampacity sufficient for the loads served and shall not be smaller than 14 AWG.

Branch Circuits

210.20 Overcurrent Protection. Branch-circuit conductors and equipment shall be protected by overcurrent protective devices that have a rating or setting that complies with 210.20(A) through (D).

(A) Continuous and Noncontinuous Loads. Where a branch circuit supplies continuous loads or any combination of continuous and noncontinuous loads, the rating of the overcurrent device shall not be less than the noncontinuous load plus 125 percent of the continuous load.

210.21 Outlet Devices. Outlet devices shall have an ampere rating that is not less than the load to be served and shall comply with 210.21(A) and (B).

(A) Lampholders. Where connected to a branch circuit having a rating in excess of 20 amperes, lampholders shall be of the heavy-duty type. A heavy-duty lampholder shall have a rating of not less than 660 watts if of the admedium type and not less than 750 watts if of any other type.

(B) Receptacles.

(1) Single Receptacle on an Individual Branch Circuit. A single receptacle installed on an individual branch circuit shall have an ampere rating not less than that of the branch circuit.

(2) Total Cord-and-Plug-Connected Load. Where connected to a branch circuit supplying two or more receptacles or outlets, a receptacle shall not supply a total cord-and-plug-connected load in excess of the maximum specified in Table 210.21(B)(2).

Table 210.21(B)(2) Maximum Cord-and-Plug-Connected Load to Receptacle

Circuit Rating (Amperes)	Receptacle Rating (Amperes)	Maximum Load (Amperes)
15 or 20	15	12
20	20	16
30	30	24

Table 210.21(B)(3) Receptacle Ratings for Various Size Circuits

Circuit Rating (Amperes)	Receptacle Rating (Amperes)
15	Not over 15
20	15 or 20
30	30
40	40 or 50
50	50

(3) Receptacle Ratings. Where connected to a branch circuit supplying two or more receptacles or outlets, receptacle ratings shall conform to the values listed in Table 210.21(B)(3), or where larger than 50 amperes, the receptacle rating shall not be less than the branch-circuit rating.

210.23 Permissible Loads. In no case shall the load exceed the branch-circuit ampere rating. An individual branch circuit shall be permitted to supply any load for which it is rated. A branch circuit supplying two or more outlets or receptacles shall supply only the loads specified according to its size as specified in 210.23(A) through (D) and as summarized in 210.24 and Table 210.24.

(A) 15- and 20-Ampere Branch Circuits. A 15- or 20-ampere branch circuit shall be permitted to supply lighting units or other utilization equipment, or a combination of both, and shall comply with 210.23(A)(1) and (A)(2).

(1) Cord-and-Plug-Connected Equipment. The rating of any one cord-and-plug-connected utilization equipment shall not exceed 80 percent of the branch-circuit ampere rating.

(2) Utilization Equipment Fastened in Place. The total rating of utilization equipment fastened in place, other than luminaires (lighting fixtures), shall not exceed 50 percent of the branch-circuit ampere rating where lighting units, cord-and-plug-connected utilization equipment not fastened in place, or both, are also supplied.

(B) 30-Ampere Branch Circuits. A 30-ampere branch circuit shall be permitted to supply fixed lighting units with

Branch Circuits

Table 210.24 Summary of Branch-Circuit Requirements

Circuit Rating	15 A	20 A	30 A	40 A	50 A
Conductors (min. size):[1]					
Circuit wires	14	12	10	8	6
Taps	14	14	14	12	12
Fixture wires and cords—See 240.5					
Overcurrent Protection	15 A	20 A	30 A	40 A	50 A
Outlet devices:					
Lampholders permitted	Any type	Any type	Heavy duty	Heavy duty	Heavy duty
Receptacle rating[2]	15 max. A	15 or 20 A	30 A	40 or 50 A	50 A
Maximum Load	15 A	20 A	30 A	40 A	50 A
Permissible load	See 210.23(A)	See 210.23(A)	See 210.23(B)	See 210.23(C)	See 210.23(C)

[1]These gauges are for copper conductors.
[2]For receptacle rating of cord-connected electric-discharge luminaires (lighting fixtures), see 410.30(C).

heavy-duty lampholders in other than a dwelling unit(s) or utilization equipment in any occupancy. A rating of any one cord-and-plug-connected utilization equipment shall not exceed 80 percent of the branch-circuit ampere rating.

(C) 40- and 50-Ampere Branch Circuits. A 40- or 50-ampere branch circuit shall be permitted to supply cooking appliances that are fastened in place in any occupancy. In other than dwelling units, such circuits shall be permitted to supply fixed lighting units with heavy-duty lampholders, infrared heating units, or other utilization equipment.

(D) Branch Circuits Larger Than 50 Amperes. Branch circuits larger than 50 amperes shall supply only nonlighting outlet loads.

210.24 Branch-Circuit Requirements—Summary. The requirements for circuits that have two or more outlets or receptacles, other than the receptacle circuits of 210.11(C)(1) and (2), are summarized in Table 210.24. This table provides only a summary of minimum requirements. See 210.19, 210.20, and 210.21 for the specific requirements applying to branch circuits.

III. REQUIRED OUTLETS

See Chapter 16 for required receptacles, and Chapter 18 for required lighting outlets.

CHAPTER 16

RECEPTACLE PROVISIONS

INTRODUCTION

This chapter contains provisions from Article 210—Branch Circuits and Article 406—Receptacles, Cord Connectors, and Attachment Plugs (Caps). The provisions from Article 210 covered in this chapter include 210.62, 210.63, and the first part of 210.50. In those sections are specifications for cord pendants, cord connections, show windows, and required receptacle outlets for heating, air-conditioning, and refrigeration equipment. While dwelling units have numerous receptacle placement provisions, commercial and industrial locations have relatively few.

Article 406 covers the rating, type, and installation of receptacles, cord connectors, and attachment plugs (cord caps). Some of the topics covered include: receptacle rating and type, general installation requirements, receptacle mounting, receptacles in damp or wet locations, and connecting the receptacle's grounding terminal to the box.

ARTICLE 210
Branch Circuits

III. REQUIRED OUTLETS

210.50 General. Receptacle outlets shall be installed as specified in 210.52 through 210.63.

(A) Cord Pendants. A cord connector that is supplied by a permanently connected cord pendant shall be considered a receptacle outlet.

(B) Cord Connections. A receptacle outlet shall be installed wherever flexible cords with attachment plugs are used. Where flexible cords are permitted to be permanently connected, receptacles shall be permitted to be omitted for such cords.

210.60 Guest Rooms.

(A) General. Guest rooms in hotels, motels, and similar occupancies shall have receptacle outlets installed in accordance with 210.52(A) and 210.52(D). Guest rooms meeting the definition of a dwelling unit shall have receptacle outlets installed in accordance with all of the applicable rules in 210.52.

(B) Receptacle Placement. In applying the provisions of 210.52(A), the total number of receptacle outlets shall not be less than the minimum number that would comply with the provisions of that section. These receptacle outlets shall be permitted to be located conveniently for permanent furniture layout. At least two receptacle outlets shall be readily accessible. Where receptacles are installed behind the bed, the receptacle shall be located to prevent the bed from contacting any attachment plug that may be installed, or the receptacle shall be provided with a suitable guard.

210.62. Show Windows. At least one receptacle outlet shall be installed directly above a show window for each 3.7 linear m (12 ft) or major fraction thereof of show window area measured horizontally at its maximum width.

210.63 Heating, Air-Conditioning, and Refrigeration Equipment Outlet. A 125-volt, single-phase, 15- or 20-

ampere-rated receptacle outlet shall be installed at an accessible location for the servicing of heating, air-conditioning, and refrigeration equipment. The receptacle shall be located on the same level and within 7.5 m (25 ft) of the heating, air-conditioning, and refrigeration equipment. The receptacle outlet shall not be connected to the load side of the equipment disconnecting means.

ARTICLE 406
Receptacles, Cord Connectors, and Attachment Plugs (Caps)

406.1 Scope. This article covers the rating, type, and installation of receptacles, cord connectors, and attachment plugs (cord caps).

406.2 Receptacle Rating and Type.

(A) Receptacles. Receptacles shall be listed for the purpose and marked with the manufacturer's name or identification and voltage and ampere ratings.

(B) Rating. Receptacles and cord connectors shall be rated not less than 15 amperes, 125 volts, or 15 amperes, 250 volts, and shall be of a type not suitable for use as lampholders.

(C) Receptacles for Aluminum Conductors. Receptacles rated 20 amperes or less and designed for the direct connection of aluminum conductors shall be marked CO/ALR.

(D) Isolated Ground Receptacles. Receptacles incorporating an isolated grounding connection intended for the reduction of electrical noise (electromagnetic interference) as permitted in 250.146(D) shall be identified by an orange triangle located on the face of the receptacle.

(1) Receptacles so identified shall be used only with grounding conductors that are isolated in accordance with 250.146(D).

(2) Isolated ground receptacles installed in nonmetallic boxes shall be covered with a nonmetallic faceplate.

Exception: Where an isolated ground receptacle is installed in a nonmetallic box, a metal faceplate shall be permitted if the box contains a feature or accessory that permits the effective grounding of the faceplate.

406.3 General Installation Requirements. Receptacle outlets shall be located in branch circuits in accordance with Part III of Article 210. General installation requirements shall be in accordance with 406.3(A) through (F).

(A) Grounding Type. Receptacles installed on 15- and 20-ampere branch circuits shall be of the grounding type. Grounding-type receptacles shall be installed only on circuits of the voltage class and current for which they are rated, except as provided in Tables 210.21(B)(2) and (B)(3).

Exception: Nongrounding-type receptacles installed in accordance with 406.3(D).

(B) To Be Grounded. Receptacles and cord connectors that have grounding contacts shall have those contacts effectively grounded.

Exception No. 2: Replacement receptacles as permitted by 406.3(D).

(C) Methods of Grounding. The grounding contacts of receptacles and cord connectors shall be grounded by connection to the equipment grounding conductor of the circuit supplying the receptacle or cord connector.

The branch-circuit wiring method shall include or provide an equipment-grounding conductor to which the grounding contacts of the receptacle or cord connector shall be connected.

(D) Replacements. Replacement of receptacles shall comply with 406.3(D)(1), (2), and (3) as applicable.

(1) Grounding-Type Receptacles. Where a grounding means exists in the receptacle enclosure or a grounding conductor is installed in accordance with 250.130(C), grounding-type receptacles shall be used and shall be connected to the grounding conductor in accordance with 406.3(C) or 250.130(C).

(2) Ground-Fault Circuit Interrupters. Ground-fault circuit-interrupter protected receptacles shall be provided

Receptacle Provisions

where replacements are made at receptacle outlets that are required to be so protected elsewhere in this *Code*.

(3) Nongrounding-Type Receptacles. Where grounding means does not exist in the receptacle enclosure, the installation shall comply with (a), (b), or (c).

(a) A nongrounding-type receptacle(s) shall be permitted to be replaced with another nongrounding-type receptacle(s).

(b) A nongrounding-type receptacle(s) shall be permitted to be replaced with a ground-fault circuit interrupter-type of receptacle(s). These receptacles shall be marked "No Equipment Ground." An equipment grounding conductor shall not be connected from the ground-fault circuit interrupter-type receptacle to any outlet supplied from the ground-fault circuit-interrupter receptacle.

(c) A nongrounding-type receptacle(s) shall be permitted to be replaced with a grounding-type receptacle(s) where supplied through a ground-fault circuit interrupter. Grounding-type receptacles supplied through the ground-fault circuit interrupter shall be marked "GFCI Protected" and "No Equipment Ground." An equipment grounding conductor shall not be connected between the grounding-type receptacles.

(E) Cord-and-Plug-Connected Equipment. The installation of grounding-type receptacles shall not be used as a requirement that all cord-and-plug-connected equipment be of the grounded type.

(F) Noninterchangeable Types. Receptacles connected to circuits that have different voltages, frequencies, or types of current (ac or dc) on the same premises shall be of such design that the attachment plugs used on these circuits are not interchangeable.

406.4 Receptacle Mounting. Receptacles shall be mounted in boxes or assemblies designed for the purpose, and such boxes or assemblies shall be securely fastened in place.

(A) Boxes That Are Set Back. Receptacles mounted in boxes that are set back of the wall surface, as permitted in

314.20, shall be installed so that the mounting yoke or strap of the receptacle is held rigidly at the surface of the wall.

(B) Boxes That Are Flush. Receptacles mounted in boxes that are flush with the wall surface or project therefrom shall be installed so that the mounting yoke or strap of the receptacle is held rigidly against the box or raised box cover.

(C) Receptacles Mounted on Covers. Receptacles mounted to and supported by a cover shall be held rigidly against the cover by more than one screw or shall be a device assembly or box cover listed and identified for securing by a single screw.

(D) Position of Receptacle Faces. After installation, receptacle faces shall be flush with or project from faceplates of insulating material and shall project a minimum of 0.4 mm (0.015 in.) from metal faceplates.

(E) Receptacles in Countertops and Similar Work Surfaces in Dwelling Units. Receptacles shall not be installed in a face-up position in countertops or similar work surfaces.

(F) Exposed Terminals. Receptacles shall be enclosed so that live wiring terminals are not exposed to contact.

406.5 Receptacle Faceplates (Cover Plates). Receptacle faceplates shall be installed so as to completely cover the opening and seat against the mounting surface.

406.6 Attachment Plugs. All attachment plugs and cord connectors shall be listed for the purpose and marked with the manufacturer's name or identification and voltage and ampere ratings.

(A) Attachment plugs and cord connectors shall be constructed so that there are no exposed current-carrying parts except the prongs, blades, or pins. The cover for wire terminations shall be a part that is essential for the operation of an attachment plug or connector (dead-front construction).

(B) Attachment plugs shall be installed so that their prongs, blades, or pins are not energized unless inserted into an energized receptacle. No receptacle shall be installed so

Receptacle Provisions

as to require an energized attachment plug as its source of supply.

406.7 Noninterchangeability. Receptacles, cord connectors, and attachment plugs shall be constructed so that receptacle or cord connectors do not accept an attachment plug with a different voltage or current rating from that for which the device is intended. However, a 20-ampere T-slot receptacle or cord connector shall be permitted to accept a 15-ampere attachment plug of the same voltage rating. Non–grounding-type receptacles and connectors shall not accept grounding-type attachment plugs.

406.8 Receptacles in Damp or Wet Locations.

(A) Damp Locations. A receptacle installed outdoors in a location protected from the weather or in other damp locations shall have an enclosure for the receptacle that is weatherproof when the receptacle is covered (attachment plug cap not inserted and receptacle covers closed).

An installation suitable for wet locations shall also be considered suitable for damp locations.

A receptacle shall be considered to be in a location protected from the weather where located under roofed open porches, canopies, marquees, and the like, and will not be subjected to a beating rain or water runoff.

(B) Wet Locations.

(1) 15- and 20-Ampere Outdoor Receptacles. 15- and 20-ampere, 125- and 250-volt receptacles installed outdoors in a wet location shall have an enclosure that is weatherproof whether or not the attachment plug cap is inserted.

(2) Other Receptacles. All other receptacles installed in a wet location shall comply with (a) or (b):

(a) A receptacle installed in a wet location where the product intended to be plugged into it is not attended while in use (e.g., sprinkler system controller, landscape lighting, holiday lights, and so forth) shall have an enclosure that is weatherproof with the attachment plug cap inserted or removed.

(b) A receptacle installed in a wet location where the product intended to be plugged into it will be attended while in use (e.g., portable tools, and so forth) shall have an enclosure that is weatherproof when the attachment plug is removed.

(C) Bathtub and Shower Space. A receptacle shall not be installed within a bathtub or shower space.

(D) Protection for Floor Receptacles. Standpipes of floor receptacles shall allow floor-cleaning equipment to be operated without damage to receptacles.

(E) Flush Mounting with Faceplate. The enclosure for a receptacle installed in an outlet box flush-mounted on a wall surface shall be made weatherproof by means of a weatherproof faceplate assembly that provides a watertight connection between the plate and the wall surface.

406.10 Connecting Receptacle Grounding Terminal to Box. The connection of the receptacle grounding terminal shall comply with 250.146.

CHAPTER 17

SWITCHES

INTRODUCTION

This chapter contains provisions from Article 404—Switches. Unlike the disconnect switches covered in Chapter 7, this chapter covers general-use snap switches. These switches are generally installed in device boxes or on box covers. Provisions from this article cover general-use snap switches (single-pole, three-way, four-way, double-pole, etc.), dimmers, and similar control switches. Specifications pertaining to the required locations for switches are in Article 210 (see Chapter 18).

ARTICLE 404
Switches

I. INSTALLATION

404.1 Scope. The provisions of this article shall apply to all switches, switching devices, and circuit breakers where used as switches.

404.2 Switch Connections.

(A) Three-Way and Four-Way Switches. Three-way and four-way switches shall be wired so that all switching is done only in the ungrounded circuit conductor. Where in metal raceways or metal-armored cables, wiring between switches and outlets shall be in accordance with 300.20(A).

Exception: Switch loops shall not require a grounded conductor.

(B) Grounded Conductors. Switches or circuit breakers shall not disconnect the grounded conductor of a circuit.

Exception: A switch or circuit breaker shall be permitted to disconnect a grounded circuit conductor where all circuit conductors are disconnected simultaneously, or where the device is arranged so that the grounded conductor cannot be disconnected until all the ungrounded conductors of the circuit have been disconnected.

404.9 Provisions for General-Use Snap Switches.

(A) Faceplates. Faceplates provided for snap switches mounted in boxes and other enclosures shall be installed so as to completely cover the opening and, where the switch is flush mounted, seat against the finished surface.

(B) Grounding. Snap switches, including dimmer and similar control switches, shall be effectively grounded and shall provide a means to ground metal faceplates, whether or not a metal faceplate is installed. Snap switches shall be considered effectively grounded if either of the following conditions is met.

Switches

(1) The switch is mounted with metal screws to a metal box or to a nonmetallic box with integral means for grounding devices.
(2) An equipment grounding conductor or equipment bonding jumper is connected to an equipment grounding termination of the snap switch.

Exception to (B): Where no grounding means exists within the snap-switch enclosure or where the wiring method does not include or provide an equipment ground, a snap switch without a grounding connection shall be permitted for replacement purposes only. A snap switch wired under the provisions of this exception and located within reach of earth, grade conducting floors, or other conducting surfaces shall be provided with a faceplate of nonconducting, noncombustible material.

404.10 Mounting of Snap Switches.

(A) Surface-Type. Snap switches used with open wiring on insulators shall be mounted on insulating material that separates the conductors at least 13 mm (1/2 in.) from the surface wired over.

(B) Box Mounted. Flush-type snap switches mounted in boxes that are set back of the wall surface as permitted in 314.20 shall be installed so that the extension plaster ears are seated against the surface of the wall. Flush-type snap switches mounted in boxes that are flush with the wall surface or project from it shall be installed so that the mounting yoke or strap of the switch is seated against the box.

404.14 Rating and Use of Snap Switches.
Snap switches shall be used within their ratings and as indicated in 404.14(A) through (D).

(A) Alternating Current General-Use Snap Switch. A form of general-use snap switch suitable only for use on ac circuits for controlling the following:

(1) Resistive and inductive loads, including electric-discharge lamps, not exceeding the ampere rating of the switch at the voltage involved
(2) Tungsten-filament lamp loads not exceeding the ampere rating of the switch at 120 volts

(3) Motor loads not exceeding 80 percent of the ampere rating of the switch at its rated voltage

(B) Alternating-Current or Direct-Current General-Use Snap Switch. A form of general-use snap switch suitable for use on either ac or dc circuits for controlling the following:

(1) Resistive loads not exceeding the ampere rating of the switch at the voltage applied.
(2) Inductive loads not exceeding 50 percent of the ampere rating of the switch at the applied voltage. Switches rated in horsepower are suitable for controlling motor loads within their rating at the voltage applied.
(3) Tungsten-filament lamp loads not exceeding the ampere rating of the switch at the applied voltage if T-rated.

(C) CO/ALR Snap Switches. Snap switches rated 20 amperes or less directly connected to aluminum conductors shall be listed and marked CO/ALR.

(D) Alternating-Current Specific-Use Snap Switches Rated for 347 Volts. Snap switches rated 347 volts ac shall be listed and shall be used only for controlling the following.

(1) Noninductive Loads. Noninductive loads other than tungsten-filament lamps not exceeding the ampere and voltage ratings of the switch.

(2) Inductive Loads. Inductive loads not exceeding the ampere and voltage ratings of the switch. Where particular load characteristics or limitations are specified as a condition of the listing, those restrictions shall be observed regardless of the ampere rating of the load.

The ampere rating of the switch shall not be less than 15 amperes at a voltage rating of 347 volts ac. Flush-type snap switches rated 347 volts ac shall not be readily interchangeable in box mounting with switches identified in 404.14(A) and (B).

(E) Dimmer Switches. General-use dimmer switches shall be used only to control permanently installed incandescent luminaires (lighting fixtures) unless listed for the control of other loads and installed accordingly.

CHAPTER 18

LIGHTING REQUIREMENTS

INTRODUCTION

This chapter contains provisions from Article 210—Branch Circuits; Article 410—Luminaires (Lighting Fixtures), Lampholders, and Lamps; and Article 411—Lighting Systems Operating at 30 Volts or Less. This chapter covers the last paragraph of Article 210. While dwelling units have numerous switch-controlled lighting outlet provisions, commercial and industrial locations have their lighting outlets arranged based on room, furniture or equipment layout, and design considerations for the necessary amount of illumination required for a specific use or operation.

Article 410 covers luminaires (lighting fixtures), lampholders, pendants, incandescent filament lamps, arc lamps, electric-discharge lamps, and the wiring and equipment forming part of such lamps, luminaires (fixtures), and lighting installations. Contained in this pocket guide are nine of the fifteen parts that are included in Article 410: I General; II Luminaire (Fixture) Locations; III Provisions at Luminaire (Fixture) Outlet Boxes, Canopies, and Pans; IV Luminaire (Fixture) Supports; V Grounding; VI Wiring of Luminaires (Fixtures); XI Special Provisions for Flush and Recessed Luminaires (Fixtures); XIII Special Provisions for Electric-Discharge Lighting Systems of 1000 Volts or less; and XV Lighting Track.

Article 411 covers lighting systems operating at 30 volts or less and their associated components.

ARTICLE 210
Branch Circuits

III. REQUIRED OUTLETS

210.70 Lighting Outlets Required.

(C) Other Than Dwelling Units. For attics and underfloor spaces containing equipment requiring servicing, such as heating, air-conditioning, and refrigeration equipment, at least one lighting outlet containing a switch or controlled by a wall switch shall be installed in such spaces. At least one point of control shall be at the usual point of entry to these spaces. The lighting outlet shall be provided at or near the equipment requiring servicing.

ARTICLE 410
Luminaires (Lighting Fixtures), Lampholders, and Lamps

I. GENERAL

410.1 Scope. This article covers luminaires (lighting fixtures), lampholders, pendants, incandescent filament lamps, arc lamps, electric-discharge lamps, the wiring and equipment forming part of such lamps, luminaires (fixtures), and lighting installations.

II. LUMINAIRE (FIXTURE) LOCATIONS

410.4 Luminaires (Fixtures) in Specific Locations.

(A) Wet and Damp Locations. Luminaires (fixtures) installed in wet or damp locations shall be installed so that water cannot enter or accumulate in wiring compartments, lampholders, or other electrical parts. All luminaires (fixtures) installed in wet locations shall be marked, "Suitable for Wet Locations." All luminaires (fixtures) installed in damp locations shall be marked, "Suitable for Wet Locations" or "Suitable for Damp Locations."

(B) Corrosive Locations. Luminaires (fixtures) installed in corrosive locations shall be of a type suitable for such locations.

(C) In Ducts or Hoods. Luminaires (fixtures) shall be permitted to be installed in commercial cooking hoods where all of the following conditions are met:

(1) The luminaire (fixture) shall be identified for use within commercial cooking hoods and installed so that the temperature limits of the materials used are not exceeded.
(2) The luminaire (fixture) shall be constructed so that all exhaust vapors, grease, oil, or cooking vapors are excluded from the lamp and wiring compartment. Diffusers shall be resistant to thermal shock.
(3) Parts of the luminaire (fixture) exposed within the hood shall be corrosion resistant or protected against corrosion, and the surface shall be smooth so as not to collect deposits and to facilitate cleaning.
(4) Wiring methods and materials supplying the luminaire(s) [fixture(s)] shall not be exposed within the cooking hood.

410.5 Luminaires (Fixtures) Near Combustible Material. Luminaires (fixtures) shall be constructed, installed, or equipped with shades or guards so that combustible material is not subjected to temperatures in excess of 90°C (194°F).

410.6 Luminaires (Fixtures) Over Combustible Material. Lampholders installed over highly combustible material shall be of the unswitched type. Unless an individual switch is provided for each luminaire (fixture), lampholders shall be located at least 2.5 m (8 ft) above the floor or shall be located or guarded so that the lamps cannot be readily removed or damaged.

410.7 Luminaires (Fixtures) in Show Windows. Chain-supported luminaires (fixtures) used in a show window shall be permitted to be externally wired. No other externally wired luminaires (fixtures) shall be used.

410.9 Space for Cove Lighting. Coves shall have adequate space and shall be located so that lamps and equipment can be properly installed and maintained.

III. PROVISIONS AT LUMINAIRE (FIXTURE) OUTLET BOXES, CANOPIES, AND PANS

410.10 Space for Conductors. Canopies and outlet boxes taken together shall provide adequate space so that luminaire (fixture) conductors and their connecting devices can be properly installed.

410.11 Temperature Limit of Conductors in Outlet Boxes. Luminaires (fixtures) shall be of such construction or installed so that the conductors in outlet boxes shall not be subjected to temperatures greater than that for which the conductors are rated.

Branch-circuit wiring, other than 2-wire or multiwire branch circuits supplying power to luminaires (fixtures) connected together, shall not be passed through an outlet box that is an integral part of a luminaire (fixture) unless the luminaire (fixture) is identified for through-wiring.

410.12 Outlet Boxes to Be Covered. In a completed installation, each outlet box shall be provided with a cover unless covered by means of a luminaire (fixture) canopy, lampholder, receptacle, or similar device.

410.13 Covering of Combustible Material at Outlet Boxes. Any combustible wall or ceiling finish exposed between the edge of a luminaire (fixture) canopy or pan and an outlet box shall be covered with noncombustible material.

410.14 Connection of Electric-Discharge Luminaires (Lighting Fixtures).

(A) Independent of the Outlet Box. Electric-discharge luminaires (lighting fixtures) supported independently of the outlet box shall be connected to the branch circuit through metal raceway, nonmetallic raceway, Type MC cable, Type AC cable, Type MI cable, nonmetallic sheathed cable, or by flexible cord as permitted in 410.30(B) or (C).

Lighting Requirements

(B) Access to Boxes. Electric-discharge luminaires (lighting fixtures) surface mounted over concealed outlet, pull, or junction boxes shall be installed with suitable openings in back of the fixture to provide access to the boxes.

IV. LUMINAIRE (FIXTURE) SUPPORTS

410.15 Supports.

(A) General. Luminaires (fixtures) and lampholders shall be securely supported. A luminaire (fixture) that weighs more than 3 kg (6 lb) or exceeds 400 mm (16 in.) in any dimension shall not be supported by the screw shell of a lampholder.

(B) Metal Poles Supporting Luminaires (Lighting Fixtures). Metal poles shall be permitted to be used to support luminaires (lighting fixtures) and as a raceway to enclose supply conductors, provided the following conditions are met:

(1) A metal pole shall have a handhole not less than 50 mm × 100 mm (2 in. × 4 in.) with a raintight cover to provide access to the supply terminations within the pole or pole base.

Exception No. 1: No handhole shall be required in a pole 2.5 m (8 ft) or less in height above grade where the supply wiring method continues without splice or pull point, and where the interior of the pole and any splices are accessible by removing the luminaire (fixture).

Exception No. 2: No handhole shall be required in a metal pole 6.0 m (20 ft) or less in height above grade that is provided with a hinged base.

(2) Where raceway risers or cable is not installed within the pole, a threaded fitting or nipple shall be brazed or welded to the pole opposite the handhole for the supply connection.

(3) A metal pole shall be provided with a grounding terminal.

   a. A pole with a handhole shall have the grounding terminal accessible from the handhole.

b. A pole with a hinged base shall have the grounding terminal accessible within the base.

Exception: No grounding terminal shall be required in a pole 2.5 m (8 ft) or less in height above grade where the supply wiring method continues without splice or pull, and where the interior of the pole and any splices are accessible by removing the luminaire (fixture).

(4) A pole with a hinged base shall have the hinged base and pole bonded together.
(5) Metal raceways or other equipment grounding conductors shall be bonded to the pole with an equipment grounding conductor recognized by 250.118 and sized in accordance with 250.122.
(6) Conductors in vertical metal poles used as raceway shall be supported as provided in 300.19.

410.16 Means of Support.

(A) Outlet Boxes. Outlet boxes or fittings installed as required by 314.23 shall be permitted to support luminaires (fixtures).

(B) Inspection. Luminaires (fixtures) shall be installed so that the connections between the luminaire (fixture) conductors and the circuit conductors can be inspected without requiring the disconnection of any part of the wiring unless the luminaires (fixtures) are connected by attachment plugs and receptacles.

(C) Suspended Ceilings. Framing members of suspended ceiling systems used to support luminaires (fixtures) shall be securely fastened to each other and shall be securely attached to the building structure at appropriate intervals. Luminaires (fixtures) shall be securely fastened to the ceiling framing member by mechanical means such as bolts, screws, or rivets. Listed clips identified for use with the type of ceiling framing member(s) and luminaire(s) [fixture(s)] shall also be permitted.

(D) Luminaire (Fixture) Studs. Luminaire (fixture) studs that are not a part of outlet boxes, hickeys, tripods, and crowfeet shall be made of steel, malleable iron, or other material suitable for the application.

Lighting Requirements

(E) Insulating Joints. Insulating joints that are not designed to be mounted with screws or bolts shall have an exterior metal casing, insulated from both screw connections.

(F) Raceway Fittings. Raceway fittings used to support a luminaire(s) [lighting fixture(s)] shall be capable of supporting the weight of the complete fixture assembly and lamp(s).

(G) Busways. Luminaires (fixtures) shall be permitted to be connected to busways in accordance with 368.12.

(H) Trees. Outdoor luminaires (lighting fixtures) and associated equipment shall be permitted to be supported by trees.

V. GROUNDING

410.17 General. Luminaires (fixtures) and lighting equipment shall be grounded as required in Article 250 and Part V of this article.

410.18 Exposed Luminaire (Fixture) Parts.

(A) Exposed Conductive Parts. Exposed metal parts shall be grounded or insulated from ground and other conducting surfaces or be inaccessible to unqualified personnel. Lamp tie wires, mounting screws, clips, and decorative bands on glass spaced at least 38 mm (1-1/2 in.) from lamp terminals shall not be required to be grounded.

(B) Made of Insulating Material. Luminaires (fixtures) directly wired or attached to outlets supplied by a wiring method that does not provide a ready means for grounding shall be made of insulating material and shall have no exposed conductive parts.

Exception: Replacement luminaires (fixtures) shall be permitted to connect an equipment grounding conductor from the outlet in compliance with 250.130(C). The luminaire (fixture) shall then be grounded in accordance with 410.18(A).

410.21 Methods of Grounding. Luminaires (fixtures) and equipment shall be considered grounded where mechanically connected to an equipment grounding conductor as specified in 250.118 and sized in accordance with 250.122.

VI. WIRING OF LUMINAIRES (FIXTURES)

410.22 Luminaire (Fixture) Wiring—General. Wiring on or within fixtures shall be neatly arranged and shall not be exposed to physical damage. Excess wiring shall be avoided. Conductors shall be arranged so that they are not subjected to temperatures above those for which they are rated.

410.23 Polarization of Luminaires (Fixtures). Luminaires (fixtures) shall be wired so that the screw shells of lampholders are connected to the same luminaire (fixture) or circuit conductor or terminal. The grounded conductor, where connected to a screw-shell lampholder, shall be connected to the screw shell.

410.24 Conductor Insulation. Luminaires (fixtures) shall be wired with conductors having insulation suitable for the environmental conditions, current, voltage, and temperature to which the conductors will be subjected.

410.28 Protection of Conductors and Insulation.

(A) Properly Secured. Conductors shall be secured in a manner that does not tend to cut or abrade the insulation.

(B) Protection Through Metal. Conductor insulation shall be protected from abrasion where it passes through metal.

(C) Luminaire (Fixture) Stems. Splices and taps shall not be located within luminaire (fixture) arms or stems.

(D) Splices and Taps. No unnecessary splices or taps shall be made within or on a luminaire (fixture).

(E) Stranding. Stranded conductors shall be used for wiring on luminaire (fixture) chains and on other movable or flexible parts.

(F) Tension. Conductors shall be arranged so that the weight of the luminaire (fixture) or movable parts does not put tension on the conductors.

Lighting Requirements

410.29 Cord-Connected Showcases. Individual showcases, other than fixed, shall be permitted to be connected by flexible cord to permanently installed receptacles, and groups of not more than six such showcases shall be permitted to be coupled together by flexible cord and separable locking-type connectors with one of the group connected by flexible cord to a permanently installed receptacle.

The installation shall comply with 410.29(A) through (E).

(A) Cord Requirements. Flexible cord shall be of the hard-service type, having conductors not smaller than the branch-circuit conductors, having ampacity at least equal to the branch-circuit overcurrent device, and having an equipment grounding conductor.

(B) Receptacles, Connectors, and Attachment Plugs. Receptacles, connectors, and attachment plugs shall be of a listed grounding type rated 15 or 20 amperes.

(C) Support. Flexible cords shall be secured to the undersides of showcases so that

(1) Wiring is not exposed to mechanical damage.
(2) A separation between cases not in excess of 50 mm (2 in.), or more than 300 mm (12 in.) between the first case and the supply receptacle, is ensured.
(3) The free lead at the end of a group of showcases has a female fitting not extending beyond the case.

(D) No Other Equipment. Equipment other than showcases shall not be electrically connected to showcases.

(E) Secondary Circuit(s). Where showcases are cord-connected, the secondary circuit(s) of each electric-discharge lighting ballast shall be limited to one showcase.

410.30 Cord-Connected Lampholders and Luminaires (Fixtures).

(A) Lampholders. Where a metal lampholder is attached to a flexible cord, the inlet shall be equipped with an insulating bushing that, if threaded, is not smaller than metric designator 12 (trade size 3/8) pipe size. The cord hole shall be of a size appropriate for the cord, and all burrs and fins shall be removed in order to provide a smooth bearing surface for the cord.

Bushing having holes 7 mm (9/32 in.) in diameter shall be permitted for use with plain pendant cord and holes 11 mm (13/32 in.) in diameter with reinforced cord.

(B) Adjustable Luminaires (Fixtures). Luminaires (fixtures) that require adjusting or aiming after installation shall not be required to be equipped with an attachment plug or cord connector, provided the exposed cord is of the hard-usage or extra-hard-usage type and is not longer than that required for maximum adjustment. The cord shall not be subject to strain or physical damage.

(C) Electric-Discharge Luminaires (Fixtures).

(1) A listed luminaire (fixture) or a listed assembly shall be permitted to be cord connected if the following conditions apply:

(1) The luminaire (fixture) is located directly below the outlet box or busway.
(2) The flexible cord meets all the following:

   a. Is visible for its entire length outside the luminaire (fixture)
   b. Is not subject to strain or physical damage
   c. Is terminated in a grounding-type attachment plug cap or busway plug or has a luminaire (fixture) assembly with a strain relief and canopy

(2) Electric-discharge luminaires (lighting fixtures) provided with mogul-base, screw-shell lampholders shall be permitted to be connected to branch circuits of 50 amperes or less by cords complying with 240.5. Receptacles and attachment plugs shall be permitted to be of lower ampere rating than the branch circuit but not less than 125 percent of the luminaire (fixture) full-load current.

(3) Electric-discharge luminaires (lighting fixtures) equipped with a flanged surface inlet shall be permitted to be supplied by cord pendants equipped with cord connectors. Inlets and connectors shall be permitted to be of lower ampere rating than the branch circuit but not less than 125 percent of the luminaire (fixture) load current.

410.31 Luminaires (Fixtures) as Raceways. Luminaires (fixtures) shall not be used as a raceway for circuit conductors unless listed and marked for use as a raceway.

410.32 Wiring Supplying Luminaires (Fixtures) Connected Together. Luminaires (fixtures) designed for end-to-end connection to form a continuous assembly, or luminaires (fixtures) connected together by recognized wiring methods, shall be permitted to contain the conductors of a 2-wire branch circuit, or one multiwire branch circuit, supplying the connected luminaires (fixtures) and need not be listed as a raceway. One additional 2-wire branch circuit separately supplying one or more of the connected luminaires (fixtures) shall also be permitted.

410.33 Branch Circuit Conductors and Ballasts. Branch-circuit conductors within 75 mm (3 in.) of a ballast shall have an insulation temperature rating not lower than 90°C (194°F) unless supplying a luminaire (fixture) listed and marked as suitable for a different insulation temperature.

XI. SPECIAL PROVISIONS FOR FLUSH AND RECESSED LUMINAIRES (FIXTURES)

410.64 General. Luminaires (fixtures) installed in recessed cavities in walls or ceilings shall comply with 410.65 through 410.72.

410.65 Temperature.

(A) Combustible Material. Luminaires (fixtures) shall be installed so that adjacent combustible material will not be subjected to temperatures in excess of 90°C (194°F).

(B) Fire-Resistant Construction. Where a luminaire (fixture) is recessed in fire-resistant material in a building of fire-resistant construction, a temperature higher than 90°C (194°F) but not higher than 150°C (302°F) shall be considered acceptable if the luminaire (fixture) is plainly marked that it is listed for that service.

(C) Recessed Incandescent Luminaires (Fixtures). Incandescent luminaires (fixtures) shall have thermal protection and shall be identified as thermally protected.

Exception No. 1: Thermal protection shall not be required in a recessed luminaire (fixture) identified for use and installed in poured concrete.

Exception No. 2: Thermal protection shall not be required in a recessed luminaire (fixture) whose design, construction, and thermal performance characteristics are equivalent to a thermally protected luminaire (fixture) and are identified as inherently protected.

410.66 Clearance and Installation.

(A) Clearance.

(1) Non-Type IC. A recessed luminaire (fixture) that is not identified for contact with insulation shall have all recessed parts spaced not less than 13 mm (1/2 in.) from combustible materials. The points of support and the trim finishing off the opening in the ceiling or wall surface shall be permitted to be in contact with combustible materials.

(2) Type IC. A recessed luminaire (fixture) that is identified for contact with insulation, Type IC, shall be permitted to be in contact with combustible materials at recessed parts, points of support, and portions passing through or finishing off the opening in the building structure.

(B) Installation. Thermal insulation shall not be installed above a recessed luminaire (fixture) or within 75 mm (3 in.) of the recessed luminaire's (fixture's) enclosure, wiring compartment, or ballast unless it is identified for contact with insulation, Type IC.

XIII. SPECIAL PROVISIONS FOR ELECTRIC-DISCHARGE LIGHTING SYSTEMS OF 1000 VOLTS OR LESS

410.73 General.

(A) Open-Circuit Voltage of 1000 Volts or Less. Equipment for use with electric-discharge lighting systems and

Lighting Requirements

designed for an open-circuit voltage of 1000 volts or less shall be of a type intended for such service.

(B) Considered as Energized. The terminals of an electric-discharge lamp shall be considered as energized where any lamp terminal is connected to a circuit of over 300 volts.

(F) High-Intensity Discharge Luminaires (Fixtures).

(1) Recessed. Recessed high-intensity luminaires (fixtures) designed to be installed in wall or ceiling cavities shall have thermal protection and be identified as thermally protected.

(2) Inherently Protected. Thermal protection shall not be required in a recessed high-intensity luminaire (fixture) whose design, construction, and thermal performance characteristics are equivalent to a thermally protected luminaire (fixture) and are identified as inherently protected.

(3) Installed in Poured Concrete. Thermal protection shall not be required in a recessed high-intensity discharge luminaire (fixture) identified for use and installed in poured concrete.

(4) Recessed Remote Ballasts. A recessed remote ballast for a high-intensity discharge luminaire (fixture) shall have thermal protection that is integral with the ballast and be identified as thermally protected.

410.76 Luminaire (Fixture) Mounting.

(A) Exposed Ballasts. Luminaires (fixtures) that have exposed ballasts or transformers shall be installed so that such ballasts or transformers will not be in contact with combustible material.

(B) Combustible Low-Density Cellulose Fiberboard. Where a surface-mounted luminaire (fixture) containing a ballast is to be installed on combustible low-density cellulose fiberboard, it shall be listed for this condition or shall be spaced not less than 38 mm (1-1/2 in.) from the surface of the fiberboard. Where such luminaires (fixtures) are partially or wholly recessed, the provisions of 410.64 through 410.72 shall apply.

XV. LIGHTING TRACK

410.100 Definition.

Lighting Track. A manufactured assembly designed to support and energize luminaires (lighting fixtures) that are capable of being readily repositioned on the track. Its length may be altered by the addition or subtraction of sections of track.

410.101 Installation.

(A) Lighting Track. Lighting track shall be permanently installed and permanently connected to a branch circuit. Only lighting track fittings shall be installed on lighting track. Lighting track fittings shall not be equipped with general-purpose receptacles.

(B) Connected Load. The connected load on lighting track shall not exceed the rating of the track. Lighting track shall be supplied by a branch circuit having a rating not more than that of the track.

(C) Locations Not Permitted. Lighting track shall not be installed in the following locations:

(1) Where likely to be subjected to physical damage
(2) In wet or damp locations
(3) Where subject to corrosive vapors
(4) In storage battery rooms
(5) In hazardous (classified) locations
(6) Where concealed
(7) Where extended through walls or partitions
(8) Less than 1.5 m (5 ft) above the finished floor except where protected from physical damage or track operating at less than 30 volts rms open-circuit voltage
(9) Within the zone measured 900 mm (3 ft) horizontally and 2.5 m (8 ft) vertically from the top of the bathtub rim

(D) Support. Fittings identified for use on lighting track shall be designed specifically for the track on which they

Lighting Requirements

are to be installed. They shall be securely fastened to the track, shall maintain polarization and grounding, and shall be designed to be suspended directly from the track.

410.103 Heavy-Duty Lighting Track. Heavy-duty lighting track is lighting track identified for use exceeding 20 amperes. Each fitting attached to a heavy-duty lighting track shall have individual overcurrent protection.

410.104 Fastening. Lighting track shall be securely mounted so that each fastening will be suitable for supporting the maximum weight of luminaires (fixtures) that can be installed. Unless identified for supports at greater intervals, a single section 1.2 m (4 ft) or shorter in length shall have two supports, and, where installed in a continuous row, each individual section of not more than 1.2 m (4 ft) in length shall have one additional support.

ARTICLE 411
Lighting Systems Operating at 30 Volts or Less

411.1 Scope. This article covers lighting systems operating at 30 volts or less and their associated components.

411.2 Definition.

Lighting Systems Operating at 30 Volts or Less. A lighting system consisting of an isolating power supply operating at 30 volts (42.4 volts peak) or less, under any load condition, with one or more secondary circuits, each limited to 25 amperes maximum, supplying luminaires (lighting fixtures) and associated equipment identified for the use.

411.3 Listing Required. Lighting systems operating at 30 volts or less shall be listed for the purpose.

411.4 Locations Not Permitted. Lighting systems operating at 30 volts or less shall not be installed (1) where concealed or extended through a building wall, unless using a wiring method specified in Chapter 3, or (2) within 3.0 m (10 ft) of pools, spas, fountains, or similar locations, except as permitted by Article 680.

411.5 Secondary Circuits.

(A) Grounding. Secondary circuits shall not be grounded.

(B) Isolation. The secondary circuit shall be insulated from the branch circuit by an isolating transformer.

(C) Bare Conductors. Exposed bare conductors and current-carrying parts shall be permitted. Bare conductors shall not be installed less than 2.1 m (7 ft) above the finished floor, unless specifically listed for a lower installation height.

411.6 Branch Circuit. Lighting systems operating at 30 volts or less shall be supplied from a maximum 20-ampere branch circuit.

CHAPTER 19

MOTORS, MOTOR CIRCUITS, AND CONTROLLERS

INTRODUCTION

This chapter contains provisions from Article 430—Motors, Motor Circuits, and Controllers. Article 430 covers motors, motor branch-circuit and feeder conductors and their protection, motor overload protection, motor control circuits, motor controllers, and motor control centers.

Contained in this *Pocket Guide* are twelve of the thirteen parts that are included in Article 430. The twelve parts are: I General; II Motor Circuit Conductors; III Motor and Branch-Circuit Overload Protection; IV Motor Branch-Circuit Short-Circuit and Ground-Fault Protection; V Motor Feeder Short-Circuit and Ground-Fault Protection; VI Motor Control Circuits; VII Motor Controllers; VIII Motor Control Centers; IX Disconnecting Means; XI Protection of Live Parts—All Voltages; XII Grounding—All Voltages; and XIII Tables.

ARTICLE 430
Motors, Motor Circuits, and Controllers

I. GENERAL

430.1 Scope. This article covers motors, motor branch-circuit and feeder conductors and their protection, motor overload protection, motor control circuits, motor controllers, and motor control centers.

FPN No. 1: Installation requirements for motor control centers are covered in 110.26(F). Air conditioning and refrigerating equipment are covered in Article 440.

430.6 Ampacity and Motor Rating Determination. The size of conductors supplying equipment covered by Article 430 shall be selected from the allowable ampacity tables in accordance with 310.15(B) or shall be calculated in accordance with 310.15(C). Where flexible cord is used, the size of the conductor shall be selected in accordance with 400.5. The required ampacity and motor ratings shall be determined as specified in 430.6(A), (B), and (C).

(A) General Motor Applications. For general motor applications, current ratings shall be determined based on (1) and (2).

(1) Table Values. The values given in Tables 430.147, 430.148, 430.149, and 430.150, including notes, shall be used to determine the ampacity of conductors or ampere ratings of switches, branch-circuit short-circuit and ground-fault protection, instead of the actual current rating marked on the motor nameplate. Where a motor is marked in amperes, but not horsepower, the horsepower rating shall be assumed to be that corresponding to the value given in Tables 430.147, 430.148, 430.149, and 430.150, interpolated if necessary.

(2) Nameplate Values. Separate motor overload protection shall be based on the motor nameplate current rating.

430.10 Wiring Space in Enclosures.

(A) General. Enclosures for motor controllers and disconnecting means shall not be used as junction boxes, auxiliary gutters, or raceways for conductors feeding through or tapping off to the other apparatus unless designs are employed that provide adequate space for this purpose.

430.14 Location of Motors.

(A) Ventilation and Maintenance. Motors shall be located so that adequate ventilation is provided and so that maintenance, such as lubrication of bearings and replacing of brushes, can be readily accomplished.

430.16 Exposure to Dust Accumulations.
In locations where dust or flying material collects on or in motors in such quantities as to seriously interfere with the ventilation or cooling of motors and thereby cause dangerous temperatures, suitable types of enclosed motors that do not overheat under the prevailing conditions shall be used.

Motors, Motor Circuits, and Controllers 253

General, 430.1 through 430.18	Part I
Motor Circuit Conductors, 430.21 through 430.29	Part II
Motor Circuit Overload Protection, 430.31 through 430.44	Part III
Motor and Branch-Circuit Short-Circuit and Ground-Fault Protection, 430.51 through 430.58	Part IV
Motor Feeder Short-Circuit and Ground-Fault Protection, 430.61 through 430.63	Part V
Motor Control Circuits, 430.71 through 430.74	Part VI
Motor Controllers, 430.81 through 430.91	Part VII
Motor Control Centers, 430.92 through 430.98	Part VIII
Disconnecting Means, 430.101 through 430.113	Part IX
Over 600 Volts, Nominal, 430.121 through 430.127	Part X
Protection of Live Parts—All Voltages, 430.131 through 430.133	Part XI
Grounding—All Voltages, 430.141 through 430.145	Part XII
Tables, Tables 430.147 through 430.151(B)	Part XIII

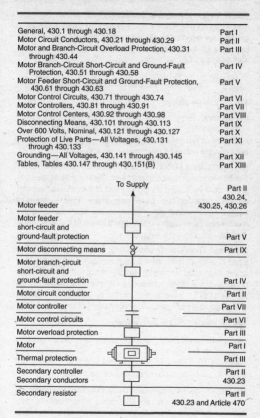

Figure 430.1 Article 430 contents.

430.17 Highest Rated or Smallest Rated Motor. In determining compliance with 430.24, 430.53(B), and 430.53(C), the highest rated or smallest rated motor shall be based on the rated full-load current as selected from Tables 430.147, 430.148, 430.149, and 430.150.

II. MOTOR CIRCUIT CONDUCTORS

430.22 Single Motor.

(A) General. Branch-circuit conductors that supply a single motor used in a continuous duty application shall have an ampacity of not less than 125 percent of the motor's full-load current rating as determined by 430.6(A)(1).

(B) Multispeed Motor. For a multispeed motor, the selection of branch-circuit conductors on the line side of the controller shall be based on the highest of the full-load current ratings shown on the motor nameplate. The selection of branch-circuit conductors between the controller and the motor shall be based on the current rating of the winding(s) that the conductors energize.

(E) Other Than Continuous Duty. Conductors for a motor used in a short-time, intermittent, periodic, or varying duty application shall have an ampacity of not less than the percentage of the motor nameplate current rating shown in Table 430.22(E), unless the authority having jurisdiction grants special permission for conductors of lower ampacity.

430.23 Wound-Rotor Secondary.

(A) Continuous Duty. For continuous duty, the conductors connecting the secondary of a wound-rotor ac motor to its controller shall have an ampacity not less than 125 percent of the full-load secondary current of the motor.

(B) Other Than Continuous Duty. For other than continuous duty, these conductors shall have an ampacity, in percent of full-load secondary current, not less than that specified in Table 430.22(E).

430.24 Several Motors or a Motor(s) and Other Load(s). Conductors supplying several motors, or a motor(s) and other load(s), shall have an ampacity not less than 125 percent of the full-load current rating of the highest rated motor plus the sum of the full-load current ratings of all the other motors in the group, as determined by 430.6(A), plus the ampacity required for the other loads.

430.28 Feeder Taps. Feeder tap conductors shall have an ampacity not less than that required by Part II, shall terminate in a branch-circuit protective device, and, in addition, shall meet one of the following requirements:

(1) Be enclosed either by an enclosed controller or by a raceway, be not more than 3.0 m (10 ft) in length, and, for field installation, be protected by an overcurrent device on the line side of the tap conductor, the rating or setting of which shall not exceed 1000 percent of the tap conductor ampacity

(2) Have an ampacity of at least one-third that of the feeder conductors, be suitably protected from physical damage or enclosed in a raceway, and be not more than 7.5 m (25 ft) in length

(3) Have the same ampacity as the feeder conductors

III. MOTOR AND BRANCH-CIRCUIT OVERLOAD PROTECTION

430.31 General. Part III specifies overload devices intended to protect motors, motor-control apparatus, and motor branch-circuit conductors against excessive heating due to motor overloads and failure to start.

Overload in electrical apparatus is an operating overcurrent that, when it persists for a sufficient length of time, would cause damage or dangerous overheating of the apparatus. It does not include short circuits or ground faults.

These provisions shall not be interpreted as requiring overload protection where it might introduce additional or increased hazards, as in the case of fire pumps.

430.32 Continuous-Duty Motors.

(A) More Than 1 Horsepower. Each continuous-duty motor rated more than 1 hp shall be protected against overload by one of the means in 430.32(A)(1) through (A)(4).

(1) Separate Overload Device. A separate overload device that is responsive to motor current. This device shall be

Table 430.22(E) Duty-Cycle Service

Classification of Service	Nameplace Current Rating Percentages			
	5-Minute Rated Motor	15-Minute Rated Motor	30- & 60-Minute Rated Motor	Continuous Rated Motor
Short-time duty operating valves, raising or lowering rolls, etc	110	120	150	—
Intermittent duty freight and passenger elevators, tool heads, pumps, drawbridges, turntables, etc. (for arc welders, see 630.11)	85	85	90	140
Periodic duty rolls, ore- and coal-handling machines, etc.	85	90	95	140
Varying duty	110	120	150	200

Note: Any motor application shall be considered as continuous duty unless the nature of the apparatus it drives is such that the motor will not operate continuously with load under any condition of use.

Motors, Motor Circuits, and Controllers

selected to trip or shall be rated at no more than the following percent of the motor nameplate full-load current rating:

Motors with a marked service factor 1.15 or greater	125%
Motors with a marked temperature rise 40°C or less	125%
All other motors	115%

Modification of this value shall be permitted as provided in 430.32(C). For a multispeed motor, each winding connection shall be considered separately.

Where a separate motor overload device is connected so that it does not carry the total current designated on the motor nameplate, such as for wye-delta starting, the proper percentage of nameplate current applying to the selection or setting of the overload device shall be clearly designated on the equipment, or the manufacturer's selection table shall take this into account.

(2) Thermal Protector. A thermal protector integral with the motor, approved for use with the motor it protects on the basis that it will prevent dangerous overheating of the motor due to overload and failure to start. The ultimate trip current of a thermally protected motor shall not exceed the following percentage of motor full-load current given in Tables 430.148, 430.149, and 430.150:

Motor full-load current 9 amperes or less	170%
Motor full-load current from 9.1 to, and including, 20 amperes	156%
Motor full-load current greater than 20 amperes	140%

If the motor current-interrupting device is separate from the motor and its control circuit is operated by a protective device integral with the motor, it shall be arranged so that the opening of the control circuit will result in interruption of current to the motor.

(B) One Horsepower or Less, Automatically Started. Any motor of 1 hp or less that is started automatically shall be protected against overload by one of the following means.

(1) Separate Overload Device. A separate overload device that is responsive to motor current. This device shall be selected to trip or shall be rated at not more than the following percentage of the motor nameplate full-load current rating:

Motors with a marked service factor 1.15 or greater	125%
Motors with a marked temperature rise 40°C or less	125%
All other motors	115%

For a multispeed motor, each winding connection shall be considered separately. Modification of this value shall be permitted as provided in 430.32(C).

(2) Thermal Protector. A thermal protector integral with the motor, approved for use with the motor that it protects on the basis that it will prevent dangerous overheating of the motor due to overload and failure to start. Where the motor current-interrupting device is separate from the motor and its control circuit is operated by a protective device integral with the motor, it shall be arranged so that the opening of the control circuit results in interruption of current to the motor.

(C) Selection of Overload Relay. Where the sensing element or setting of the overload relay selected in accordance with 430.32(A)(1) and 430.32(B)(1) is not sufficient to start the motor or to carry the load, higher size sensing elements or incremental settings shall be permitted to be used, provided the trip current of the overload relay does not exceed the following percentage of motor nameplate full-load current rating:

Motors with a marked service factor 1.15 or greater	140%
Motors with a marked temperature rise 40°C or less	140%
All other motors	130%

If not shunted during the starting period of the motor as provided in 430.35, the overload device shall have sufficient time delay to permit the motor to start and accelerate its load.

FPN: A Class 20 or 30 overload relay will provide a longer motor acceleration time than a Class 10 or 20, respectively. Use of a higher class overload relay may preclude the need for selection of a higher trip current.

(D) One Horsepower or Less, Nonautomatically Started.

(1) Within Sight from Controller. Each continuous-duty motor rated at 1 hp or less that is not permanently installed, is nonautomatically started, and is within sight from the controller location shall be permitted to be protected against overload by the branch-circuit short-circuit and ground-fault protective device. This branch-circuit protective device shall not be larger than that specified in Part IV of Article 430.

(2) Not Within Sight from Controller. Any such motor that is not in sight from the controller location shall be protected as specified in 430.32(B). Any motor rated at 1 hp or less that is permanently installed shall be protected in accordance with 430.32(B).

430.36 Fuses—In Which Conductor. Where fuses are used for motor overload protection, a fuse shall be inserted in each ungrounded conductor and also in the grounded conductor if the supply system is 3-wire, 3-phase ac with one conductor grounded.

430.37 Devices Other Than Fuses—In Which Conductor. Where devices other than fuses are used for motor overload protection, Table 430.37 shall govern the minimum allowable number and location of overload units such as trip coils or relays.

430.38 Number of Conductors Opened by Overload Device. Motor overload devices, other than fuses or thermal protectors, shall simultaneously open a sufficient number of ungrounded conductors to interrupt current flow to the motor.

430.39 Motor Controller as Overload Protection. A motor controller shall also be permitted to serve as an overload device if the number of overload units complies with

Table 430.37 and if these units are operative in both the starting and running position in the case of a dc motor, and in the running position in the case of an ac motor.

430.40 Overload Relays. Overload relays and other devices for motor overload protection that are not capable of opening short circuits or ground faults shall be protected by fuses or circuit breakers with ratings or settings in accordance with 430.52 or by a motor short-circuit protector in accordance with 430.52.

430.42 Motors on General-Purpose Branch Circuits. Overload protection for motors used on general-purpose branch circuits as permitted in Article 210 shall be provided as specified in 430.42(A), (B), (C), or (D).

(A) Not Over 1 Horsepower. One or more motors without individual overload protection shall be permitted to be connected to a general-purpose branch circuit only where the installation complies with the limiting conditions specified in 430.32(B) and (D) and 430.53(A)(1) and (A)(2).

(B) Over 1 Horsepower. Motors of ratings larger than specified in 430.53(A) shall be permitted to be connected to general-purpose branch circuits only where each motor is protected by overload protection selected to protect the motor as specified in 430.32. Both the controller and the motor overload device shall be approved for group installation with the short-circuit and ground-fault protective device selected in accordance with 430.53.

430.43 Automatic Restarting. A motor overload device that can restart a motor automatically after overload tripping shall not be installed unless approved for use with the motor it protects. A motor overload device that can restart a motor automatically after overload tripping shall not be installed if automatic restarting of the motor can result in injury to persons.

430.44 Orderly Shutdown. If immediate automatic shutdown of a motor by a motor overload protective device(s) would introduce additional or increased hazard(s) to a person(s) and continued motor operation is necessary for safe shutdown of equipment or process, a motor overload sensing device(s) conforming with the provisions of Part III of

Motors, Motor Circuits, and Controllers

Table 430.37 Overload Units

Kind of Motor	Supply System	Number and Location of Overload Units, Such as Trip Coils or Relays
1-phase ac or dc	2-wire, 1 phase ac or dc ungrounded	1 in either conductor
1-phase ac or dc	2-wire, 1-phase ac or dc, one conductor grounded	1 in ungrounded conductor
1-phase ac or dc	3-wire, 1-phase ac or dc, grounded neutral	1 in either ungrounded conductor
1-phase ac	Any 3-phase	1 in ungrounded conductor
2-phase ac	3-wire, 2-phase ac, ungrounded	2, one in each phase
2-phase ac	3-wire, 2-phase ac, one conductor grounded	2 in ungrounded conductors
2-phase ac	4-wire, 2-phase ac, grounded or ungrounded	2, one per phase in ungrounded conductors
2-phase ac	Grounded neutral or 5-wire, 2-phase ac, ungrounded	2, one per phase in ungrounded phase wire
3-phase ac	Any 3-phase	3, one in each phase*

*Exception: An overload unit in each phase shall not be required where overload protection is provided by other approved means.

this article shall be permitted to be connected to a supervised alarm instead of causing immediate interruption of the motor circuit, so that corrective action or an orderly shutdown can be initiated.

IV. MOTOR BRANCH-CIRCUIT SHORT-CIRCUIT AND GROUND-FAULT PROTECTION

430.51 General. Part IV specifies devices intended to protect the motor branch-circuit conductors, the motor control apparatus, and the motors against overcurrent due to short circuits or grounds. These rules add to or amend the provisions of Article 240. The devices specified in Part IV do not include the types of devices required by 210.8, 230.95, and 527.6.

430.52 Rating or Setting for Individual Motor Circuit.

(A) General. The motor branch-circuit short-circuit and ground-fault protective device shall comply with 430.52(B) and either 430.52(C) or (D), as applicable.

(B) All Motors. The motor branch-circuit short-circuit and ground-fault protective device shall be capable of carrying the starting current of the motor.

(C) Rating or Setting.

(1) In Accordance with Table 430.52. A protective device that has a rating or setting not exceeding the value calculated according to the values given in Table 430.52 shall be used.

Exception No. 1: Where the values for branch-circuit short-circuit and ground-fault protective devices determined by Table 430.52 do not correspond to the standard sizes or ratings of fuses, nonadjustable circuit breakers, thermal protective devices, or possible settings of adjustable circuit breakers, a higher size, rating, or possible setting that does not exceed the next higher standard ampere rating shall be permitted.

(2) Overload Relay Table. Where maximum branch-circuit short-circuit and ground-fault protective device ratings are shown in the manufacturer's overload relay table for use with a motor controller or are otherwise marked on the equipment, they shall not be exceeded even if higher values are allowed as shown above.

430.53 Several Motors or Loads on One Branch Circuit. Two or more motors or one or more motors and other loads shall be permitted to be connected to the same branch circuit under conditions specified in 430.53(D) and in 430.53(A), (B), or (C).

(A) Not Over 1 Horsepower. Several motors, each not exceeding 1 hp in rating, shall be permitted on a nominal 120-volt branch circuit protected at not over 20 amperes or a branch circuit of 600 volts, nominal, or less, protected at not over 15 amperes, if all of the following conditions are met:

Motors, Motor Circuits, and Controllers

Table 430.52 Maximum Rating or Setting of Motor Branch-Circuit Short-Circuit and Ground-Fault Protective Devices

	Percentage of Full-Load Current			
Type of Motor	Nontime Delay Fuse[1]	Dual Element (Time-Delay) Fuse[1]	Instantaneous Trip Breaker	Inverse Time Breaker[2]
Single-phase motors	300	175	800	250
AC polyphase motors other than wound-rotor				
Squirrel cage—other than Design E or Design B energy efficient	300	175	800	250
Design E or Design B energy efficient	300	175	1100	250
Synchronous[3]	300	175	800	250
Wound rotor	150	150	800	150
Direct current (constant voltage)	150	150	250	150

Note: For certain exceptions to the values specified, see 430.54.

[1] The values in the Nontime Delay Fuse column apply to Time-Delay Class CC fuses.

[2] The values given in the last column also cover the ratings of nonadjustable inverse time types of circuit breakers that may be modified as in 430.52(C), Exception No. 1 and No. 2.

[3] Synchronous motors of the low-torque, low-speed type (usually 450 rpm or lower), such as are used to drive reciprocating compressors, pumps, and so forth, that start unloaded, do not require a fuse rating or circuit-breaker setting in excess of 200 percent of full-load current.

(1) The full-load rating of each motor does not exceed 6 amperes.
(2) The rating of the branch-circuit short-circuit and ground-fault protective device marked on any of the controllers is not exceeded.
(3) Individual overload protection conforms to 430.32.

(B) If Smallest Rated Motor Protected. If the branch-circuit short-circuit and ground-fault protective device is selected not to exceed that allowed by 430.52 for the smallest rated motor, two or more motors or one or more motors and other load(s), with each motor having individual overload protection, shall be permitted to be connected to a branch circuit where it can be determined that the branch-circuit short-circuit and ground-fault protective device will not open under the most severe normal conditions of service that might be encountered.

(C) Other Group Installations. Two or more motors of any rating or one or more motors and other load(s), with each motor having individual overload protection, shall be permitted to be connected to one branch circuit where the motor controller(s) and overload device(s) are (1) installed as a listed factory assembly and the motor branch-circuit short-circuit and ground-fault protective device either is provided as part of the assembly or is specified by a marking on the assembly, or (2) the motor branch-circuit short-circuit and ground-fault protective device, the motor controller(s), and overload device(s) are field-installed as separate assemblies listed for such use and provided with manufacturers' instructions for use with each other, and (3) all of the following conditions are complied with:

(1) Each motor overload device is listed for group installation with a specified maximum rating of fuse, inverse time circuit breaker, or both.
(2) Each motor controller is listed for group installation with a specified maximum rating of fuse, circuit breaker, or both.
(3) Each circuit breaker is one of the inverse time type and listed for group installation.
(4) The branch circuit shall be protected by fuses or inverse time circuit breakers having a rating not exceeding that specified in 430.52 for the highest rated motor connected to the branch circuit plus an amount equal to the sum of the full-load current ratings of all other motors and the ratings of other loads connected to the circuit. Where this calculation results in a rating less than the ampacity of the supply conductors, it shall be permitted to increase the maximum rating of the fuses

Motors, Motor Circuits, and Controllers 265

or circuit breaker to a value not exceeding that permitted by 240.4(B).

(5) The branch-circuit fuses or inverse time circuit breakers are not larger than allowed by 430.40 for the overload relay protecting the smallest rated motor of the group.

(D) Single Motor Taps. For group installations described above, the conductors of any tap supplying a single motor shall not be required to have an individual branch-circuit short-circuit and ground-fault protective device, provided they comply with one of the following:

(1) No conductor to the motor shall have an ampacity less than that of the branch-circuit conductors.

(2) No conductor to the motor shall have an ampacity less than one-third that of the branch-circuit conductors, with a minimum in accordance with 430.22, the conductors to the motor overload device being not more than 7.5 m (25 ft) long and being protected from physical damage.

(3) Conductors from the branch-circuit short-circuit and ground-fault protective device to a listed manual motor controller additionally marked "Suitable for Tap Conductor Protection in Group Installations" shall be permitted to have an ampacity not less than 1/10 the rating or setting of the branch-circuit short-circuit and ground-fault protective device. The conductors from the controller to the motor shall have an ampacity in accordance with 430.22. The conductors from the branch-circuit short-circuit and ground-fault protective device to the controller shall (1) be suitably protected from physical damage and enclosed either by an enclosed controller or by a raceway and shall be not more than 3 m (10 ft) long or (2) shall have an ampacity not less than that of the branch circuit conductors.

V. MOTOR FEEDER SHORT-CIRCUIT AND GROUND-FAULT PROTECTION

430.61 General. Part V specifies protective devices intended to protect feeder conductors supplying motors against overcurrents due to short circuits or grounds.

430.62 Rating or Setting—Motor Load.

(A) Specific Load. A feeder supplying a specific fixed motor load(s) and consisting of conductor sizes based on 430.24 shall be provided with a protective device having a rating or setting not greater than the largest rating or setting of the branch-circuit short-circuit and ground-fault protective device for any motor supplied by the feeder [based on the maximum permitted value for the specific type of a protective device in accordance with 430.52, or 440.22(A) for hermetic refrigerant motor-compressors], plus the sum of the full-load currents of the other motors of the group.

Where the same rating or setting of the branch-circuit short-circuit and ground-fault protective device is used on two or more of the branch circuits supplied by the feeder, one of the protective devices shall be considered the largest for the above calculations.

(B) Other Installations. Where feeder conductors have an ampacity greater than required by 430.24, the rating or setting of the feeder overcurrent protective device shall be permitted to be based on the ampacity of the feeder conductors.

430.63 Rating or Setting—Power and Lighting Loads.

Where a feeder supplies a motor load and, in addition, a lighting or a lighting and appliance load, the feeder protective device shall have a rating sufficient to carry the lighting or lighting and appliance load, plus the following:

(1) For a single motor, the rating permitted by 430.52
(2) For a single hermetic refrigerant motor-compressor, the rating permitted by 440.22
(3) For two or more motors, the rating permitted by 430.62

VI. MOTOR CONTROL CIRCUITS

430.71 General.
Part VI contains modifications of the general requirements and applies to the particular conditions of motor control circuits.

430.72 Overcurrent Protection.

(A) General. A motor control circuit tapped from the load side of a motor branch-circuit short-circuit and ground-fault protective device(s) and functioning to control the motor(s) connected to that branch circuit shall be protected against overcurrent in accordance with 430.72. Such a tapped control circuit shall not be considered to be a branch circuit and shall be permitted to be protected by either a supplementary or branch-circuit overcurrent protective device(s). A motor control circuit other than such a tapped control circuit shall be protected against overcurrent in accordance with 725.23 or the notes to Tables 11(A) and (B) in Chapter 9, as applicable.

(B) Conductor Protection. The overcurrent protection for conductors shall be provided as specified in 430.72(B)(1) or (B)(2).

Exception No. 1: Where the opening of the control circuit would create a hazard as, for example, the control circuit of a fire pump motor, and the like, conductors of control circuits shall require only short-circuit and ground-fault protection and shall be permitted to be protected by the motor branch-circuit short-circuit and ground-fault protective device(s).

(1) Separate Overcurrent Protection. Where the motor branch-circuit short-circuit and ground-fault protective device does not provide protection in accordance with 430.72(B)(2), separate overcurrent protection shall be provided. The overcurrent protection shall not exceed the values specified in Column A of Table 430.72(B).

(2) Branch-Circuit Overcurrent Protective Device. Conductors shall be permitted to be protected by the motor branch-circuit short-circuit and ground-fault protective device and shall require only short-circuit and ground-fault protection. Where the conductors do not extend beyond the motor control equipment enclosure, the rating of the protective device(s) shall not exceed the value specified in Column B of Table 430.72(B). Where the conductors extend beyond the motor control equipment enclosure, the rating of the protective device(s) shall not exceed the value specified in Column C of Table 430.72(B).

(C) Control Circuit Transformer. Where a motor control circuit transformer is provided, the transformer shall be protected in accordance with 430.72(C)(1), (2), (3), (4), or (5).

(1) Compliance with Article 725. Where the transformer supplies a Class 1 power-limited circuit, Class 2, or Class 3

Table 430.72(B) Maximum Rating of Overcurrent Protective Device in Amperes

Control Circuit Conductor Size (AWG)	Column A Separate Protection Provided		Protection Provided by Motor Branch-Circuit Protection Device(s)			
			Column B Conductors Within Enclosure		Column C Conductors Extend Beyond Enclosure	
	Copper	Aluminum or Copper-Clad Aluminum	Copper	Aluminum or Copper-Clad Aluminum	Copper	Aluminum or Copper-Clad Aluminum
18	7	—	25	—	7	—
16	10	—	40	—	10	—
14	(Note 1)	—	100	—	45	—
12	(Note 1)	(Note 1)	120	100	60	45
10	(Note 1)	(Note 1)	160	140	90	75
Larger than 10	(Note 1)	(Note 1)	(Note 2)	(Note 2)	(Note 3)	(Note 3)

Notes:
1. Value specified in 310.15 as applicable.
2. 400 percent of value specified in Table 310.17 for 60°C conductors.
3. 300 percent of value specified in Table 310.16 for 60°C conductors.

Motors, Motor Circuits, and Controllers

remote-control circuit conforming with the requirements of Article 725, the protection shall comply with Article 725.

(2) Compliance with Article 450. Protection shall be permitted to be provided in accordance with 450.3.

(3) Less Than 50 Volt-Amperes. Control circuit transformers rated less than 50 volt-amperes (VA) and that are an integral part of the motor controller and located within the motor controller enclosure shall be permitted to be protected by primary overcurrent devices, impedance limiting means, or other inherent protective means.

(4) Primary Less Than 2 Amperes. Where the control circuit transformer rated primary current is less than 2 amperes, an overcurrent device rated or set at not more than 500 percent of the rated primary current shall be permitted in the primary circuit.

(5) Other Means. Protection shall be permitted to be provided by other approved means.

430.73 Mechanical Protection of Conductor. Where damage to a motor control circuit would constitute a hazard, all conductors of such a remote motor control circuit that are outside the control device itself shall be installed in a raceway or be otherwise suitably protected from physical damage.

Where one side of the motor control circuit is grounded, the motor control circuit shall be arranged so that an accidental ground in the control circuit remote from the motor controller will (1) not start the motor and (2) not bypass manually operated shutdown devices or automatic safety shutdown devices.

430.74 Disconnection.

(A) General. Motor control circuits shall be arranged so that they will be disconnected from all sources of supply when the disconnecting means is in the open position. The disconnecting means shall be permitted to consist of two or more separate devices, one of which disconnects the motor and the controller from the source(s) of power supply for the motor, and the other(s), the motor control circuit(s) from its power supply. Where separate devices are used, they shall be located immediately adjacent to each other.

(B) Control Transformer in Controller Enclosure. Where a transformer or other device is used to obtain a reduced voltage for the motor control circuit and is located in the controller enclosure, such transformer or other device shall be connected to the load side of the disconnecting means for the motor control circuit.

VII. MOTOR CONTROLLERS

430.81 General. Part VII is intended to require suitable controllers for all motors.

430.83 Ratings. The controller shall have a rating as specified in 430.83(A), unless otherwise permitted in 430.83(B) or (C), or as specified in 430.83(D), under the conditions specified.

(A) General.

(1) Horsepower Ratings. Controllers, other than inverse time circuit breakers and molded case switches, shall have horsepower ratings at the application voltage not lower than the horsepower rating of the motor. A controller for a Design E motor rated more than 2 hp shall (1) be marked as rated for use with a Design E motor or (2) have a horsepower rating not less than 1.4 times the rating of a motor rated 3 through 100 hp or not less than 1.3 times the rating of a motor rated over 100 hp.

(2) Circuit Breaker. A branch-circuit inverse time circuit breaker rated in amperes shall be permitted as a controller for all motors, including Design E. Where this circuit breaker is also used for overload protection, it shall conform to the appropriate provisions of this article governing overload protection.

(3) Molded Case Switch. A molded case switch rated in amperes shall be permitted as a controller for all motors, including Design E.

(B) Small Motors. Devices as specified in 430.81(B) and (C) shall be permitted as a controller.

(C) Stationary Motors of 2 Horsepower or Less. For stationary motors rated at 2 hp or less and 300 volts or less, the controller shall be permitted to be either of the following:

Motors, Motor Circuits, and Controllers

(1) A general-use switch having an ampere rating not less than twice the full-load current rating of the motor

(2) On ac circuits, a general-use snap switch suitable only for use on ac (not general-use ac–dc snap switches) where the motor full-load current rating is not more than 80 percent of the ampere rating of the switch

(D) Torque Motors. For torque motors, the controller shall have a continuous-duty, full-load current rating not less than the nameplate current rating of the motor. For a motor controller rated in horsepower but not marked with the foregoing current rating, the equivalent current rating shall be determined from the horsepower rating by using Tables 430.147, 430.148, 430.149, or 430.150.

(E) Voltage Rating. A controller with a straight voltage rating, for example, 240 volts or 480 volts, shall be permitted to be applied in a circuit in which the nominal voltage between any two conductors does not exceed the controller's voltage rating. A controller with a slash rating, for example, 120/240 volts or 480/277 volts, shall only be applied in a circuit in which the nominal voltage to ground from any conductor does not exceed the lower of the two values of the controller's voltage rating and the nominal voltage between any two conductors does not exceed the higher value of the controller's voltage rating.

430.91. Motor Controller Enclosure Types. Table 430.91 provides the basis for selecting enclosures for use in specific locations other than hazardous (classified) locations. The enclosures are not intended to protect against conditions such as condensation, icing, corrosion, or contamination that may occur within the enclosure or enter via the conduit or unsealed openings. These internal conditions shall require special consideration by the installer and user.

VIII. MOTOR CONTROL CENTERS

430.92 General. Part VIII covers motor control centers installed for the control of motors, lighting, and power circuits.

430.94 Overcurrent Protection. Motor control centers shall be provided with overcurrent protection in accordance with Parts I, II, and IX of Article 240. The ampere rating or

Table 430.91 Motor Controller Enclosure Selection

Provides a Degree of Protection Against the Following Environmental Conditions	For Outdoor Use									
	Enclosure Type Number[1]									
	3	3R	3S	3X	3RX	3SX	4	4X	6	6P
Incidental contact with the enclosed equipment	X	X	X	X	X	X	X	X	X	X
Rain, snow, and sleet	X	X	X	X	X	X	X	X	X	X
Sleet[2]	—	—	X	—	—	X	—	—	—	—
Windblown dust	X	—	X	X	—	X	X	X	—	X
Hosedown	—	—	—	—	—	—	X	X	X	X
Corrosive agents	—	—	—	—	X	X	—	X	—	X
Temporary submersion	—	—	—	—	—	—	—	—	X	X
Prolonged submersion	—	—	—	—	—	—	—	—	—	X

Motors, Motor Circuits, and Controllers

Provides a Degree of Protection Against the Following Environmental Conditions	For Indoor Use — Enclosure Type Number[1]									
	1	2	4	4X	5	6	6P	12	12K	13
Incidental contact with the enclosed equipment	X	X	X	X	X	X	X	X	X	X
Falling dirt	X	X	X	X	X	X	X	X	X	X
Falling liquids, and light splashing	—	X	X	X	X	X	X	X	X	X
Circulating dust, lint, fibers, and flyings[2]	—	—	X	X	—	X	X	X	X	X
Settling airborne dust, lint, fibers, and flyings	—	—	X	X	X	X	X	X	X	X
Hosedown and splashing water	—	—	X	X	—	X	X	—	—	—
Oil and coolant seepage	—	—	—	—	—	—	—	X	X	X
Oil or coolant spraying and splashing	—	—	—	—	—	—	—	—	—	X
Corrosive agents	—	—	—	X	—	—	X	—	—	—
Temporary submersion	—	—	—	—	—	X	X	—	—	—
Prolonged submersion	—	—	—	—	—	—	X	—	—	—

[1] Enclosure type number shall be marked on the motor controller enclosure.
[2] Mechanism shall be operable when ice covered.

setting of the overcurrent protective device shall not exceed the rating of the common power bus. This protection shall be provided by (1) an overcurrent protective device located ahead of the motor control center or (2) a main overcurrent protective device located within the motor control center.

430.96 Grounding. Multisection motor control centers shall be bonded together with an equipment grounding conductor or an equivalent grounding bus sized in accordance with Table 250.122. Equipment grounding conductors shall terminate on this grounding bus or to a grounding termination point provided in a single-section motor control center.

IX. DISCONNECTING MEANS

430.101 General. Part IX is intended to require disconnecting means capable of disconnecting motors and controllers from the circuit.

430.102 Location.

(A) Controller. An individual disconnecting means shall be provided for each controller and shall disconnect the controller. The disconnecting means shall be located in sight from the controller location.

Exception No. 2: A single disconnecting means shall be permitted for a group of coordinated controllers that drive several parts of a single machine or piece of apparatus. The disconnecting means shall be located in sight from the controllers, and both the disconnecting means and the controllers shall be located in sight from the machine or apparatus.

(B) Motor. A disconnecting means shall be located in sight from the motor location and the driven machinery location. The disconnecting means required in accordance with 430.102(A) shall be permitted to serve as the disconnecting means for the motor if it is located in sight from the motor location and the driven machinery location.

Exception: The disconnecting means shall not be required to be in sight from the motor and the driven machinery location under either condition (a) or (b), provided the disconnecting means required in accordance with 430.102(A)

Motors, Motor Circuits, and Controllers

is individually capable of being locked in the open position. The provision for locking or adding a lock to the disconnecting means shall be permanently installed on or at the switch or circuit breaker used as the disconnecting means.

(a) Where such a location of the disconnecting means is impracticable or introduces additional or increased hazards to persons or property

(b) In industrial installations, with written safety procedures, where conditions of maintenance and supervision ensure that only qualified persons service the equipment

> FPN No. 1: Some examples of increased or additional hazards include, but are not limited to, motors rated in excess of 100 hp, multimotor equipment, submersible motors, motors associated with variable frequency drives, and motors located in hazardous (classified) locations.
>
> FPN No. 2: For information on lockout/tagout procedures, see NFPA 70E-2000, *Standard for Electrical Safety Requirements for Employee Workplaces*.

430.103 Operation. The disconnecting means shall open all ungrounded supply conductors and shall be designed so that no pole can be operated independently. The disconnecting means shall be permitted in the same enclosure with the controller.

430.104 To Be Indicating. The disconnecting means shall plainly indicate whether it is in the open (off) or closed (on) position.

430.107 Readily Accessible. At least one of the disconnecting means shall be readily accessible.

430.108 Every Switch. Every disconnecting means in the motor circuit between the point of attachment to the feeder and the point of connection to the motor shall comply with the requirements of 430.109 and 430.110.

430.109 Type. The disconnecting means shall be a type specified in 430.109(A), unless otherwise permitted in 430.109(B) through (G), under the conditions specified.

(A) General.

(1) Motor Circuit Switch. A listed motor-circuit switch rated in horsepower. For Design E motors rated greater than 2 hp, the motor circuit switch shall be either (1) marked as rated for use with Design E motors or (2) have a horsepower rating not less than 1.4 times the rating of a motor rated 3–100 hp, or not less than 1.3 times the rating of a motor rated over 100 hp.

(2) Molded Case Circuit Breaker. A listed molded case circuit breaker.

(3) Molded Case Switch. A listed molded case switch.

(4) Instantaneous Trip Circuit Breaker. An instantaneous trip circuit breaker that is part of a listed combination motor controller.

(5) Self-Protected Combination Controller. Listed self-protected combination controller.

(6) Manual Motor Controller. Listed manual motor controllers additionally marked "Suitable as Motor Disconnect" shall be permitted as a disconnecting means where installed between the final motor branch-circuit short-circuit and ground-fault protective device and the motor.

(B) Stationary Motors of 1/8 Horsepower or Less. For stationary motors of 1/8 hp or less, the branch-circuit overcurrent device shall be permitted to serve as the disconnecting means.

(C) Stationary Motors of 2 Horsepower or Less. For stationary motors rated at 2 hp or less and 300 volts or less, the disconnecting means shall be permitted to be one of the devices specified in (1), (2), or (3):

(1) A general-use switch having an ampere rating not less than twice the full-load current rating of the motor
(2) On ac circuits, a general-use snap switch suitable only for use on ac (not general-use ac–dc snap switches) where the motor full-load current rating is not more than 80 percent of the ampere rating of the switch

Motors, Motor Circuits, and Controllers

(3) A listed manual motor controller having a horsepower rating not less than the rating of the motor and marked "Suitable as Motor Disconnect"

(D) Autotransformer-Type Controlled Motors. For motors of over 2 hp to and including 100 hp, the separate disconnecting means required for a motor with an autotransformer-type controller shall be permitted to be a general-use switch where all of the following provisions are met:

(1) The motor drives a generator that is provided with overload protection.
(2) The controller is capable of interrupting the locked-rotor current of the motors, is provided with a no voltage release, and is provided with running overload protection not exceeding 125 percent of the motor full-load current rating.
(3) Separate fuses or an inverse time circuit breaker rated or set at not more than 150 percent of the motor full-load current are provided in the motor branch circuit.

(E) Isolating Switches. For stationary motors rated at more than 40 hp dc or 100 hp ac, the disconnecting means shall be permitted to be a general-use or isolating switch where plainly marked "Do not operate under load."

(F) Cord-and-Plug-Connected Motors. For a cord-and-plug-connected motor, a horsepower-rated attachment plug and receptacle having ratings no less than the motor ratings shall be permitted to serve as the disconnecting means for other than a Design E motor and for a Design E motor rated 2 hp or less. For a Design E motor rated more than 2 hp, an attachment plug and receptacle used as the disconnecting means shall have a horsepower rating not less than 1.4 times the motor rating. A horsepower-rated attachment plug and receptacle shall not be required for a cord-and-plug-connected appliance in accordance with 422.33, a room air conditioner in accordance with 440.63, or a portable motor rated 1/3 hp or less.

(G) Torque Motors. For torque motors, the disconnecting means shall be permitted to be a general-use switch.

430.110 Ampere Rating and Interrupting Capacity.

(A) General. The disconnecting means for motor circuits rated 600 volts, nominal, or less shall have an ampere rating of at least 115 percent of the full-load current rating of the motor.

(B) For Torque Motors. Disconnecting means for a torque motor shall have an ampere rating of at least 115 percent of the motor nameplate current.

(C) For Combination Loads. Where two or more motors are used together or where one or more motors are used in combination with other loads, such as resistance heaters, and where the combined load may be simultaneous on a single disconnecting means, the ampere and horsepower ratings of the combined load shall be determined as follows.

(1) Horsepower Rating. The rating of the disconnecting means shall be determined from the sum of all currents, including resistance loads, at the full-load condition and also at the locked-rotor condition. The combined full-load current and the combined locked-rotor current so obtained shall be considered as a single motor for the purpose of this requirement as follows.

The full-load current equivalent to the horsepower rating of each motor shall be selected from Tables 430.147, 430.148, 430.149, or 430.150. These full-load currents shall be added to the rating in amperes of other loads to obtain an equivalent full-load current for the combined load.

The locked-rotor current equivalent to the horsepower rating of each motor shall be selected from Tables 430.151(A) or 430.151(B). The locked-rotor currents shall be added to the rating in amperes of other loads to obtain an equivalent locked-rotor current for the combined load. Where two or more motors or other loads cannot be started simultaneously, the largest sum of locked rotor currents of a motor or group of motors that can be started simultaneously and the full load currents of other concurrent loads shall be permitted to be used to determine the equivalent locked-rotor current for the simultaneous combined loads.

(2) Ampere Rating. The ampere rating of the disconnecting means shall not be less than 115 percent of the sum of

Motors, Motor Circuits, and Controllers

all currents at the full-load condition determined in accordance with 430.110(C)(1).

(3) Small Motors. For small motors not covered by Tables 430.147, 430.148, 430.149, or 430.150, the locked-rotor current shall be assumed to be six times the full-load current.

430.111 Switch or Circuit Breaker as Both Controller and Disconnecting Means. A switch or circuit breaker shall be permitted to be used as both the controller and disconnecting means if it complies with 430.111(A) and is one of the types specified in 430.111(B).

(A) General. The switch or circuit breaker complies with the requirements for controllers specified in 430.83, opens all ungrounded conductors to the motor, and is protected by an overcurrent device in each ungrounded conductor (which shall be permitted to be the branch-circuit fuses). The overcurrent device protecting the controller shall be permitted to be part of the controller assembly or shall be permitted to be separate. An autotransformer-type controller shall be provided with a separate disconnecting means.

(B) Type. The device shall be one of the types specified in 430.111(B)(1), (2), or (3).

(1) Air-Break Switch. An air-break switch, operable directly by applying the hand to a lever or handle.

(2) Inverse Time Circuit Breaker. An inverse time circuit breaker operable directly by applying the hand to a lever or handle. The circuit breaker shall be permitted to be both power and manually operable.

(3) Oil Switch. An oil switch used on a circuit whose rating does not exceed 600 volts or 100 amperes, or by special permission on a circuit exceeding this capacity where under expert supervision. The oil switch shall be permitted to be both power and manually operable.

430.112 Motors Served by Single Disconnecting Means. Each motor shall be provided with an individual disconnecting means.

XI. PROTECTION OF LIVE PARTS—ALL VOLTAGES

430.131 General. Part XI specifies that live parts shall be protected in a manner judged adequate for the hazard involved.

430.132 Where Required. Exposed live parts of motors and controllers operating at 50 volts or more between terminals shall be guarded against accidental contact by enclosure or by location as follows:

(1) By installation in a room or enclosure that is accessible only to qualified persons
(2) By installation on a suitable balcony, gallery, or platform, elevated and arranged so as to exclude unqualified persons
(3) By elevation 2.5 m (8 ft) or more above the floor

430.133 Guards for Attendants. Where live parts of motors or controllers operating at over 150 volts to ground are guarded against accidental contact only by location as specified in 430.132, and where adjustment or other attendance may be necessary during the operation of the apparatus, suitable insulating mats or platforms shall be provided so that the attendant cannot readily touch live parts unless standing on the mats or platforms.

XII. GROUNDING—ALL VOLTAGES

430.141 General. Part XII specifies the grounding of exposed non–current-carrying metal parts, likely to become energized, of motor and controller frames to prevent a voltage above ground in the event of accidental contact between energized parts and frames. Insulation, isolation, or guarding are suitable alternatives to grounding of motors under certain conditions.

430.142 Stationary Motors. The frames of stationary motors shall be grounded under any of the following conditions:

(1) Where supplied by metal-enclosed wiring
(2) Where in a wet location and not isolated or guarded
(3) If in a hazardous (classified) location as covered in Articles 500 through 517

Motors, Motor Circuits, and Controllers

(4) If the motor operates with any terminal at over 150 volts to ground

Where the frame of the motor is not grounded, it shall be permanently and effectively insulated from the ground.

430.143 Portable Motors. The frames of portable motors that operate at over 150 volts to ground shall be guarded or grounded.

430.144 Controllers. Controller enclosures shall be grounded regardless of voltage. Controller enclosures shall have means for attachment of an equipment grounding conductor termination in accordance with 250.8.

430.145 Method of Grounding. Where required, grounding shall be done in the manner specified in Part V of Article 250.

(A) Grounding Through Terminal Housings. Where the wiring to fixed motors is metal-enclosed cable or in metal raceways, junction boxes to house motor terminals shall be provided, and the armor of the cable or the metal raceways shall be connected to them in the manner specified in Article 250.

(B) Separation of Junction Box from Motor. The junction box required by 430.145(A) shall be permitted to be separated from the motor by not more than 1.8 m (6 ft), provided the leads to the motor are Type AC cable or armored cord or are stranded leads enclosed in liquidtight flexible metal conduit, flexible metal conduit, intermediate metal conduit, rigid metal conduit, or electrical metallic tubing not smaller than metric designator 12 (trade size 3/8), the armor or raceway being connected both to the motor and to the box.

Liquidtight flexible nonmetallic conduit and rigid nonmetallic conduit shall be permitted to enclose the leads to the motor, provided the leads are stranded and the required equipment grounding conductor is connected to both the motor and to the box.

Where stranded leads are used, protected as specified above, each strand within the conductor shall be not larger than 10 AWG and shall comply with other requirements of this *Code* for conductors to be used in raceways.

(C) Grounding of Controller-Mounted Devices. Instrument transformer secondaries and exposed non–current-carrying metal or other conductive parts or cases of instrument transformers, meters, instruments, and relays shall be grounded as specified in 250.170 through 250.178.

XIII. TABLES

Table 430.147 Full-Load Current in Amperes, Direct-Current Motors

The following values of full-load currents* are for motors running at base speed.

Horsepower	Armature Voltage Rating*					
	90 Volts	120 Volts	180 Volts	240 Volts	500 Volts	550 Volts
1/4	4.0	3.1	2.0	1.6	—	—
1/3	5.2	4.1	2.6	2.0	—	—
1/2	6.8	5.4	3.4	2.7	—	—
3/4	9.6	7.6	4.8	3.8	—	—
1	12.2	9.5	6.1	4.7	—	—
1-1/2	—	13.2	8.3	6.6	—	—
2	—	17	10.8	8.5	—	—
3	—	25	16	12.2	—	—
5	—	40	27	20	—	—
7-1/2	—	58	—	29	13.6	12.2
10	—	76	—	38	18	16
15	—	—	—	55	27	24
20	—	—	—	72	34	31
25	—	—	—	89	43	38
30	—	—	—	106	51	46
40	—	—	—	140	67	61
50	—	—	—	173	83	75
60	—	—	—	206	99	90
75	—	—	—	255	123	111
100	—	—	—	341	164	148
125	—	—	—	425	205	185
150	—	—	—	506	246	222
200	—	—	—	675	330	294

*These are average dc quantities.

Motors, Motor Circuits, and Controllers

Table 430.148 Full-Load Currents in Amperes, Single-Phase Alternating-Current Motors

The following values of full-load currents are for motors running at usual speeds and motors with normal torque characteristics. Motors built for especially low speeds or high torques may have higher full-load currents, and multispeed motors will have full-load current varying with speed, in which case the nameplate current ratings shall be used.

The voltages listed are rated motor voltages. The currents listed shall be permitted for system voltage ranges of 110 to 120 and 220 to 240 volts.

Horsepower	115 Volts	200 Volts	208 Volts	230 Volts
1/6	4.4	2.5	2.4	2.2
1/4	5.8	3.3	3.2	2.9
1/3	7.2	4.1	4.0	3.6
1/2	9.8	5.6	5.4	4.9
3/4	13.8	7.9	7.6	6.9
1	16	9.2	8.8	8.0
1-1/2	20	11.5	11.0	10
2	24	13.8	13.2	12
3	34	19.6	18.7	17
5	56	32.2	30.8	28
7-1/2	80	46.0	44.0	40
10	100	57.5	55.0	50

Table 430.150 Full-Load Current, Three-Phase Alternating-Current Motors

The following values of full-load currents are typical for motors running at speeds usual for belted motors and motors with normal torque characteristics.

Motors built for low speeds (1200 rpm or less) or high torques may require more running current, and multispeed motors will have full-load current varying with speed. In these cases, the nameplate current rating shall be used.

The voltages listed are rated motor voltages. The currents listed shall be permitted for system voltage ranges of 110 to 120, 220 to 240, 440 to 480, and 550 to 600 volts.

Horsepower	Induction-Type Squirrel Cage and Wound Rotor (Amperes)							Synchronous-Type Unity Power Factor* (Amperes)				
	115 Volts	200 Volts	208 Volts	230 Volts	460 Volts	575 Volts	2300 Volts	230 Volts	460 Volts	575 Volts	2300 Volts
1/2	4.4	2.5	2.4	2.2	1.1	0.9	—	—	—	—	—
3/4	6.4	3.7	3.5	3.2	1.6	1.3	—	—	—	—	—
1	8.4	4.8	4.6	4.2	2.1	1.7	—	—	—	—	—
1-1/2	12.0	6.9	6.6	6.0	3.0	2.4	—	—	—	—	—
2	13.6	7.8	7.5	6.8	3.4	2.7	—	—	—	—	—
3	—	11.0	10.6	9.6	4.8	3.9	—	—	—	—	—
5	—	17.5	16.7	15.2	7.6	6.1	—	—	—	—	—
7-1/2	—	25.3	24.2	22	11	9	—	—	—	—	—

Motors, Motor Circuits, and Controllers

HP											
10	—	32.2	30.8	28	14	11	—	—	—	—	—
15	—	48.3	46.2	42	21	17	—	—	—	—	—
20	—	62.1	59.4	54	27	22	—	—	—	—	—
25	—	78.2	74.8	68	34	27	—	53	26	21	12
30	—	92	88	80	40	32	—	63	32	26	15
40	—	120	114	104	52	41	—	83	41	33	20
50	—	150	143	130	65	52	16	104	52	42	25
60	—	177	169	154	77	62	20	123	61	49	30
75	—	221	211	192	96	77	26	155	78	62	40
100	—	285	273	248	124	99	31	202	101	81	—
125	—	359	343	312	156	125	37	253	126	101	—
150	—	414	396	360	180	144	49	302	151	121	—
200	—	552	528	480	240	192	49	400	201	161	—
250	—	—	—	—	302	242	60	—	—	—	—
300	—	—	—	—	361	289	72	—	—	—	—
350	—	—	—	—	414	336	83	—	—	—	—
400	—	—	—	—	477	382	95	—	—	—	—
450	—	—	—	—	515	412	103	—	—	—	—
500	—	—	—	—	590	472	118	—	—	—	—

*For 90 and 80 percent power factor, the figures shall be multiplied by 1.1 and 1.25, respectively.

Table 430.151(A) Conversion Table of Single-Phase Locked-Rotor Currents for Selection of Disconnecting Means and Controllers as Determined from Horsepower and Voltage Rating

For use only with 430.110, 440.12, 440.41, and 455.8(C).

Rated Horsepower	Maximum Locked-Rotor Current in Amperes, Single Phase		
	115 Volts	208 Volts	230 Volts
1/2	58.8	32.5	29.4
3/4	82.8	45.8	41.4
1	96	53	48
1-1/2	120	66	60
2	144	80	72
3	204	113	102
5	336	186	168
7-1/2	480	265	240
10	600	332	300

Motors, Motor Circuits, and Controllers

Table 430.151(B) Conversion Table of Polyphase Design B, C, D, and E Maximum Locked-Rotor Currents for Selection of Disconnecting Means and Controllers as Determined from Horsepower and Voltage Rating and Design Letter

For use only with 430.110, 440.12*, 440.41*, and 455.8(C).

	Maximum Motor Locked-Rotor Current in Amperes Two- and Three-Phase, Design B, C, D, and E									
	115 Volts		200 Volts		208 Volts		230 Volts		460 Volts		575 Volts	
Rated Horsepower	B, C, D	E	B, C, D	E	B, C, D	E	B, C, D	E	B, C, D	E	B, C, D	E
1/2	40	40	23	23	22.1	22.1	20	20	10	10	8	8
3/4	50	50	28.8	28.8	27.6	27.6	25	25	12.5	12.5	10	10
1	60	60	34.5	34.5	33	33	30	30	15	15	12	12
1-1/2	80	80	46	46	44	44	40	40	20	20	16	16
2	100	100	57.5	57.5	55	55	50	50	25	25	20	20
3	—	—	73.6	84	71	81	64	73	32	36.5	25.6	29.2
5	—	—	105.8	140	102	135	92	122	46	61	36.8	48.8
7-1/2	—	—	146	210	140	202	127	183	63.5	91.5	50.8	73.2
10	—	—	186.3	259	179	249	162	225	81	113	64.8	90
15	—	—	267	388	257	373	232	337	116	169	93	135

Continued

Table 430.151(B) Conversion Table of Polyphase Design B, C, D, and E Maximum Locked-Rotor Currents for Selection of Disconnecting Means and Controllers as Determined from Horsepower and Voltage Rating and Design Letter *Continued*

For use only with 430.110, 440.12*, 440.41*, and 455.8(C).

Rated Horsepower	Maximum Motor Locked-Rotor Current in Amperes Two- and Three-Phase, Design B, C, D, and E										
	115 Volts		200 Volts		208 Volts		230 Volts		460 Volts		575 Volts	
	B, C, D	E	B, C, D	E	B, C, D	E	B, C, D	E	B, C, D	E	B, C, D	E
20	—	—	334	516	321	497	290	449	145	225	116	180
25	—	—	420	646	404	621	365	562	183	281	146	225
30	—	—	500	775	481	745	435	674	218	337	174	270
40	—	—	667	948	641	911	580	824	290	412	232	330
50	—	—	834	1185	802	1139	725	1030	363	515	290	412
60	—	—	1001	1421	962	1367	870	1236	435	618	348	494
75	—	—	1248	1777	1200	1708	1085	1545	543	773	434	618
100	—	—	1668	2154	1603	2071	1450	1873	725	937	580	749
125	—	—	2087	2692	2007	2589	1815	2341	908	1171	726	936

Motors, Motor Circuits, and Controllers

150	—	—	2496	—	—	—	—	1085	1405	868	1124
200	—	—	3335	—	—	—	—	1450	1873	1160	1498
250	—	—	3230	2400	3106	2170	2809	1825	2344	1460	1875
300	—	—	4307	3207	4141	2900	3745	2220	2809	1760	2247
350	—	—	—	—	—	—	—	2550	3277	2040	2622
400	—	—	—	—	—	—	—	2900	3745	2320	2996
450	—	—	—	—	—	—	—	3250	4214	2600	3371
500	—	—	—	—	—	—	—	3625	4682	2900	3746

*In determining compliance with 440.12 and 440.41, the values in the B, C, D columns shall be used.

CHAPTER 20

AIR-CONDITIONING AND REFRIGERATING EQUIPMENT

INTRODUCTION

This chapter contains provisions from Article 440—Air-Conditioning and Refrigerating Equipment. Article 440 covers electric motor–driven air-conditioning and refrigerating equipment, including their branch circuits and controllers. Provided in Article 440 are special considerations necessary for circuits supplying hermetic refrigerant motor-compressors and for any air-conditioning or refrigerating equipment that is supplied from a branch circuit that supplies a hermetic refrigerant motor-compressor.

Contained in this *Pocket Guide* are six of the seven parts that are included in Article 440. The six parts are: I General; II Disconnecting Means; III Branch-Circuit Short-Circuit and Ground-Fault Protection; IV Branch-Circuit Conductors; V Controllers for Motor-Compressors; and VI Motor-Compressor and Branch-Circuit Overload Protection.

ARTICLE 440
Air-Conditioning and Refrigerating Equipment

I. GENERAL

440.1 Scope. The provisions of this article apply to electric motor-driven air-conditioning and refrigerating equipment and to the branch circuits and controllers for such equipment. It provides for the special considerations necessary for circuits supplying hermetic refrigerant motor-compressors and for any air-conditioning or refrigerating equipment that is supplied from a branch circuit that supplies a hermetic refrigerant motor-compressor.

440.6 Ampacity and Rating. The size of conductors for equipment covered by this article shall be selected from Tables 310.16 through 310.19 or calculated in accordance with 310.15 as applicable. The required ampacity of conductors and rating of equipment shall be determined according to 440.6(A) and (B).

(A) Hermetic Refrigerant Motor-Compressor. For a hermetic refrigerant motor-compressor, the rated-load current marked on the nameplate of the equipment in which the motor-compressor is employed shall be used in determining the rating or ampacity of the disconnecting means, the branch-circuit conductors, the controller, the branch-circuit short-circuit and ground-fault protection, and the separate motor overload protection. Where no rated-load current is shown on the equipment nameplate, the rated-load current shown on the compressor nameplate shall be used.

Exception No. 1: Where so marked, the branch-circuit selection current shall be used instead of the rated-load current to determine the rating or ampacity of the disconnecting means, the branch-circuit conductors, the controller, and the branch-circuit short-circuit and ground-fault protection.

Exception No. 2: For cord-and-plug-connected equipment, the nameplate marking shall be used in accordance with 440.22(B), Exception No. 2.

(B) Multimotor Equipment. For multimotor equipment employing a shaded-pole or permanent split-capacitor-

Air-Conditioning and Refrigerating Equipment

type fan or blower motor, the full-load current for such motor marked on the nameplate of the equipment in which the fan or blower motor is employed shall be used instead of the horsepower rating to determine the ampacity or rating of the disconnecting means, the branch-circuit conductors, the controller, the branch-circuit short-circuit and ground-fault protection, and the separate overload protection. This marking on the equipment nameplate shall not be less than the current marked on the fan or blower motor nameplate.

440.7 Highest Rated (Largest) Motor. In determining compliance with this article and with 430.24, 430.53(B) and (C), and 430.62(A), the highest rated (largest) motor shall be considered to be the motor that has the highest rated-load current. Where two or more motors have the same highest rated-load current, only one of them shall be considered as the highest rated (largest) motor. For other than hermetic refrigerant motor-compressors, and fan or blower motors as covered in 440.6(B), the full-load current used to determine the highest rated motor shall be the equivalent value corresponding to the motor horsepower rating selected from Tables 430.148, 430.149, or 430.150.

440.8 Single Machine. An air-conditioning or refrigerating system shall be considered to be a single machine under the provisions of 430.87, Exception, and 430.112, Exception. The motors shall be permitted to be located remotely from each other.

II. DISCONNECTING MEANS

440.11 General. The provisions of Part II are intended to require disconnecting means capable of disconnecting air-conditioning and refrigerating equipment, including motor-compressors and controllers from the circuit conductors.

440.12 Rating and Interrupting Capacity.

(A) Hermetic Refrigerant Motor-Compressor. A disconnecting means serving a hermetic refrigerant motor-compressor shall be selected on the basis of the nameplate rated-load current or branch-circuit selection current, whichever is greater, and locked-rotor current, respectively, of the motor-compressor as follows:

(1) Ampere Rating. The ampere rating shall be at least 115 percent of the nameplate rated-load current or branch-circuit selection current, whichever is greater.

Exception: A listed nonfused motor circuit switch having a horsepower rating not less than the equivalent horsepower determined in accordance with 440.12(A)(2) shall be permitted have an ampere rating less than 115 percent of the specified current.

(2) Equivalent Horsepower. To determine the equivalent horsepower in complying with the requirements of 430.109, the horsepower rating shall be selected from Tables 430.148, 430.149, or 430.150 corresponding to the rated-load current or branch-circuit selection current, whichever is greater, and also the horsepower rating from Tables 430.151(A) or 430.151(B) corresponding to the locked-rotor current. In case the nameplate rated-load current or branch-circuit selection current and locked-rotor current do not correspond to the currents shown in Tables 430.148, 430.149, 430.150, 430.151(A), or 430.151(B), the horsepower rating corresponding to the next higher value shall be selected. In case different horsepower ratings are obtained when applying these tables, a horsepower rating at least equal to the larger of the values obtained shall be selected.

(B) Combination Loads. Where the combined load of two or more hermetic refrigerant motor-compressors or one or more hermetic refrigerant motor-compressor with other motors or loads may be simultaneous on a single disconnecting means, the rating for the disconnecting means shall be determined in accordance with 440.12(B)(1) and (B)(2).

(1) Horsepower Rating. The horsepower rating of the disconnecting means shall be determined from the sum of all currents, including resistance loads, at the rated-load condition and also at the locked-rotor condition. The combined rated-load current and the combined locked-rotor current so obtained shall be considered as a single motor for the purpose of this requirement as follows:

(a) The full-load current equivalent to the horsepower rating of each motor, other than a hermetic refrigerant motor-compressor, and fan or blower motors as covered in 440.6(B) shall be selected from Tables 430.148, 430.149, or 430.150. These full-load currents shall be added to the motor-compressor rated-load current(s) or branch-circuit selection current(s), whichever is greater, and to the rating in amperes of other loads to obtain an equivalent full-load current for the combined load.

(b) The locked-rotor current equivalent to the horsepower rating of each motor, other than a hermetic refrigerant motor-compressor, shall be selected from Tables 430.151(A) or 430.151(B), and, for fan and blower motors of the shaded-pole or permanent split-capacitor type marked with the locked-rotor current, the marked value shall be used. The locked-rotor currents shall be added to the motor-compressor locked-rotor current(s) and to the rating in amperes of other loads to obtain an equivalent locked-rotor current for the combined load. Where two or more motors or other loads such as resistance heaters, or both, cannot be started simultaneously, appropriate combinations of locked-rotor and rated-load current or branch-circuit selection current, whichever is greater, shall be an acceptable means of determining the equivalent locked-rotor current for the simultaneous combined load.

(2) Full-Load Current Equivalent. The ampere rating of the disconnecting means shall be at least 115 percent of the sum of all currents at the rated-load condition determined in accordance with 440.12(B)(1).

(C) Small Motor-Compressors. For small motor-compressors having the locked-rotor current marked on the nameplate, or for small motors not covered by Tables 430.147, 430.148, 430.149, or 430.150, the locked-rotor current shall be assumed to be six times the rated-load current.

(D) Every Switch. Every disconnecting means in the refrigerant motor-compressor circuit between the point of attachment to the feeder and the point of connection to the refrigerant motor-compressor shall comply with the requirements of 440.12.

(E) Disconnecting Means Rated in Excess of 100 Horsepower. Where the rated-load or locked-rotor current as determined above would indicate a disconnecting means rated in excess of 100 hp, the provisions of 430.109(E) shall apply.

440.13 Cord-Connected Equipment. For cord-connected equipment such as room air conditioners, household refrigerators and freezers, drinking water coolers, and beverage dispensers, a separable connector or an attachment plug and receptacle shall be permitted to serve as the disconnecting means.

440.14 Location. Disconnecting means shall be located within sight from and readily accessible from the air-conditioning or refrigerating equipment. The disconnecting means shall be permitted to be installed on or within the air-conditioning or refrigerating equipment.

The disconnecting means shall not be located on panels that are designed to allow access to the air-conditioning or refrigeration equipment.

III. BRANCH-CIRCUIT SHORT-CIRCUIT AND GROUND-FAULT PROTECTION

440.21 General. The provisions of Part III specify devices intended to protect the branch-circuit conductors, control apparatus, and motors in circuits supplying hermetic refrigerant motor-compressors against overcurrent due to short circuits and grounds. They are in addition to or amendatory of the provisions of Article 240.

440.22 Application and Selection.

(A) Rating or Setting for Individual Motor Compressor. The motor-compressor branch-circuit short-circuit and ground-fault protective device shall be capable of carrying the starting current of the motor. A protective device having a rating or setting not exceeding 175 percent of the motor-compressor rated-load current or branch-circuit selection current, whichever is greater, shall be permitted, provided that, where the protection specified is not sufficient for the starting current of the motor, the rating or setting shall be

Air-Conditioning and Refrigerating Equipment

permitted to be increased but shall not exceed 225 percent of the motor rated-load current or branch-circuit selection current, whichever is greater.

Exception: The rating of the branch-circuit short-circuit and ground-fault protective device shall not be required to be less than 15 amperes.

(B) Rating or Setting for Equipment. The equipment branch-circuit short-circuit and ground-fault protective device shall be capable of carrying the starting current of the equipment. Where the hermetic refrigerant motor-compressor is the only load on the circuit, the protection shall conform with 440.22(A). Where the equipment incorporates more than one hermetic refrigerant motor-compressor or a hermetic refrigerant motor-compressor and other motors or other loads, the equipment short-circuit and ground-fault protection shall conform with 430.53 and the following.

(1) Motor-Compressor Largest Load. Where a hermetic refrigerant motor-compressor is the largest load connected to the circuit, the rating or setting of the branch-circuit short-circuit and ground-fault protective device shall not exceed the value specified in 440.22(A) for the largest motor-compressor plus the sum of the rated-load current or branch-circuit selection current, whichever is greater, of the other motor-compressor(s) and the ratings of the other loads supplied.

(2) Motor-Compressor Not Largest Load. Where a hermetic refrigerant motor-compressor is not the largest load connected to the circuit, the rating or setting of the branch-circuit short-circuit and ground-fault protective device shall not exceed a value equal to the sum of the rated-load current or branch-circuit selection current, whichever is greater, rating(s) for the motor-compressor(s) plus the value specified in 430.53(C)(4) where other motor loads are supplied, or the value specified in 240.4 where only nonmotor loads are supplied in addition to the motor-compressor(s).

Exception No. 1: Equipment that starts and operates on a 15- or 20-ampere 120-volt, or 15-ampere 208- or 240-volt single-phase branch circuit, shall be permitted to be protected by the 15- or 20-ampere overcurrent device protecting the branch circuit, but if the maximum branch-circuit short-circuit and ground-fault protective device rating marked on the equipment is less than these values, the circuit protective device shall not exceed the value marked on the equipment nameplate.

Exception No. 2: The nameplate marking of cord-and-plug-connected equipment rated not greater than 250 volts, single-phase, such as household refrigerators and freezers, drinking water coolers, and beverage dispensers, shall be used in determining the branch-circuit requirements, and each unit shall be considered as a single motor unless the nameplate is marked otherwise.

(C) Protective Device Rating Not to Exceed the Manufacturer's Values. Where maximum protective device ratings shown on a manufacturer's overload relay table for use with a motor controller are less than the rating or setting selected in accordance with 440.22(A) and (B), the protective device rating shall not exceed the manufacturer's values marked on the equipment.

IV. BRANCH-CIRCUIT CONDUCTORS

440.31 General. The provisions of Part IV and Article 310 specify ampacities of conductors required to carry the motor current without overheating under the conditions specified, except as modified in 440.6(A), Exception No. 1.

The provisions of these articles shall not apply to integral conductors of motors, motor controllers and the like, or to conductors that form an integral part of approved equipment.

FPN: See 300.1(B) and 310.1 for similar requirements.

440.32 Single Motor-Compressor. Branch-circuit conductors supplying a single motor-compressor shall have an ampacity not less than 125 percent of either the motor-compressor rated-load current or the branch-circuit selection current, whichever is greater.

Air-Conditioning and Refrigerating Equipment

For a wye-start, delta-run connected motor-compressor, the selection of branch-circuit conductors between the controller and the motor-compressor shall be permitted to be based on 58 percent of either the motor-compressor rated-load current or the branch-circuit selection current, whichever is greater.

440.33 Motor-Compressor(s) With or Without Additional Motor Loads. Conductors supplying one or more motor-compressor(s) with or without an additional load(s) shall have an ampacity not less than the sum of the rated-load or branch-circuit selection current ratings, whichever is larger, of all the motor-compressors plus the full-load currents of the other motors, plus 25 percent of the highest motor or motor-compressor rating in the group.

Exception No. 1: Where the circuitry is interlocked so as to prevent the starting and running of a second motor-compressor or group of motor-compressors, the conductor size shall be determined from the largest motor-compressor or group of motor-compressors that is to be operated at a given time.

Exception No. 2: The branch circuit conductors for room air conditioners shall be in accordance with Part VII of Article 440.

440.34 Combination Load. Conductors supplying a motor-compressor load in addition to a lighting or appliance load as computed from Article 220 and other applicable articles shall have an ampacity sufficient for the lighting or appliance load plus the required ampacity for the motor-compressor load determined in accordance with 440.33, or, for a single motor-compressor, in accordance with 440.32.

V. CONTROLLERS FOR MOTOR-COMPRESSORS

440.41 Rating.

(A) Motor-Compressor Controller. A motor-compressor controller shall have both a continuous-duty full-load current rating and a locked-rotor current rating not less than the nameplate rated-load current or branch-circuit selection current, whichever is greater, and locked-rotor current, respectively, of the compressor. In case the motor controller is rated in horsepower but is without one or both of the foregoing current ratings, equivalent currents shall be determined from the ratings as follows. Tables 430.148, 430.149, and 430.150 shall be used to determine the equivalent full-load current rating. Tables 430.151(A) and 430.151(B) shall be used to determine the equivalent locked-rotor current ratings.

(B) Controller Serving More Than One Load. A controller serving more than one motor-compressor or a motor-compressor and other loads shall have a continuous-duty full-load current rating and a locked-rotor current rating not less than the combined load as determined in accordance with 440.12(B).

VI. MOTOR-COMPRESSOR AND BRANCH-CIRCUIT OVERLOAD PROTECTION

440.51 General. The provisions of Part VI specify devices intended to protect the motor-compressor, the motor-control apparatus, and the branch-circuit conductors against excessive heating due to motor overload and failure to start.

440.52 Application and Selection.

(A) Protection of Motor-Compressor. Each motor-compressor shall be protected against overload and failure to start by one of the following means:

(1) A separate overload relay that is responsive to motor-compressor current. This device shall be selected to trip at not more than 140 percent of the motor-compressor rated-load current.

(2) A thermal protector integral with the motor-compressor, approved for use with the motor-compressor that it protects on the basis that it will prevent dangerous overheating of the motor-compressor due to overload and failure to start. If the current-interrupting device is separate from the motor-compressor and its control circuit is operated by a protective device integral with the motor-compressor, it shall be arranged so that the

opening of the control circuit will result in interruption of current to the motor-compressor.

(3) A fuse or inverse time circuit breaker responsive to motor current, which shall also be permitted to serve as the branch-circuit short-circuit and ground-fault protective device. This device shall be rated at not more than 125 percent of the motor-compressor rated-load current. It shall have sufficient time delay to permit the motor-compressor to start and accelerate its load. The equipment or the motor-compressor shall be marked with this maximum branch-circuit fuse or inverse time circuit breaker rating.

(4) A protective system, furnished or specified and approved for use with the motor-compressor that it protects on the basis that it will prevent dangerous overheating of the motor-compressor due to overload and failure to start. If the current-interrupting device is separate from the motor-compressor and its control circuit is operated by a protective device that is not integral with the current-interrupting device, it shall be arranged so that the opening of the control circuit will result in interruption of current to the motor-compressor.

(B) Protection of Motor-Compressor Control Apparatus and Branch-Circuit Conductors. The motor-compressor controller(s), the disconnecting means, and the branch-circuit conductors shall be protected against overcurrent due to motor overload and failure to start by one of the following means, which shall be permitted to be the same device or system protecting the motor-compressor in accordance with 440.52(A):

(1) An overload relay selected in accordance with 440.52(A)(1)
(2) A thermal protector applied in accordance with 440.52(A)(2), that will not permit a continuous current in excess of 156 percent of the marked rated-load current or branch-circuit selection current
(3) A fuse or inverse time circuit breaker selected in accordance with 440.52(A)(3)
(4) A protective system, in accordance with 440.52(A)(4), that will not permit a continuous current in excess of

156 percent of the marked rated-load current or branch-circuit selection current

440.53 Overload Relays. Overload relays and other devices for motor overload protection that are not capable of opening short circuits shall be protected by fuses or inverse time circuit breakers with ratings or settings in accordance with Part III unless approved for group installation or for part-winding motors and marked to indicate the maximum size of fuse or inverse time circuit breaker by which they shall be protected.

440.54 Motor-Compressors and Equipment on 15- or 20-Ampere Branch Circuits—Not Cord-and-Attachment-Plug-Connected. Overload protection for motor-compressors and equipment used on 15- or 20-ampere 120-volt, or 15-ampere 208- or 240-volt single-phase branch circuits as permitted in Article 210 shall be permitted as indicated in 440.54(A) and (B).

(A) Overload Protection. The motor-compressor shall be provided with overload protection selected as specified in 440.52(A). Both the controller and motor overload protective device shall be approved for installation with the short-circuit and ground-fault protective device for the branch circuit to which the equipment is connected.

(B) Time Delay. The short-circuit and ground-fault protective device protecting the branch circuit shall have sufficient time delay to permit the motor-compressor and other motors to start and accelerate their loads.

440.55 Cord-and-Attachment-Plug-Connected Motor-Compressors and Equipment on 15- or 20-Ampere Branch Circuits. Overload protection for motor-compressors and equipment that are cord-and-attachment-plug-connected and used on 15- or 20-ampere 120-volt, or 15-ampere 208- or 240-volt single-phase branch circuits as permitted in Article 210 shall be permitted as indicated in 440.55(A), (B), and (C).

(A) Overload Protection. The motor-compressor shall be provided with overload protection as specified in 440.52(A). Both the controller and the motor overload pro-

Air-Conditioning and Refrigerating Equipment

tective device shall be approved for installation with the short-circuit and ground-fault protective device for the branch circuit to which the equipment is connected.

(B) Attachment Plug and Receptacle Rating. The rating of the attachment plug and receptacle shall not exceed 20 amperes at 125 volts or 15 amperes at 250 volts.

(C) Time Delay. The short-circuit and ground-fault protective device protecting the branch circuit shall have sufficient time delay to permit the motor-compressor and other motors to start and accelerate their loads.

CHAPTER 21

GENERATORS

INTRODUCTION

This chapter contains provisions from Article 445—Generators. Article 445 covers the installation of generators. Some of the provisions covered in this article include: location, overcurrent protection, ampacities of conductors, bushings, and disconnecting means required for generators. A generator installed as a separately derived system must be grounded in accordance with the provisions of 250.30. An alternating ac power source such as an on-site generator is not a separately derived system if the neutral is solidly interconnected to a service-supplied neutral. Portable and vehicle-mounted generators must be grounded per 250.34.

ARTICLE 445
Generators

445.1 Scope. This article covers the installation of generators.

445.10 Location. Generators shall be of a type suitable for the locations in which they are installed. They shall also meet the requirements for motors in 430.14. Generators installed in hazardous (classified) locations as described in Articles 500 through 503 and Article 505, or in other locations as described in Articles 510 through 517, and in Articles 518, 520, 525, 530, 665, and 695 shall also comply with the applicable provisions of those articles.

445.12 Overcurrent Protection.

(A) Constant-Voltage Generators. Constant-voltage generators, except ac generator exciters, shall be protected from overloads by inherent design, circuit breakers, fuses, or other acceptable overcurrent protective means suitable for the conditions of use.

(B) Two-Wire Generators. Two-wire, dc generators shall be permitted to have overcurrent protection in one conductor only if the overcurrent device is actuated by the entire current generated other than the current in the shunt field. The overcurrent device shall not open the shunt field.

(C) 65 Volts or Less. Generators operating at 65 volts or less and driven by individual motors shall be considered as protected by the overcurrent device protecting the motor if these devices will operate when the generators are delivering not more than 150 percent of their full-load rated current.

(D) Balancer Sets. Two-wire, dc generators used in conjunction with balancer sets to obtain neutrals for 3-wire systems shall be equipped with overcurrent devices that will disconnect the 3-wire system in case of excessive unbalancing of voltages or currents.

(E) Three-Wire, Direct-Current Generators. Three-wire, dc generators, whether compound or shunt wound, shall be equipped with overcurrent devices, one in each arma-

Generators

ture lead, and connected so as to be actuated by the entire current from the armature. Such overcurrent devices shall consist either of a double-pole, double-coil circuit breaker or of a 4-pole circuit breaker connected in the main and equalizer leads and tripped by two overcurrent devices, one in each armature lead. Such protective devices shall be interlocked so that no one pole can be opened without simultaneously disconnecting both leads of the armature from the system.

Exception to (A) through (E): Where deemed by the authority having jurisdiction, a generator is vital to the operation of an electrical system and the generator should operate to failure to prevent a greater hazard to persons. The overload sensing device(s) shall be permitted to be connected to an annunciator or alarm supervised by authorized personnel instead of interrupting the generator circuit.

445.13 Ampacity of Conductors. The ampacity of the conductors from the generator terminals to the first distribution device(s) containing overcurrent protection shall not be less than 115 percent of the nameplate current rating of the generator. It shall be permitted to size the neutral conductors in accordance with 220.22. Conductors that must carry ground-fault currents shall not be smaller than required by 250.24(B). Neutral conductors of dc generators that must carry ground-fault currents shall not be smaller than the minimum required size of the largest conductor.

445.14 Protection of Live Parts. Live parts of generators operated at more than 50 volts to ground shall not be exposed to accidental contact where accessible to unqualified persons.

445.15 Guards for Attendants. Where necessary for the safety of attendants, the requirements of 430.133 shall apply.

445.16 Bushings. Where wires pass through an opening in an enclosure, conduit box, or barrier, a bushing shall be used to protect the conductors from the edges of an opening having sharp edges. The bushing shall have smooth, well-rounded surfaces where it may be in contact with the conductors. If used where oils, grease, or other contaminants may be present, the bushing shall be made of a material not deleteriously affected.

445.17 Generator Terminal Housings. Generator terminal housings shall comply with 430.12. Where a horsepower rating is required to determine the required minimum size of the generator terminal housing, the full-load current of the generator shall be compared with comparable motors in Tables 430.147 through 430.150. The higher horsepower rating of Tables 430.147 and 430.150 shall be used whenever the generator selection is between two ratings.

445.18 Disconnecting Means Required for Generators. Generators shall be equipped with a disconnect by means of which the generator and all protective devices and control apparatus are able to be disconnected entirely from the circuits supplied by the generator except where:

(1) The driving means for the generator can be readily shut down; and
(2) The generator is not arranged to operate in parallel with another generator or other source of voltage.

CHAPTER 22

TRANSFORMERS

INTRODUCTION

This chapter contains provisions from Article 450—Transformers. This article covers the installation of all transformers unless the transformer is specifically exempted by one of the eight exceptions in 450.1. The word transformer is intended to mean an individual transformer, single- or polyphase, identified by a single nameplate, unless otherwise indicated in Article 450. This article is divided into three parts: I General Provisions, II Specific Provisions Applicable to Different Types of Transformers, and III Transformer Vaults.

ARTICLE 450

Transformers and Transformer Vaults (Including Secondary Ties)

450.1 Scope. This article covers the installation of all transformers.

Exception No. 1: Current transformers.

Exception No. 2: Dry-type transformers that constitute a component part of other apparatus and comply with the requirements for such apparatus.

Exception No. 3: Transformers that are an integral part of an X-ray, high-frequency, or electrostatic-coating apparatus.

Exception No. 4: Transformers used with Class 2 and Class 3 circuits that comply with Article 725.

Exception No. 5: Transformers for sign and outline lighting that comply with Article 600.

Exception No. 6: Transformers for electric-discharge lighting that comply with Article 410.

Exception No. 7: Transformers used for power-limited fire alarm circuits that comply with Part III of Article 760.

Exception No. 8: Transformers used for research, development, or testing, where effective arrangements are provided to safeguard persons from contacting energized parts.

This article covers the installation of transformers dedicated to supplying power to a fire pump installation as modified by Article 695.

This article also covers the installation of transformers in hazardous (classified) locations as modified by Articles 501 through 504.

I. GENERAL PROVISIONS

450.3 Overcurrent Protection. Overcurrent protection of transformers shall comply with 450.3(A), (B), or (C). As used in this section, the word *transformer* shall mean a transformer or polyphase bank of two or more single-phase transformers operating as a unit.

Transformers

(B) Transformers 600 Volts, Nominal, or Less. Overcurrent protection shall be provided in accordance with Table 450.3(B).

(C) Voltage Transformers. Voltage transformers installed indoors or enclosed shall be protected with primary fuses.

450.4 Autotransformers 600 Volts, Nominal, or Less.

(A) Overcurrent Protection. Each autotransformer 600 volts, nominal, or less shall be protected by an individual overcurrent device installed in series with each ungrounded input conductor. Such overcurrent device shall be rated or set at not more than 125 percent of the rated full-load input current of the autotransformer. Where this calculation does not correspond to a standard rating of a fuse or nonadjustable circuit breaker and the rated input current is 9 amperes or more, the next higher standard rating described in 240.6 shall be permitted. An overcurrent device shall not be installed in series with the shunt winding (the winding common to both the input and the output circuits) of the autotransformer between Points A and B as shown in Figure 450.4.

Exception: Where the rated input current of the autotransformer is less than 9 amperes, an overcurrent device rated or set at not more than 167 percent of the input current shall be permitted.

(B) Transformer Field-Connected as an Auto-transformer. A transformer field-connected as an autotransformer shall be identified for use at elevated voltage.

450.5 Grounding Autotransformers.
Grounding autotransformers covered in this section are zigzag or T-connected transformers connected to 3-phase, 3-wire ungrounded systems for the purpose of creating a 3-phase, 4-wire distribution system or providing a neutral reference for grounding purposes. Such transformers shall have a continuous per-phase current rating and a continuous neutral current rating.

(A) Three-Phase, 4-Wire System. A grounding autotransformer used to create a 3-phase, 4-wire distribution system

Table 450.3(B) Maximum Rating or Setting of Overcurrent Protection for Transformers 600 Volts and Less (as a Percentage of Transformer-Rated Current)

Protection Method	Primary Protection			Secondary Protection (see Note 2.)	
	Currents of 9 Amperes or More	Currents Less Than 9 Amperes	Currents Less Than 2 Amperes	Currents of 9 Amperes or More	Currents Less Than 9 Amperes
Primary only protection	125% (see Note 1.)	167%	300%	Not required	Not required
Primary and secondary protection	250% (see Note 3.)	250% (see Note 3.)	250% (see Note 3.)	125% (see Note 1.)	167%

Notes:

1. Where 125 percent of this current does not correspond to a standard rating of a fuse or nonadjustable circuit breaker, a higher rating that does not exceed the next higher standard rating shall be permitted.

2. Where secondary overcurrent protection is required, the secondary overcurrent device shall be permitted to consist of not more than six circuit breakers or six sets of fuses grouped in one location. Where multiple overcurrent devices are utilized, the total of all the device ratings shall not exceed the allowed value of a single overcurrent device. If both breakers and fuses are utilized as the overcurrent device, the total of the device ratings shall not exceed that allowed for fuses.

3. A transformer equipped with coordinated thermal overload protection by the manufacturer and arranged to interrupt the primary current shall be permitted to have primary overcurrent protection rated or set at a value that is not more than six times the rated current of the transformer for transformers having not more than 6 percent impedance and not more than four times the rated current of the transformer for transformers having more than 6 percent but not more than 10 percent impedance.

See also Table 450.3(A) in NEC for transformers with primary over 600 volts.

from a 3-phase, 3-wire ungrounded system shall conform to 450.5(A)(1) through (A)(4).

(1) Connections. The transformer shall be directly connected to the ungrounded phase conductors and shall not be switched or provided with overcurrent protection that is independent of the main switch and common-trip overcurrent protection for the 3-phase, 4-wire system.

(2) Overcurrent Protection. An overcurrent sensing device shall be provided that will cause the main switch or common-trip overcurrent protection referred to in 450.5(A)(1) to open if the load on the autotransformer reaches or exceeds 125 percent of its continuous current per-phase or neutral rating. Delayed tripping for temporary overcurrents sensed at the autotransformer overcurrent device shall be permitted for the purpose of allowing proper operation of branch or feeder protective devices on the 4-wire system.

(3) Transformer Fault Sensing. A fault-sensing system that causes the opening of a main switch or common-trip overcurrent device for the 3-phase, 4-wire system shall be provided to guard against single-phasing or internal faults.

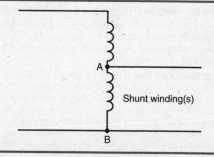

Figure 450.4 Autotransformer.

(4) Rating. The autotransformer shall have a continuous neutral-current rating that is sufficient to handle the maximum possible neutral unbalanced load current of the 4-wire system.

(B) Ground Reference for Fault Protection Devices. A grounding autotransformer used to make available a specified magnitude of ground-fault current for operation of a ground-responsive protective device on a 3-phase, 3-wire ungrounded system shall conform to 450.5(B)(1) and (2).

(1) Rating. The autotransformer shall have a continuous neutral-current rating sufficient for the specified ground-fault current.

(2) Overcurrent Protection. An overcurrent protective device of adequate short-circuit rating that will open simultaneously all ungrounded conductors when it operates shall be applied in the grounding autotransformer branch circuit and shall be rated or set at a current not exceeding 125 percent of the autotransformer continuous per-phase current rating or 42 percent of the continuous-current rating of any series connected devices in the autotransformer neutral connection. Delayed tripping for temporary overcurrents to permit the proper operation of ground-responsive tripping devices on the main system shall be permitted but shall not exceed values that would be more than the short-time current rating of the grounding autotransformer or any series connected devices in the neutral connection thereto.

(C) Ground Reference for Damping Transitory Overvoltages. A grounding autotransformer used to limit transitory overvoltages shall be of suitable rating and connected in accordance with 450.5(A)(1).

450.6 Secondary Ties. A secondary tie is a circuit operating at 600 volts, nominal, or less between phases that connects two power sources or power supply points, such as the secondaries of two transformers. The tie shall be permitted to consist of one or more conductors per phase.

As used in this section, the word *transformer* means a transformer or a bank of transformers operating as a unit.

Transformers

(A) Tie Circuits. Tie circuits shall be provided with overcurrent protection at each end as required in Article 240.

Under the conditions described in 450.6(A)(1) and 450.6(A)(2), the overcurrent protection shall be permitted to be in accordance with 450.6(A)(3).

(1) Loads at Transformer Supply Points Only. Where all loads are connected at the transformer supply points at each end of the tie and overcurrent protection is not provided in accordance with Article 240, the rated ampacity of the tie shall not be less than 67 percent of the rated secondary current of the largest transformer connected to the secondary tie system.

(2) Loads Connected Between Transformer Supply Points. Where load is connected to the tie at any point between transformer supply points and overcurrent protection is not provided in accordance with Article 240, the rated ampacity of the tie shall not be less than 100 percent of the rated secondary current of the largest transformer connected to the secondary tie system.

(3) Tie Circuit Protection. Under the conditions described in 450.6(A)(1) and (A)(2), both ends of each tie conductor shall be equipped with a protective device that opens at a predetermined temperature of the tie conductor under short-circuit conditions. This protection shall consist of one of the following: (1) a fusible link cable connector, terminal, or lug, commonly known as a limiter, each being of a size corresponding with that of the conductor and of construction and characteristics according to the operating voltage and the type of insulation on the tie conductors or (2) automatic circuit breakers actuated by devices having comparable current-time characteristics.

(4) Interconnection of Phase Conductors Between Transformer Supply Points. Where the tie consists of more than one conductor per phase, the conductors of each phase shall comply with one of the following provisions.

(a) *Interconnected.* The conductors shall be interconnected in order to establish a load supply point, and the protection specified in 450.6(A)(3) shall be provided in each tie conductor at this point.

(b) *Not Interconnected.* The loads shall be connected to one or more individual conductors of a paralleled conductor tie without interconnecting the conductors of each phase and without the protection specified in 450.6(A)(3) at load connection points. Where this is done, the tie conductors of each phase shall have a combined capacity of not less than 133 percent of the rated secondary current of the largest transformer connected to the secondary tie system, the total load of such taps shall not exceed the rated secondary current of the largest transformer, and the loads shall be equally divided on each phase and on the individual conductors of each phase as far as practicable.

(5) Tie Circuit Control. Where the operating voltage exceeds 150 volts to ground, secondary ties provided with limiters shall have a switch at each end that, when open, de-energizes the associated tie conductors and limiters. The current rating of the switch shall not be less than the rated current of the conductors connected to the switch. It shall be capable of opening its rated current, and it shall be constructed so that it will not open under the magnetic forces resulting from short-circuit current.

(B) Overcurrent Protection for Secondary Connections. Where secondary ties are used, an overcurrent device rated or set at not more than 250 percent of the rated secondary current of the transformers shall be provided in the secondary connections of each transformer. In addition, an automatic circuit breaker actuated by a reverse-current relay set to open the circuit at not more than the rated secondary current of the transformer shall be provided in the secondary connection of each transformer.

450.8 Guarding. Transformers shall be guarded as specified in 450.8(A) through (D).

(A) Mechanical Protection. Appropriate provisions shall be made to minimize the possibility of damage to transformers from external causes where the transformers are exposed to physical damage.

(B) Case or Enclosure. Dry-type transformers shall be provided with a noncombustible moisture-resistant case or enclosure that provides protection against the accidental insertion of foreign objects.

(C) Exposed Energized Parts. Switches or other equipment operating at 600 volts, nominal, or less and serving only equipment within a transformer enclosure shall be permitted to be installed in the transformer enclosure if accessible to qualified persons only. All energized parts shall be guarded in accordance with 110.27 and 110.34.

(D) Voltage Warning. The operating voltage of exposed live parts of transformer installations shall be indicated by signs or visible markings on the equipment or structures.

450.9 Ventilation. The ventilation shall be adequate to dispose of the transformer full-load losses without creating a temperature rise that is in excess of the transformer rating.

Transformers with ventilating openings shall be installed so that the ventilating openings are not blocked by walls or other obstructions. The required clearances shall be clearly marked on the transformer.

450.10 Grounding. Exposed non–current-carrying metal parts of transformer installations, including fences, guards, and so forth, shall be grounded where required under the conditions and in the manner specified for electric equipment and other exposed metal parts in Article 250.

450.12 Terminal Wiring Space. The minimum wire-bending space at fixed, 600-volt and below terminals of transformer line and load connections shall be as required in 312.6. Wiring space for pigtail connections shall conform to Table 314.16(B).

450.13 Accessibility. All transformers and transformer vaults shall be readily accessible to qualified personnel for inspection and maintenance or shall meet the requirements of 450.13(A) or (B).

(A) Open Installations. Dry-type transformers 600 volts, nominal, or less, located in the open on walls, columns, or structures, shall not be required to be readily accessible.

(B) Hollow Space Installations. Dry-type transformers 600 volts, nominal, or less and not exceeding 50 kVA shall be permitted in hollow spaces of buildings not permanently closed in by structure, provided they meet the ventilation requirements of 450.9 and separation from combustible materials requirements of 450.21(A). Transformers so installed shall not be required to be readily accessible.

II. SPECIFIC PROVISIONS APPLICABLE TO DIFFERENT TYPES OF TRANSFORMERS

450.21 Dry-Type Transformers Installed Indoors.

(A) Not Over 112-1/2 kVA. Dry-type transformers installed indoors and rated 112-1/2 kVA or less shall have a separation of at least 305 mm (12 in.) from combustible material unless separated from the combustible material by a fire-resistant, heat-insulated barrier.

(B) Over 112-1/2 kVA. Individual dry-type transformers of more than 112-1/2 kVA rating shall be installed in a transformer room of fire-resistant construction. Unless specified otherwise in this article, the term *fire resistant* means a construction having a minimum fire rating of 1 hour.

450.22 Dry-Type Transformers Installed Outdoors.

Dry-type transformers installed outdoors shall have a weatherproof enclosure.

Transformers exceeding 112-1/2 kVA shall not be located within 305 mm (12 in.) of combustible materials of buildings unless the transformer has Class 155 insulation systems or higher and is completely enclosed except for ventilating openings.

III. TRANSFORMER VAULTS

450.41 Location. Vaults shall be located where they can be ventilated to the outside air without using flues or ducts wherever such an arrangement is practicable.

450.42 Walls, Roofs, and Floors. The walls and roofs of vaults shall be constructed of materials that have adequate structural strength for the conditions with a minimum fire re-

Transformers

sistance of 3 hours. The floors of vaults in contact with the earth shall be of concrete that is not less than 100 mm (4 in.) thick, but where the vault is constructed with a vacant space or other stories below it, the floor shall have adequate structural strength for the load imposed thereon and a minimum fire resistance of 3 hours. For the purposes of this section, studs and wallboard construction shall not be acceptable.

Exception: Where transformers are protected with automatic sprinkler, water spray, carbon dioxide, or halon, construction of 1-hour rating shall be permitted.

450.43 Doorways. Vault doorways shall be protected in accordance with 450.43(A), (B), and (C).

(A) Type of Door. Each doorway leading into a vault from the building interior shall be provided with a tight-fitting door that has a minimum fire rating of 3 hours. The authority having jurisdiction shall be permitted to require such a door for an exterior wall opening where conditions warrant.

Exception: Where transformers are protected with automatic sprinkler, water spray, carbon dioxide, or halon, construction of 1-hour rating shall be permitted.

(B) Sills. A door sill or curb that is of sufficient height to confine the oil from the largest transformer within the vault shall be provided, and in no case shall the height be less than 100 mm (4 in.).

(C) Locks. Doors shall be equipped with locks, and doors shall be kept locked, access being allowed only to qualified persons. Personnel doors shall swing out and be equipped with panic bars, pressure plates, or other devices that are normally latched but open under simple pressure.

450.45 Ventilation Openings. Where required by 450.9, openings for ventilation shall be provided in accordance with 450.45(A) through (F).

(A) Location. Ventilation openings shall be located as far as possible from doors, windows, fire escapes, and combustible material.

(B) Arrangement. A vault ventilated by natural circulation of air shall be permitted to have roughly half of the total area of openings required for ventilation in one or more openings near the floor and the remainder in one or more openings in the roof or in the sidewalls near the roof, or all of the area required for ventilation shall be permitted in one or more openings in or near the roof.

(C) Size. For a vault ventilated by natural circulation of air to an outdoor area, the combined net area of all ventilating openings, after deducting the area occupied by screens, gratings, or louvers, shall be not less than 1900 mm^2 (3 in.^{2}) per kVA of transformer capacity in service, and in no case shall the net area be less than 0.1 m^2 (1 ft^2) for any capacity under 50 kVA.

(D) Covering. Ventilation openings shall be covered with durable gratings, screens, or louvers, according to the treatment required in order to avoid unsafe conditions.

(E) Dampers. All ventilation openings to the indoors shall be provided with automatic closing fire dampers that operate in response to a vault fire. Such dampers shall possess a standard fire rating of not less than 1-1/2 hours.

(F) Ducts. Ventilating ducts shall be constructed of fire-resistant material.

450.46 Drainage. Where practicable, vaults containing more than 100 kVA transformer capacity shall be provided with a drain or other means that will carry off any accumulation of oil or water in the vault unless local conditions make this impracticable. The floor shall be pitched to the drain where provided.

450.47 Water Pipes and Accessories. Any pipe or duct system foreign to the electrical installation shall not enter or pass through a transformer vault. Piping or other facilities provided for vault fire protection, or for transformer cooling, shall not be considered foreign to the electrical installation.

450.48 Storage in Vaults. Materials shall not be stored in transformer vaults.

CHAPTER 23

SWIMMING POOLS, FOUNTAINS, AND SIMILAR INSTALLATIONS

INTRODUCTION

This chapter contains provisions from Article 680—Swimming Pools, Fountains, and Similar Installations. The provisions of this article apply to the construction and installation of electric wiring for and equipment in (or adjacent to) all swimming, wading, and decorative pools, fountains, hot tubs, spas, and hydromassage bathtubs, whether permanently installed or storable, as well as to metallic auxiliary equipment (such as pumps, filters, etc.) The requirements of Article 680 pertain to grounding, bonding, conductor clearances, receptacles, lighting, equipment, and ground-fault circuit-interrupter protection for personnel.

Contained in this *Pocket Guide* are four of the seven parts that are included in Article 680: I General; II Permanently Installed Pools; IV Spas and Hot Tubs; and V Fountains.

ARTICLE 680
Swimming Pools, Fountains, and Similar Installations

I. GENERAL

680.1 Scope. The provisions of this article apply to the construction and installation of electrical wiring for and equipment in or adjacent to all swimming, wading, therapeutic, and decorative pools, fountains, hot tubs, spas, and hydromassage bathtubs, whether permanently installed or storable, and to metallic auxiliary equipment, such as pumps, filters, and similar equipment. The term *body of water* used throughout Part I applies to all bodies of water covered in this scope unless otherwise amended.

680.6 Grounding. Electrical equipment shall be grounded in accordance with Parts V, VI, and VII of Article 250 and connected by wiring methods of Chapter 3, except as modified by this article. The following equipment shall be grounded:

(1) Through-wall lighting assemblies and underwater luminaires (lighting fixtures), other than those low-voltage systems listed for the application without a grounding conductor
(2) All electrical equipment located within 1.5 m (5 ft) of the inside wall of the specified body of water
(3) All electrical equipment associated with the recirculating system of the specified body of water
(4) Junction boxes
(5) Transformer enclosures
(6) Ground-fault circuit interrupters
(7) Panelboards that are not part of the service equipment and that supply any electrical equipment associated with the specified body of water

680.7 Cord-and-Plug-Connected Equipment. Fixed or stationary equipment other than an underwater luminaire (lighting fixture) for a permanently installed pool shall be permitted to be connected with a flexible cord to facilitate the removal or disconnection for maintenance or repair.

Swimming Pools, Fountains, and Similar Installations 323

(A) Length. For other than storable pools, the flexible cord shall not exceed 900 mm (3 ft) in length.

(B) Equipment Grounding. The flexible cord shall have a copper equipment grounding conductor sized in accordance with 250.122 but not smaller than 12 AWG. The cord shall terminate in a grounding-type attachment plug.

(C) Construction. The equipment grounding conductors shall be connected to a fixed metal part of the assembly. The removable part shall be mounted on or bonded to the fixed metal part.

680.8 Overhead Conductor Clearances.

(A) Power. With respect to service drop conductors and open overhead wiring, swimming pool and similar installations shall comply with the minimum clearances given in Table 680.8 and illustrated in Figure 680.8.

680.10 Underground Wiring Location.
Underground wiring shall not be permitted under the pool or within the area extending 1.5 m (5 ft) horizontally from the inside wall of the pool unless this wiring is necessary to supply pool equipment permitted by this article. Where space limitations prevent wiring from being routed a distance 1.5 m (5 ft) or more from the pool, such wiring shall be permitted where installed in rigid metal conduit, intermediate metal conduit,

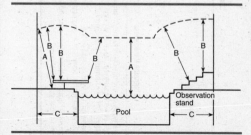

Figure 680.8 Clearances from pool structures.

Table 680.8 Overhead Conductor Clearances

Clearance Parameters	Insulated Cables, 0–750 Volts to Ground, Supported on and Cabled Together with an Effectively Grounded Bare Messenger or Effectively Grounded Neutral Conductor		All Other Conductors Voltage to Ground			
			0 through 15 kV		Over 15 through 50 kV	
	m	ft	m	ft	m	ft
A. Clearance in any direction to the water level, edge of water surface, base of diving platform, or permanently anchored raft	6.9	22.5	7.5	25	8.0	27
B. Clearance in any direction to the observation stand, tower, or diving platform	4.4	14.5	5.2	17	5.5	18
C. Horizontal limit of clearance measured from inside wall of the pool	This limit shall extend to the outer edge of the structures listed in A and B of this table but not less than 3 m (10 ft).					

Swimming Pools, Fountains, and Similar Installations 325

Table 680.10 Minimum Burial Depths

Wiring Method	Minimum Burial	
	mm	in.
Rigid metal conduit	150	6
Intermediate metal conduit	150	6
Nonmetallic raceways listed for direct burial without concrete encasement	450	18
Other approved raceways*	450	18

*Raceways approved for burial only where concrete encased shall require a concrete envelope not less than 50 mm. (2 in.) thick.

or a nonmetallic raceway system. All metal conduit shall be corrosion resistant and suitable for the location. The minimum burial depth shall be as given in Table 680.10.

680.11 Equipment Rooms and Pits. Electric equipment shall not be installed in rooms or pits that do not have drainage that adequately prevents water accumulation during normal operation or filter maintenance.

680.12 Maintenance Disconnecting Means. One or more means to disconnect all ungrounded conductors shall be provided for all utilization equipment other than lighting. Each means shall be accessible and within sight from its equipment.

II. PERMANENTLY INSTALLED POOLS

680.20 General. Electrical installations at permanently installed pools shall comply with the provisions of Part I and Part II of this article.

680.21 Motors.

(A) Wiring Methods.

(1) General. The branch circuits for pool-associated motors shall be installed in rigid metal conduit, intermediate metal conduit, rigid nonmetallic conduit, or Type MC cable listed for the location. Other wiring methods and materials shall be permitted in specific locations or applications as covered in this section. Any wiring method employed shall contain a copper equipment grounding conductor sized in accordance with 250.122 but not smaller than 12 AWG.

(2) On or Within Buildings. Where installed on or within buildings, electrical metallic tubing shall be permitted.

(3) Flexible Connections. Where necessary to employ flexible connections at or adjacent to the motor, liquidtight flexible metal or nonmetallic conduit with approved fittings shall be permitted.

(5) Cord-and-Plug Connections. Pool-associated motors shall be permitted to employ cord-and-plug connections. The flexible cord shall not exceed 900 mm (3 ft) in length. The flexible cord shall include an equipment grounding conductor sized in accordance with 250.122 and shall terminate in a grounding-type attachment plug.

(B) Double Insulated Pool Pumps. A listed cord-and-plug-connected pool pump incorporating an approved system of double insulation that provides a means for grounding only the internal and nonaccessible, non–current-carrying metal parts of the pump shall be connected to any wiring method recognized in Chapter 3 that is suitable for the location.

680.22 Area Lighting, Receptacles, and Equipment.

(A) Receptacles.

(1) Circulation and Sanitation System, Location. Receptacles that provide power for water-pump motors or for other loads directly related to the circulation and sanitation system shall be located at least 3.0 m (10 ft) from the inside walls of the pool, or not less than 1.5 m (5 ft) from the inside walls of the pool if they meet all of the following conditions:

(1) Consist of single receptacles
(2) Employ a locking configuration
(3) Are of the grounding type
(4) Have GFCI protection

Swimming Pools, Fountains, and Similar Installations 327

(2) Other Receptacles, Location. Other receptacles shall be not less than 3.0 m (10 ft) from the inside walls of a pool.

(5) GFCI Protection. All 125-volt receptacles located within 6.0 m (20 ft) of the inside walls of a pool or fountain shall be protected by a ground-fault circuit interrupter. Receptacles that supply pool pump motors and that are rated 15 or 20 amperes, 120 volt through 240 volts, single phase, shall be provided with GFCI protection.

(6) Measurements. In determining the dimensions in this section addressing receptacle spacings, the distance to be measured shall be the shortest path the supply cord of an appliance connected to the receptacle would follow without piercing a floor, wall, ceiling, doorway with hinged or sliding door, window opening, or other effective permanent barrier.

(B) Luminaires (Lighting Fixtures), Lighting Outlets, and Ceiling-Suspended (Paddle) Fans.

(1) New Outdoor Installation Clearances. In outdoor pool areas, luminaires (lighting fixtures), lighting outlets, and ceiling-suspended (paddle) fans installed above the pool or the area extending 1.5 m (5 ft) horizontally from the inside walls of the pool shall be installed at a height not less than 3.7 m (12 ft) above the maximum water level of the pool.

(2) Indoor Clearances. For installations in indoor pool areas, the clearances shall be the same as for outdoor areas unless modified as provided in this paragraph. If the branch circuit supplying the equipment is protected by a ground-fault circuit interrupter, the following equipment shall be permitted at a height not less than 2.3 m (7 ft 6 in.) above the maximum pool water level:

(1) Totally enclosed luminaires (fixtures)
(2) Ceiling-suspended (paddle) fans identified for use beneath ceiling structures such as provided on porches or patios

(3) Existing Installations. Existing luminaires (lighting fixtures) and lighting outlets located less than 1.5 m (5 ft) measured horizontally from the inside walls of a pool shall be not less than 1.5 m (5 ft) above the surface of the maximum water level, shall be rigidly attached to the existing structure, and shall be protected by a ground-fault circuit interrupter.

(4) GFCI Protection in Adjacent Areas. Luminaires (lighting fixtures), lighting outlets, and ceiling-suspended (paddle) fans installed in the area extending between 1.5 m (5 ft) and 3.0 m (10 ft) horizontally from the inside walls of a pool shall be protected by a ground-fault circuit interrupter unless installed not less than 1.5 m (5 ft) above the maximum water level and rigidly attached to the structure adjacent to or enclosing the pool.

(5) Cord-and-Plug-Connected Luminaires (Lighting Fixtures). Cord-and-plug-connected luminaires (lighting fixtures) shall comply with the requirements of 680.7 where installed within 4.9 m (16 ft) of any point on the water surface, measured radially.

(C) Switching Devices. Switching devices shall be located at least 1.5 m (5 ft) horizontally from the inside walls of a pool unless separated from the pool by a solid fence, wall, or other permanent barrier. Alternatively, a switch that is listed as being acceptable for use within 1.5 m (5 ft) shall be permitted.

680.23 Underwater Luminaires (Lighting Fixtures). This section covers all luminaires (lighting fixtures) installed below the normal water level of the pool.

(A) General.

(1) Luminaire (Fixture) Design, Normal Operation. The design of an underwater luminaire (lighting fixture) supplied from a branch circuit either directly or by way of a transformer meeting the requirements of this section shall be such that, where the luminaire (fixture) is properly installed without a ground-fault circuit interrupter, there is no shock hazard with any likely combination of fault conditions during normal use (not relamping).

(2) Transformers. Transformers used for the supply of underwater luminaires (fixtures), together with the transformer enclosure, shall be listed for the purpose. The transformer shall be an isolated winding type with an ungrounded sec-

Swimming Pools, Fountains, and Similar Installations 329

ondary that has a grounded metal barrier between the primary and secondary windings.

(3) GFCI Protection, Relamping. A ground-fault circuit interrupter shall be installed in the branch circuit supplying luminaires (fixtures) operating at more than 15 volts, so that there is no shock hazard during relamping. The installation of the ground-fault circuit interrupter shall be such that there is no shock hazard with any likely fault-condition combination that involves a person in a conductive path from any ungrounded part of the branch circuit or the luminaire (fixture) to ground.

(4) Voltage Limitation. No luminaires (lighting fixtures) shall be installed for operation on supply circuits over 150 volts between conductors.

(5) Location, Wall-Mounted Luminaires (Fixtures). Luminaires (lighting fixtures) mounted in walls shall be installed with the top of the luminaire (fixture) lens not less than 450 mm (18 in.) below the normal water level of the pool, unless the luminaire (lighting fixture) is listed and identified for use at lesser depths. No luminaire (fixture) shall be installed less than 100 mm (4 in.) below the normal water level of the pool.

(6) Bottom-Mounted Luminaires (Fixtures). A luminaire (lighting fixture) facing upward shall have the lens adequately guarded to prevent contact by any person.

(7) Dependence on Submersion. Luminaires (fixtures) that depend on submersion for safe operation shall be inherently protected against the hazards of overheating when not submerged.

(8) Compliance. Compliance with these requirements shall be obtained by the use of a listed underwater luminaire (lighting fixture) and by installation of a listed ground-fault circuit interrupter in the branch circuit or a listed transformer for luminaires (fixtures) operating at not more than 15 volts.

(B) Wet-Niche Luminaires (Fixtures).

(1) Forming Shells. Forming shells shall be installed for the mounting of all wet-niche underwater luminaires (fixtures) and shall be equipped with provisions for conduit entries. Metal parts of the luminaire (fixture) and forming shell in contact with the pool water shall be of brass or other approved corrosion-resistant metal. All forming shells used with nonmetallic conduit systems, other than those that are part of a listed low-voltage lighting system not requiring grounding, shall include provisions for terminating an 8 AWG copper conductor.

(2) Wiring Extending Directly to the Forming Shell. Conduit shall be installed from the forming shell to a suitable junction box or other enclosure located as provided in 680.24. Conduit shall be rigid metal, intermediate metal, liquidtight flexible nonmetallic, or rigid nonmetallic.

(a) Metal Conduit. Metal conduit shall be approved and shall be of brass or other approved corrosion-resistant metal.

(b) Nonmetallic Conduit. Where a nonmetallic conduit is used, an 8 AWG insulated solid or stranded copper equipment grounding conductor shall be installed in this conduit unless a listed low-voltage lighting system not requiring grounding is used. The equipment grounding conductor shall be terminated in the forming shell, junction box or transformer enclosure, or ground-fault circuit-interrupter enclosure. The termination of the 8 AWG equipment grounding conductor in the forming shell shall be covered with, or encapsulated in, a listed potting compound to protect the connection from the possible deteriorating effect of pool water.

(3) Equipment Grounding Provisions for Cords. Wet-niche luminaires (lighting fixtures) that are supplied by a flexible cord or cable shall have all exposed non–current-carrying metal parts grounded by an insulated copper equipment grounding conductor that is an integral part of the cord or cable. This grounding conductor shall be connected to a grounding terminal in the supply junction box, transformer enclosure, or other enclosure. The grounding conductor shall not be smaller than the supply conductors and not smaller than 16 AWG.

(4) Luminaire (Fixture) Grounding Terminations. The end of the flexible-cord jacket and the flexible-cord conduc-

Swimming Pools, Fountains, and Similar Installations

tor terminations within a luminaire (fixture) shall be covered with, or encapsulated in, a suitable potting compound to prevent the entry of water into the luminaire (fixture) through the cord or its conductors. In addition, the grounding connection within a luminaire (fixture) shall be similarly treated to protect such connection from the deteriorating effect of pool water in the event of water entry into the luminaire (fixture).

(5) Luminaire (Fixture) Bonding. The luminaire (fixture) shall be bonded to and secured to the forming shell by a positive locking device that ensures a low-resistance contact and requires a tool to remove the luminaire (fixture) from the forming shell. Bonding shall not be required for luminaires (fixtures) that are listed for the application and have no non–current-carrying metal parts.

(C) Dry-Niche Luminaires (Fixtures).

(1) Construction. A dry-niche luminaire (lighting fixture) shall be provided with a provision for drainage of water and a means for accommodating one equipment grounding conductor for each conduit entry.

(2) Junction Box. A junction box shall not be required but, if used, shall not be required to be elevated or located as specified in 680.24(A)(2) if the luminaire (fixture) is specifically identified for the purpose.

(D) No-Niche Luminaires (Fixtures). A no-niche luminaire (fixture) shall meet the construction requirements of 680.23(B)(3) and be installed in accordance with the requirements of 680.23(B). Where connection to a forming shell is specified, the connection shall be to the mounting bracket.

(E) Through-Wall Lighting Assembly. A through-wall lighting assembly shall be equipped with a threaded entry or hub, or a nonmetallic hub listed for the purpose, for the purpose of accommodating the termination of the supply conduit. A through-wall lighting assembly shall meet the construction requirements of 680.23(B)(3) and be installed in accordance with the requirements of 680.23. Where connection to a forming shell is specified, the connection shall be to the conduit termination point.

(F) Branch-Circuit Wiring.

(1) Wiring Methods. Branch-circuit wiring on the supply side of enclosures and junction boxes connected to conduits run to wet-niche and no-niche luminaires (fixtures), and the field wiring compartments of dry-niche luminaires (fixtures), shall be installed using rigid metal conduit, intermediate metal conduit, liquidtight flexible nonmetallic conduit, or rigid nonmetallic conduit. Where installed on buildings, electrical metallic tubing shall be permitted, and where installed within buildings, electrical nonmetallic tubing or electrical metallic tubing shall be permitted.

Exception: Where connecting to transformers for pool lights, liquidtight flexible metal conduit or liquidtight flexible nonmetallic conduit shall be permitted. The length shall not exceed 1.8 m (6 ft) for any one length or exceed 3.0 m (10 ft) in total length used. Liquidtight flexible nonmetallic conduit, Type B (LFNC-B), shall be permitted in lengths longer than 1.8 m (6 ft).

(2) Equipment Grounding. Through-wall lighting assemblies, wet-niche, dry-niche, or no-niche luminaires (lighting fixtures) shall be connected to an insulated copper equipment grounding conductor installed with the circuit conductors. The equipment grounding conductor shall be installed without joint or splice except as permitted in (a) and (b). The equipment grounding conductor shall be sized in accordance with Table 250.122 but shall not be smaller than 12 AWG.

Exception: An equipment grounding conductor between the wiring chamber of the secondary winding of a transformer and a junction box shall be sized in accordance with the overcurrent device in this circuit.

(a) If more than one underwater luminaire (lighting fixture) is supplied by the same branch circuit, the equipment grounding conductor, installed between the junction boxes, transformer enclosures, or other enclosures in the supply circuit to wet-niche luminaires (fixtures), or between the field-wiring compartments of dry-niche luminaires (fixtures), shall be permitted to be terminated on grounding terminals.

Swimming Pools, Fountains, and Similar Installations 333

(b) If the underwater luminaire (lighting fixture) is supplied from a transformer, ground-fault circuit interrupter, clock-operated switch, or a manual snap switch that is located between the panelboard and a junction box connected to the conduit that extends directly to the underwater luminaire (lighting fixture), the equipment grounding conductor shall be permitted to terminate on grounding terminals on the transformer, ground-fault circuit interrupter, clock-operated switch enclosure, or an outlet box used to enclose a snap switch.

(3) Conductors. Conductors on the load side of a ground-fault circuit interrupter or of a transformer, used to comply with the provisions of 680.23(A)(8), shall not occupy raceways, boxes, or enclosures containing other conductors unless one of the following conditions applies:

(1) The other conductors are protected by ground-fault circuit interrupters.
(2) The other conductors are grounding conductors.
(3) The other conductors are supply conductors to a feed-through type ground-fault circuit interrupter.
(4) Ground-fault circuit interrupters shall be permitted in a panelboard that contains circuits protected by other than ground-fault circuit interrupters.

680.24 Junction Boxes and Enclosures for Transformers or Ground-Fault Circuit Interrupters.

(A) Junction Boxes. A junction box connected to a conduit that extends directly to a forming shell or mounting bracket of a no-niche luminaire (fixture) shall meet the requirements of this section.

(1) Construction. The junction box shall be listed and labeled for the purpose and shall comply with the following conditions:

(1) Be equipped with threaded entries or hubs or a non-metallic hub listed for the purpose
(2) Be comprised of copper, brass, suitable plastic, or other approved corrosion-resistant material

(3) Be provided with electrical continuity between every connected metal conduit and the grounding terminals by means of copper, brass, or other approved corrosion-resistant metal that is integral with the box

(2) Installation. Where the luminaire (fixture) operates over 15 volts, the junction box location shall comply with (a) and (b). Where the luminaire (fixture) operates at less than 15 volts, the junction box location shall be permitted to comply with (c).

(a) Vertical Spacing. The junction box shall be located not less than 100 mm (4 in.), measured from the inside of the bottom of the box, above the ground level, or pool deck, or not less than 200 mm (8 in.) above the maximum pool water level, whichever provides the greater elevation.

(b) Horizontal Spacing. The junction box shall be located not less than 1.2 m (4 ft) from the inside wall of the pool, unless separated from the pool by a solid fence, wall, or other permanent barrier.

(c) Flush Deck Box. If used on a lighting system operating at 15 volts or less, a flush deck box shall be permitted if both of the following apply:

(1) An approved potting compound is used to fill the box to prevent the entrance of moisture.
(2) The flush deck box is located not less than 1.2 m (4 ft) from the inside wall of the pool.

(B) Other Enclosures. An enclosure for a transformer, ground-fault circuit interrupter, or a similar device connected to a conduit that extends directly to a forming shell or mounting bracket of a no-niche luminaire (fixture) shall meet the requirements of this section.

(1) Construction. The enclosure shall be listed and labeled for the purpose and meet the following requirements:

(1) Equipped with threaded entries or hubs or a nonmetallic hub listed for the purpose
(2) Comprised of copper, brass, suitable plastic, or other approved corrosion-resistant material

Swimming Pools, Fountains, and Similar Installations

(3) Provided with an approved seal, such as duct seal at the conduit connection, that prevents circulation of air between the conduit and the enclosures

(4) Provided with electrical continuity between every connected metal conduit and the grounding terminals by means of copper, brass, or other approved corrosion-resistant metal that is integral with the box

(2) Installation.

(a) Vertical Spacing. The enclosure shall be located not less than 100 mm (4 in.), measured from the inside of the bottom of the box, above the ground level, or pool deck, or not less than 200 mm (8 in.) above the maximum pool water level, whichever provides the greater elevation.

(b) Horizontal Spacing. The enclosure shall be located not less than 1.2 m (4 ft) from the inside wall of the pool, unless separated from the pool by a solid fence, wall, or other permanent barrier.

(C) Protection. Junction boxes and enclosures mounted above the grade of the finished walkway around the pool shall not be located in the walkway unless afforded additional protection, such as by location under diving boards, adjacent to fixed structures, and the like.

(D) Grounding Terminals. Junction boxes, transformer enclosures, and ground-fault circuit-interrupter enclosures connected to a conduit that extends directly to a forming shell or mounting bracket of a no-niche luminaire (fixture) shall be provided with a number of grounding terminals that shall be no fewer than one more than the number of conduit entries.

(E) Strain Relief. The termination of a flexible cord of an underwater luminaire (lighting fixture) within a junction box, transformer enclosure, ground-fault circuit interrupter, or other enclosure shall be provided with a strain relief.

(F) Grounding. The junction box, transformer enclosure, or other enclosure in the supply circuit to a wet-niche or no-niche luminaire (lighting fixture) and the field-wiring chamber of a dry-niche luminaire (lighting fixture) shall be grounded to the equipment grounding terminal of the panelboard. This terminal shall be directly connected to the panelboard enclosure.

680.25 Feeders. These provisions shall apply to any feeder on the supply side of panelboards supplying branch circuits for pool equipment covered in Part II of this article and on the load side of the service equipment or the source of a separately derived system.

(A) Wiring Methods. Feeders shall be installed in rigid metal conduit, intermediate metal conduit, liquidtight flexible nonmetallic conduit, or rigid nonmetallic conduit. Electrical metallic tubing shall be permitted where installed on or within a building, and electrical nonmetallic tubing shall be permitted where installed within a building.

(B) Grounding. An equipment grounding conductor shall be installed with the feeder conductors between the grounding terminal of the pool equipment panelboard and the grounding terminal of the applicable service equipment or source of a separately derived system. For other than (1) existing feeders covered in 680.25(A), Exception or (2) feeders to separate buildings that do not utilize an insulated equipment grounding conductor in accordance with 680.25(B)(2), this equipment grounding conductor shall be insulated.

(1) Size. This conductor shall be sized in accordance with 250.122 but not smaller than 12 AWG. On separately derived systems, this conductor shall be sized in accordance with Table 250.66 but not smaller than 8 AWG.

(2) Separate Buildings. A feeder to a separate building shall be permitted to supply swimming pool equipment branch circuits, or feeders supplying swimming pool equipment branch circuits, if the grounding arrangements in the separate building meet the requirements in 250.32. Where installed, a separate equipment grounding conductor shall be an insulated conductor.

680.26 Bonding.

(A) Performance. The bonding required by this section shall be installed to eliminate voltage gradients in the pool area as prescribed.

Swimming Pools, Fountains, and Similar Installations 337

FPN: This section does not require that the 8 AWG or larger solid copper bonding conductor be extended or attached to any remote panelboard, service equipment, or any electrode.

(B) Bonded Parts. The parts specified in 680.26(B)(1) through (B)(5) shall be bonded together.

(1) Metallic Structural Components. All metallic parts of the pool structure, including the reinforcing metal of the pool shell, coping stones, and deck, shall be bonded. The usual steel tie wires shall be considered suitable for bonding the reinforcing steel together, and welding or special clamping shall not be required. These tie wires shall be made tight. If reinforcing steel is effectively insulated by an encapsulating nonconductive compound at the time of manufacture and installation, it shall not be required to be bonded. Where reinforcing steel is encapsulated with a nonconductive compound, provisions shall be made for an alternate means to eliminate voltage gradients that would otherwise be provided by unencapsulated, bonded reinforcing steel.

(2) Underwater Lighting. All forming shells and mounting brackets of no-niche luminaires (fixtures) shall be bonded unless a listed low-voltage lighting system with nonmetallic forming shells not requiring bonding is used.

(3) Metal Fittings. All metal fittings within or attached to the pool structure shall be bonded. Isolated parts that are not over 100 mm (4 in.) in any dimension and do not penetrate into the pool structure more than 25 mm (1 in.) shall not require bonding.

(4) Electrical Equipment. Metal parts of electrical equipment associated with the pool water circulating system, including pump motors and metal parts of equipment associated with pool covers, including electric motors, shall be bonded. Metal parts of listed equipment incorporating an approved system of double insulation and providing a means for grounding internal nonaccessible, non–current-carrying metal parts shall not be bonded.

Where a double-insulated water-pump motor is installed under the provisions of this rule, a solid 8 AWG copper conductor that is of sufficient length to make a bonding connection to a replacement motor shall be extended from the bonding grid to an accessible point in the motor vicinity. Where there is no connection between the swimming pool bonding grid and the equipment grounding system for the premises, this bonding conductor shall be connected to the equipment grounding conductor of the motor circuit.

(5) Metal Wiring Methods and Equipment. Metal-sheathed cables and raceways, metal piping, and all fixed metal parts except those separated from the pool by a permanent barrier shall be bonded that are within the following distances of the pool:

(1) Within 1.5 m (5 ft) horizontally of the inside walls of the pool
(2) Within 3.7 m (12 ft) measured vertically above the maximum water level of the pool, or any observation stands, towers, or platforms, or any diving structures

(C) Common Bonding Grid. The parts specified in 680.26(B) shall be connected to a common bonding grid with a solid copper conductor, insulated, covered, or bare, not smaller than 8 AWG. Connection shall be made by exothermic welding or by pressure connectors or clamps that are labeled as being suitable for the purpose and are of stainless steel, brass, copper, or copper alloy. The common bonding grid shall be permitted to be any of the following:

(1) The structural reinforcing steel of a concrete pool where the reinforcing rods are bonded together by the usual steel tie wires or the equivalent
(2) The wall of a bolted or welded metal pool
(3) A solid copper conductor, insulated, covered, or bare, not smaller than 8 AWG
(4) Rigid metal conduit or intermediate metal conduit of brass or other identified corrosion-resistant metal conduit

(D) Connections. Where structural reinforcing steel or the walls of bolted or welded metal pool structures are used as a common bonding grid for nonelectrical parts, the connections shall be made in accordance with 250.8.

Swimming Pools, Fountains, and Similar Installations 339

(E) Pool Water Heaters. For pool water heaters rated at more than 50 amperes that have specific instructions regarding bonding and grounding, only those parts designated to be bonded shall be bonded, and only those parts designated to be grounded shall be grounded.

IV. SPAS AND HOT TUBS

680.40 General. Electrical installations at spas and hot tubs shall comply with the provisions of Part I and Part IV of this article.

680.41 Emergency Switch for Spas and Hot Tubs. A clearly labeled emergency shutoff or control switch for the purpose of stopping the motor(s) that provide power to the recirculation system and jet system shall be installed at a point readily accessible to the users and not less than 1.5 m (5 ft) away, adjacent to, and within sight of the spa or hot tub. This requirement shall not apply to single-family dwellings.

680.42 Outdoor Installations. A spa or hot tub installed outdoors shall comply with the provisions of Parts I and II of this article, except as permitted in 680.42(A) and (B), that would otherwise apply to pools installed outdoors.

(A) Flexible Connections. Listed packaged spa or hot tub equipment assemblies or self-contained spas or hot tubs utilizing a factory-installed or assembled control panel or panelboard shall be permitted to use flexible connections as covered in 680.42(A)(1) and (A)(2).

(1) Flexible Conduit. Liquidtight flexible metal conduit or liquidtight flexible nonmetallic conduit shall be permitted in lengths of not more than 1.8 m (6 ft).

(2) Cord-and-Plug Connections. Cord-and-plug connections with a cord not longer than 4.6 m (15 ft) shall be permitted where protected by a ground-fault circuit interrupter.

(B) Bonding. Bonding by metal-to-metal mounting on a common frame or base shall be permitted. The metal bands or hoops used to secure wooden staves shall not be required to be bonded as required in 680.26.

(C) Interior Wiring to Outdoor Installations. In the interior of a one-family dwelling or in the interior of another building or structure associated with a one-family dwelling, any of the wiring methods recognized in Chapter 3 of this *Code* that contain a copper equipment grounding conductor that is insulated or enclosed within the outer sheath of the wiring method and not smaller than 12 AWG shall be permitted to be used for the connection to motor, heating, and control loads that are part of a self-contained spa or hot tub, or a packaged spa or hot tub equipment assembly. Wiring to an underwater light shall comply with 680.23 or 680.33.

680.43 Indoor Installations. A spa or hot tub installed indoors shall comply with the provisions of Parts I and II of this article except as modified by this section, and shall be connected by the wiring methods of Chapter 3.

(A) Receptacles. At least one 125-volt, 15- or 20-ampere receptacle on a general-purpose branch circuit shall be located not less than 1.5 m (5 ft) from and not exceeding 3.0 m (10 ft) from the inside wall of the spa or hot tub.

(1) Location. Receptacles shall be located at least 1.5 m (5 ft) measured horizontally from the inside walls of the spa or hot tub.

(2) Protection, General. Receptacles rated 125 volts and 30 amperes or less and located within 3.0 m (10 ft) of the inside walls of a spa or hot tub shall be protected by a ground-fault circuit interrupter.

(3) Protection, Spa or Hot Tub Supply Receptacle. Receptacles that provide power for a spa or hot tub shall be ground-fault circuit-interrupter protected.

(4) Measurements. In determining the dimensions in this section addressing receptacle spacings, the distance to be measured shall be the shortest path the supply cord of an appliance connected to the receptacle would follow without piercing a floor, wall, ceiling, doorway with hinged or sliding door, window opening, or other effective permanent barrier.

(B) Installation of Luminaires (Lighting Fixtures), Lighting Outlets, and Ceiling-Suspended (Paddle) Fans.

Swimming Pools, Fountains, and Similar Installations 341

(1) Elevation. Luminaires (lighting fixtures), except as covered in 680.43(B)(2), lighting outlets, and ceiling-suspended (paddle) fans located over the spa or hot tub or within 1.5 m (5 ft) from the inside walls of the spa or hot tub shall comply with the clearances specified in (a), (b), and (c) above the maximum water level.

(a) Without GFCI. Where no GFCI protection is provided, the mounting height shall be not less than 3.7 m (12 ft).

(b) With GFCI. Where GFCI protection is provided, the mounting height shall be permitted to be not less than 2.3 m (7 ft 6 in.).

(c) Below 2.3 m (7 ft 6 in.). Luminaires (lighting fixtures) meeting the requirements of item (1) or (2) and protected by a ground-fault circuit interrupter shall be permitted to be installed less than 2.3 m (7 ft 6 in.) over a spa or hot tub.

(1) Recessed luminaires (fixtures) with a glass or plastic lens, nonmetallic or electrically isolated metal trim, and suitable for use in damp locations
(2) Surface-mounted luminaires (fixtures) with a glass or plastic globe, a nonmetallic body, or a metallic body isolated from contact, and suitable for use in damp locations

(C) Wall Switches. Switches shall be located at least 1.5 m (5 ft), measured horizontally, from the inside walls of the spa or hot tub.

(D) Bonding. The following parts shall be bonded together:

(1) All metal fittings within or attached to the spa or hot tub structure
(2) Metal parts of electrical equipment associated with the spa or hot tub water circulating system, including pump motors
(3) Metal conduit and metal piping that are within 1.5 m (5 ft) of the inside walls of the spa or hot tub and that are not separated from the spa or hot tub by a permanent barrier
(4) All metal surfaces that are within 1.5 m (5 ft) of the inside walls of the spa or hot tub and that are not separated from the spa or hot tub area by a permanent barrier

Exception: Small conductive surfaces not likely to become energized, such as air and water jets and drain fittings, where not connected to metallic piping, towel bars, mirror frames, and similar nonelectrical equipment, shall not be required to be bonded.

(5) Electrical devices and controls that are not associated with the spas or hot tubs and that are located not less than 1.5 m (5 ft) from such units; otherwise they shall be bonded to the spa or hot tub system

(E) Methods of Bonding. All metal parts associated with the spa or hot tub shall be bonded by any of the following methods:

(1) The interconnection of threaded metal piping and fittings
(2) Metal-to-metal mounting on a common frame or base
(3) The provisions of a copper bonding jumper, insulated, covered, or bare, not smaller than 8 AWG solid

(F) Grounding. The following equipment shall be grounded:

(1) All electric equipment located within 1.5 m (5 ft) of the inside wall of the spa or hot tub
(2) All electric equipment associated with the circulating system of the spa or hot tub

680.44 Protection. Except as otherwise provided in this section, the outlet(s) that supplies a self-contained spa or hot tub, a packaged hot tub equipment assembly, or a field-assembled spa or hot tub shall be protected by a ground-fault circuit interrupter.

(A) Listed Units. If so marked, a listed self-contained unit or listed packaged equipment assembly that includes integral ground-fault circuit-interrupter protection for all electrical parts within the unit or assembly (pumps, air blowers, heaters, lights, controls, sanitizer generators, wiring, and so forth) shall be permitted without additional GFCI protection.

(B) Other Units. A field assembled spa or hot tub rated 3 phase or rated over 250 volts or with a heater load of more than 50 amperes shall not require the supply to be protected by a ground-fault circuit interrupter.

Swimming Pools, Fountains, and Similar Installations 343

(C) Combination Pool and Spa or Hot Tub. A combination pool/hot tub or spa assembly commonly bonded need not be protected by a ground-fault circuit interrupter.

V. FOUNTAINS

680.50 General. The provisions of Part I and Part V of this article shall apply to all permanently installed fountains as defined in 680.2. Fountains that have water common to a pool shall additionally comply with the requirements in Part II of this article. Part V does not cover self-contained, portable fountains not larger than 1.5 m (5 ft) in any dimension. Portable fountains shall comply with Parts II and III of Article 422.

680.51 Luminaires (Lighting Fixtures), Submersible Pumps, and Other Submersible Equipment.

(A) Ground-Fault Circuit Interrupter. Fountain equipment, unless listed for operation at 15 volts or less and supplied by a transformer that complies with 680.23(A)(2), shall be protected by a ground-fault circuit interrupter.

(B) Operating Voltage. No luminaires (lighting fixtures) shall be installed for operation on supply circuits over 150 volts between conductors. Submersible pumps and other submersible equipment shall operate at 300 volts or less between conductors.

(C) Luminaire (Lighting Fixture) Lenses. Luminaires (lighting fixtures) shall be installed with the top of the luminaire (fixture) lens below the normal water level of the fountain unless listed for above-water locations. A luminaire (lighting fixture) facing upward shall have the lens adequately guarded to prevent contact by any person.

(D) Overheating Protection. Electrical equipment that depends on submersion for safe operation shall be protected against overheating by a low-water cutoff or other approved means when not submerged.

(E) Wiring. Equipment shall be equipped with provisions for threaded conduit entries or be provided with a suitable flexible cord. The maximum length of exposed cord in the fountain shall be limited to 3.0 m (10 ft). Cords extending beyond the fountain perimeter shall be enclosed in approved wiring enclosures. Metal parts of equipment in contact with water shall be of brass or other approved corrosion-resistant metal.

(F) Servicing. All equipment shall be removable from the water for relamping or normal maintenance. Luminaires (fixtures) shall not be permanently embedded into the fountain structure such that the water level must be reduced or the fountain drained for relamping, maintenance, or inspection.

(G) Stability. Equipment shall be inherently stable or be securely fastened in place.

680.52 Junction Boxes and Other Enclosures.

(A) General. Junction boxes and other enclosures used for other than underwater installation shall comply with 680.24.

(B) Underwater Junction Boxes and Other Underwater Enclosures. Junction boxes and other underwater enclosures shall meet the requirements of 680.52(B)(1) and (B)(2).

(1) Construction.

(a) Underwater enclosures shall be equipped with provisions for threaded conduit entries or compression glands or seals for cord entry.

(b) Underwater enclosures shall be submersible, and made of copper, brass, or other approved corrosion-resistant material.

(2) Installation. Underwater enclosure installations shall comply with (a) and (b).

(a) Underwater enclosures shall be filled with an approved potting compound to prevent the entry of moisture.

(b) Underwater enclosures shall be firmly attached to the supports or directly to the fountain surface and bonded as required. Where the junction box is supported only by the conduit, the conduit shall be of copper, brass, or other approved corrosion-resistant metal. Where the box is fed by

Swimming Pools, Fountains, and Similar Installations 345

nonmetallic conduit, it shall have additional supports and fasteners of copper, brass, or other approved corrosion-resistant material.

680.53 Bonding. All metal piping systems associated with the fountain shall be bonded to the equipment grounding conductor of the branch circuit supplying the fountain.

680.54 Grounding. The following equipment shall be grounded:

(1) All electrical equipment located within the fountain or within 1.5 m (5 ft) of the inside wall of the fountain
(2) All electrical equipment associated with the recirculating system of the fountain
(3) Panelboards that are not part of the service equipment and that supply any electrical equipment associated with the fountain

680.55 Methods of Grounding.

(A) Applied Provisions. The provisions of 680.21(A), 680.23(B)(3), 680.23(F)(1) and (2), 680.24(F), and 680.25 shall apply.

(B) Supplied by a Flexible Cord. Electrical equipment that is supplied by a flexible cord shall have all exposed non–current-carrying metal parts grounded by an insulated copper equipment grounding conductor that is an integral part of this cord. The grounding conductor shall be connected to a grounding terminal in the supply junction box, transformer enclosure, or other enclosure.

680.56 Cord-and-Plug-Connected Equipment.

(A) Ground-Fault Circuit Interrupter. All electrical equipment, including power-supply cords, shall be protected by ground-fault circuit interrupters.

(B) Cord Type. Flexible cord immersed in or exposed to water shall be of a type for extra-hard usage, as designated in Table 400.4 and shall be a listed type with a "W" suffix.

(C) Sealing. The end of the flexible cord jacket and the flexible cord conductor termination within equipment shall be covered with, or encapsulated in, a suitable potting compound to prevent the entry of water into the equipment through the cord or its conductors. In addition, the ground connection within equipment shall be similarly treated to protect such connections from the deteriorating effect of water that may enter into the equipment.

(D) Terminations. Connections with flexible cord shall be permanent, except that grounding-type attachment plugs and receptacles shall be permitted to facilitate removal or disconnection for maintenance, repair, or storage of fixed or stationary equipment not located in any water-containing part of a fountain.

CHAPTER 24

EMERGENCY SYSTEMS

INTRODUCTION

This chapter contains provisions from Article 700—Emergency Systems. Article 700 contains provisions that pertain to the installation, operation, and maintenance of emergency systems. These systems consist of circuits and equipment intended to supply, distribute, and control electricity for illumination or power, or both, to required facilities when the normal electrical supply or system is interrupted.

Article 700 is divided into six parts: I General; II Circuit Wiring; III Sources of Power; IV Emergency System Circuits for Lighting and Power; V Control—Emergency Lighting Circuits; and VI Overcurrent Protection.

ARTICLE 700
Emergency Systems

I. GENERAL

700.1 Scope. The provisions of this article apply to the electrical safety of the installation, operation, and maintenance of emergency systems consisting of circuits and equipment intended to supply, distribute, and control electricity for illumination, power, or both, to required facilities when the normal electrical supply or system is interrupted.

Emergency systems are those systems legally required and classed as emergency by municipal, state, federal, or other codes, or by any governmental agency having jurisdiction. These systems are intended to automatically supply illumination, power, or both, to designated areas and equipment in the event of failure of the normal supply or in the event of accident to elements of a system intended to supply, distribute, and control power and illumination essential for safety to human life.

700.4 Tests and Maintenance.

(A) Conduct or Witness Test. The authority having jurisdiction shall conduct or witness a test of the complete system upon installation and periodically afterward.

(B) Tested Periodically. Systems shall be tested periodically on a schedule acceptable to the authority having jurisdiction to ensure the systems are maintained in proper operating condition.

(C) Battery Systems Maintenance. Where battery systems or unit equipments are involved, including batteries used for starting, control, or ignition in auxiliary engines, the authority having jurisdiction shall require periodic maintenance.

(D) Written Record. A written record shall be kept of such tests and maintenance.

(E) Testing Under Load. Means for testing all emergency lighting and power systems during maximum anticipated load conditions shall be provided.

Emergency Systems

700.5 Capacity.

(A) Capacity and Rating. An emergency system shall have adequate capacity and rating for all loads to be operated simultaneously. The emergency system equipment shall be suitable for the maximum available fault current at its terminals.

(B) Selective Load Pickup, Load Shedding, and Peak Load Shaving. The alternate power source shall be permitted to supply emergency, legally required standby, and optional standby system loads where automatic selective load pickup and load shedding is provided as needed to ensure adequate power to (1) the emergency circuits, (2) the legally required standby circuits, and (3) the optional standby circuits, in that order of priority. The alternate power source shall be permitted to be used for peak load shaving, provided the above conditions are met.

Peak load-shaving operation shall be permitted for satisfying the test requirement of 700.4(B), provided all other conditions of 700.4 are met.

A portable or temporary alternate source shall be available whenever the emergency generator is out of service for major maintenance or repair.

700.6 Transfer Equipment.

(A) General. Transfer equipment, including automatic transfer switches, shall be automatic, identified for emergency use, and approved by the authority having jurisdiction. Transfer equipment shall be designed and installed to prevent the inadvertent interconnection of normal and emergency sources of supply in any operation of the transfer equipment. Transfer equipment and electric power production systems installed to permit operation in parallel with the normal source shall meet the requirements of Article 705.

(B) Bypass Isolation Switches. Means shall be permitted to bypass and isolate the transfer equipment. Where bypass isolation switches are used, inadvertent parallel operation shall be avoided.

(C) Automatic Transfer Switches. Automatic transfer switches shall be electrically operated and mechanically held.

(D) Use. Transfer equipment shall supply only emergency loads.

700.7 Signals. Audible and visual signal devices shall be provided, where practicable, for the purpose described in 700.7(A) through (D).

(A) Derangement. To indicate derangement of the emergency source.

(B) Carrying Load. To indicate that the battery is carrying load.

(C) Not Functioning. To indicate that the battery charger is not functioning.

(D) Ground Fault. To indicate a ground fault in solidly grounded wye emergency systems of more than 150 volts to ground and circuit-protective devices rated 1000 amperes or more. The sensor for the ground-fault signal devices shall be located at, or ahead of, the main system disconnecting means for the emergency source, and the maximum setting of the signal devices shall be for a ground-fault current of 1200 amperes. Instructions on the course of action to be taken in event of indicated ground fault shall be located at or near the sensor location.

700.8 Signs.

(A) Emergency Sources. A sign shall be placed at the service entrance equipment indicating type and location of on-site emergency power sources.

(B) Grounding. Where the grounded circuit conductor connected to the emergency source is connected to a grounding electrode conductor at a location remote from the emergency source, there shall be a sign at the grounding location that shall identify all emergency and normal sources connected at that location.

II. CIRCUIT WIRING

700.9 Wiring, Emergency System.

(A) Identification. All boxes and enclosures (including transfer switches, generators, and power panels) for emer-

Emergency Systems

gency circuits shall be permanently marked so they will be readily identified as a component of an emergency circuit or system.

(B) Wiring. Wiring of two or more emergency circuits supplied from the same source shall be permitted in the same raceway, cable, box, or cabinet. Wiring from an emergency source or emergency source distribution overcurrent protection to emergency loads shall be kept entirely independent of all other wiring and equipment, unless otherwise permitted in (1) through (4):

(1) Wiring from the normal power source located in transfer equipment enclosures
(2) Wiring supplied from two sources in exit or emergency luminaires (lighting fixtures)
(3) Wiring from two sources in a common junction box, attached to exit or emergency luminaires (lighting fixtures)
(4) Wiring within a common junction box attached to unit equipment, containing only the branch circuit supplying the unit equipment and the emergency circuit supplied by the unit equipment

(C) Wiring Design and Location. Emergency wiring circuits shall be designed and located so as to minimize the hazards that might cause failure due to flooding, fire, icing, vandalism, and other adverse conditions.

(D) Fire Protection. Emergency systems shall meet the following additional requirements in assembly occupancies for not less than 1000 persons or in buildings above 23 m (75 ft) in height with any of the following occupancy classes: assembly, educational, residential, detention and correctional, business, and mercantile.

(1) Feeder-Circuit Wiring. Feeder-circuit wiring shall meet one of the following conditions:

(1) Be installed with buildings that are fully protected by an approved automatic fire suppression system
(2) Be a listed electrical circuit protective system with a minimum 1-hour fire rating
(3) Be protected by a listed thermal barrier system for electrical system components

(4) Be protected by a fire-rated assembly listed to achieve a minimum fire rating of 1 hour
(5) Be embedded in not less than 50 mm (2 in.) of concrete
(6) Be a cable listed to maintain circuit integrity for not less than 1 hour when installed in accordance with the listing requirements

(2) Feeder-Circuit Equipment. Equipment for feeder circuits (including transfer switches, transformers, and panelboards) shall be located either in spaces fully protected by approved automatic fire suppression systems (including sprinklers, carbon dioxide systems) or in spaces with a 1-hour fire resistance rating.

III. SOURCES OF POWER

700.12 General Requirements. Current supply shall be such that, in the event of failure of the normal supply to, or within, the building or group of buildings concerned, emergency lighting, emergency power, or both shall be available within the time required for the application but not to exceed 10 seconds. The supply system for emergency purposes, in addition to the normal services to the building and meeting the general requirements of this section, shall be one or more of the types of systems described in 700.12(A) through (D). Unit equipment in accordance with 700.12(E) shall satisfy the applicable requirements of this article.

In selecting an emergency source of power, consideration shall be given to the occupancy and the type of service to be rendered, whether of minimum duration, as for evacuation of a theater, or longer duration, as for supplying emergency power and lighting due to an indefinite period of current failure from trouble either inside or outside the building.

Equipment shall be designed and located to minimize the hazards that might cause complete failure due to flooding, fires, icing, and vandalism.

Equipment for sources of power as described in 700.12(A) through (D) where located within assembly occupancies for greater than 1000 persons or in buildings above 23 m (75 ft) in height with any of the following occu-

Emergency Systems

pancy classes—assembly, educational, residential, detention and correctional, business, and mercantile—shall be installed either in spaces fully protected by approved automatic fire suppression systems (sprinklers, carbon dioxide systems, and so forth), or in spaces with a 1-hour fire rating.

(A) Storage Battery. Storage batteries used as a source of power for emergency systems shall be of suitable rating and capacity to supply and maintain the total load for a period of 1-1/2 hours minimum, without the voltage applied to the load falling below 87-1/2 percent of normal.

Batteries, whether of the acid or alkali type, shall be designed and constructed to meet the requirements of emergency service and shall be compatible with the charger for that particular installation.

For a sealed battery, the container shall not be required to be transparent. However, for the lead acid battery that requires water additions, transparent or translucent jars shall be furnished. Automotive-type batteries shall not be used.

An automatic battery charging means shall be provided.

(B) Generator Set.

(1) Prime Mover-Driven. For a generator set driven by a prime mover acceptable to the authority having jurisdiction and sized in accordance with 700.5, means shall be provided for automatically starting the prime mover on failure of the normal service and for automatic transfer and operation of all required electrical circuits. A time-delay feature permitting a 15-minute setting shall be provided to avoid retransfer in case of short-time reestablishment of the normal source.

(2) Internal Combustion as Prime Movers. Where internal combustion engines are used as the prime mover, an on-site fuel supply shall be provided with an on-premise fuel supply sufficient for not less than 2 hours' full-demand operation of the system. Where power is needed for the operation of the fuel transfer pumps to deliver fuel to a generator set day tank, this pump shall be connected to the emergency power system.

(3) Dual Supplies. Prime movers shall not be solely dependent on a public utility gas system for their fuel supply or municipal water supply for their cooling systems. Means shall be provided for automatically transferring from one fuel supply to another where dual fuel supplies are used.

(4) Battery Power and Dampers. Where a storage battery is used for control or signal power or as the means of starting the prime mover, it shall be suitable for the purpose and shall be equipped with an automatic charging means independent of the generator set. Where the battery charger is required for the operation of the generator set, it shall be connected to the emergency system. Where power is required for the operation of dampers used to ventilate the generator set, the dampers shall be connected to the emergency system.

(5) Auxiliary Power Supply. Generator sets that require more than 10 seconds to develop power shall be permitted if an auxiliary power supply energizes the emergency system until the generator can pick up the load.

(6) Outdoor Generator Sets. Where an outdoor housed generator set is equipped with a readily accessible disconnecting means located within sight of the building or structure supplied, an additional disconnecting means shall not be required where ungrounded conductors pass through the building or structure.

(C) Uninterruptible Power Supplies. Uninterruptible power supplies used to provide power for emergency systems shall comply with the applicable provisions of 700.12(A) and (B).

(D) Separate Service. Where acceptable to the authority having jurisdiction as suitable for use as an emergency source, a second service shall be permitted. This service shall be in accordance with Article 230, with separate service drop or lateral, widely separated electrically and physically from the normal service to minimize the possibility of simultaneous interruption of supply.

(E) Unit Equipment. Individual unit equipment for emergency illumination shall consist of the following:

Emergency Systems

(1) A rechargeable battery
(2) A battery charging means
(3) Provisions for one or more lamps mounted on the equipment, or shall be permitted to have terminals for remote lamps, or both
(4) A relaying device arranged to energize the lamps automatically upon failure of the supply to the unit equipment

The batteries shall be of suitable rating and capacity to supply and maintain at not less than 87-1/2 percent of the nominal battery voltage for the total lamp load associated with the unit for a period of at least 1-1/2 hours, or the unit equipment shall supply and maintain not less than 60 percent of the initial emergency illumination for a period of at least 1-1/2 hours. Storage batteries, whether of the acid or alkali type, shall be designed and constructed to meet the requirements of emergency service.

Unit equipment shall be permanently fixed in place (i.e., not portable) and shall have all wiring to each unit installed in accordance with the requirements of any of the wiring methods in Chapter 3. Flexible cord-and-plug connection shall be permitted, provided that the cord does not exceed 900 mm (3 ft) in length. The branch circuit feeding the unit equipment shall be the same branch circuit as that serving the normal lighting in the area and connected ahead of any local switches. The branch circuit that feeds unit equipment shall be clearly identified at the distribution panel. Emergency luminaires (illumination fixtures) that obtain power from a unit equipment and are not part of the unit equipment shall be wired to the unit equipment as required by 700.9 and by one of the wiring methods of Chapter 3.

IV. EMERGENCY SYSTEM CIRCUITS FOR LIGHTING AND POWER

700.15 Loads on Emergency Branch Circuits. No appliances and no lamps, other than those specified as required for emergency use, shall be supplied by emergency lighting circuits.

700.16 Emergency Illumination. Emergency illumination shall include all required means of egress lighting, illuminated exit signs, and all other lights specified as necessary to provide required illumination.

Emergency lighting systems shall be designed and installed so that the failure of any individual lighting element, such as the burning out of a light bulb, cannot leave in total darkness any space that requires emergency illumination.

Where high-intensity discharge lighting such as high- and low-pressure sodium, mercury vapor, and metal halide is used as the sole source of normal illumination, the emergency lighting system shall be required to operate until normal illumination has been restored.

700.17 Circuits for Emergency Lighting. Branch circuits that supply emergency lighting shall be installed to provide service from a source complying with 700.12 when the normal supply for lighting is interrupted. Such installations shall provide either one of the following:

(1) An emergency lighting supply, independent of the general lighting supply, with provisions for automatically transferring the emergency lights upon the event of failure of the general lighting system supply
(2) Two or more separate and complete systems with independent power supply, each system providing sufficient current for emergency lighting purposes

Unless both systems are used for regular lighting purposes and are both kept lighted, means shall be provided for automatically energizing either system upon failure of the other. Either or both systems shall be permitted to be a part of the general lighting system of the protected occupancy if circuits supplying lights for emergency illumination are installed in accordance with other sections of this article.

700.18 Circuits for Emergency Power. For branch circuits that supply equipment classed as emergency, there shall be an emergency supply source to which the load will be transferred automatically upon the failure of the normal supply.

V. CONTROL—EMERGENCY LIGHTING CIRCUITS

700.20 Switch Requirements. The switch or switches installed in emergency lighting circuits shall be arranged so that only authorized persons have control of emergency lighting.

Switches connected in series or 3- and 4-way switches shall not be used.

700.21 Switch Location. All manual switches for controlling emergency circuits shall be in locations convenient to authorized persons responsible for their actuation. In places of assembly, such as theaters, a switch for controlling emergency lighting systems shall be located in the lobby or at a place conveniently accessible thereto.

In no case shall a control switch for emergency lighting in a theater, motion-picture theater, or place of assembly be placed in a motion-picture projection booth or on a stage or platform.

700.22 Exterior Lights. Those lights on the exterior of a building that are not required for illumination when there is sufficient daylight shall be permitted to be controlled by an automatic light-actuated device.

VI. OVERCURRENT PROTECTION

700.25 Accessibility. The branch-circuit overcurrent devices in emergency circuits shall be accessible to authorized persons only.

> FPN: Fuses and circuit breakers for emergency circuit overcurrent protection, where coordinated to ensure selective clearing of fault currents, increase overall reliability of the system.

700.26 Ground-Fault Protection of Equipment. The alternate source for emergency systems shall not be required to have ground-fault protection of equipment with automatic disconnecting means. Ground-fault indication of the emergency source shall be provided per 700.7(D).

CHAPTER 25

CONDUIT AND TUBING FILL TABLES

INTRODUCTION

This chapter contains tables and notes from Chapter 9—Tables and Annex C—Conduit and Tubing Fill Tables for Conductors and Fixture Wires of the Same Size. These tables are used to determine the maximum number of conductors permitted in raceways (conduit and tubing). While Tables 4 and 5 of Chapter 9 are used when different size conductors are installed in the same conduit, Annex C is used when all the conductors are the same size. Table 8—Conductor Properties (of Chapter 9) provides useful information such as size (AWG or kcmil), circular mils, overall area, and the resistance of copper and aluminum conductors.

CHAPTER 9 TABLES

Notes to Tables

(1) See Annex C for the maximum number of conductors and fixture wires, all of the same size (total cross-sectional area including insulation) permitted in trade sizes of the applicable conduit or tubing.

(2) Table 1 applies only to complete conduit or tubing systems and is not intended to apply to sections of conduit or tubing used to protect exposed wiring from physical damage.

(3) Equipment grounding or bonding conductors, where installed, shall be included when calculating conduit or tubing fill. The actual dimensions of the equipment grounding or bonding conductor (insulated or bare) shall be used in the calculation.

(4) Where conduit or tubing nipples having a maximum length not to exceed 600 mm (24 in.) are installed between boxes, cabinets, and similar enclosures, the nipples shall be permitted to be filled to 60 percent of their total cross-sectional area, and 310.15(B)(2)(a) adjustment factors need not apply to this condition.

Table 1 Percent of Cross Section of Conduit and Tubing for Conductors

Number of Conductors	All Conductor Types
1	53
2	31
Over 2	40

FPN No. 1: Table 1 is based on common conditions of proper cabling and alignment of conductors where the length of the pull and the number of bends are within reasonable limits. It should be recognized that, for certain conditions, a larger size conduit or a lesser conduit fill should be considered.

FPN No. 2: When pulling three conductors or cables into a raceway, if the ratio of the raceway (inside diameter) to the conductor or cable (outside diameter) is between 2.8 and 3.2, jamming can occur. While jamming can occur when pulling four or more conductors or cables into a raceway, the probability is very low.

(5) For conductors not included in Chapter 9, such as multiconductor cables, the actual dimensions shall be used.

(6) For combinations of conductors of different sizes, use Table 5 and Table 5A for dimensions of conductors and Table 4 for the applicable conduit or tubing dimensions.

(7) When calculating the maximum number of conductors permitted in a conduit or tubing, all of the same size (total cross-sectional area including insulation), the next higher whole number shall be used to determine the maximum number of conductors permitted when the calculation results in a decimal of 0.8 or larger.

(8) Where bare conductors are permitted by other sections of this *Code*, the dimensions for bare conductors in Table 8 shall be permitted.

(9) A multiconductor cable of two or more conductors shall be treated as a single conductor for calculating percentage conduit fill area. For cables that have elliptical cross sections, the cross-sectional area calculation shall be based on using the major diameter of the ellipse as a circle diameter.

Table 4 Dimensions and Percent Area of Conduit and Tubing
(Areas of Conduit or Tubing for the Combinations of Wires Permitted in Table 1)

Metric Designator	Trade Size	Nominal Internal Diameter		Total Area 100%		2 Wires 31%		Over 2 Wires 40%		1 Wire 53%		60%	
		mm	in.	mm²	in.²	mm²	in.²	mm²	in.²	mm²	in.²	mm²	in.²
Article 358—Electrical Metallic Tubing (EMT)													
16	1/2	15.8	0.622	196	0.304	61	0.094	78	0.122	104	0.161	118	0.182
21	3/4	20.9	0.824	343	0.533	106	0.165	137	0.213	182	0.283	206	0.320
27	1	26.6	1.049	556	0.864	172	0.268	222	0.346	295	0.458	333	0.519
35	1-1/4	35.1	1.380	968	1.496	300	0.464	387	0.598	513	0.793	581	0.897
41	1-1/2	40.9	1.610	1314	2.036	407	0.631	526	0.814	696	1.079	788	1.221
53	2	52.5	2.067	2165	3.356	671	1.040	866	1.342	1147	1.778	1299	2.013
63	2-1/2	69.4	2.731	3783	5.858	1173	1.816	1513	2.343	2005	3.105	2270	3.515
78	3	85.2	3.356	5701	8.846	1767	2.742	2280	3.538	3022	4.688	3421	5.307
91	3-1/2	97.4	3.834	7451	11.545	2310	3.579	2980	4.618	3949	6.119	4471	6.927
103	4	110.1	4.334	9521	14.753	2951	4.573	3808	5.901	5046	7.819	5712	8.852

Conduit and Tubing Fill Tables

Article 362—Electrical Nonmetallic Tubing (ENT)

16	1/2	14.2	0.560	158	0.246	49	0.076	63	0.099	84	0.131	95	0.148
21	3/4	19.3	0.760	293	0.454	91	0.141	117	0.181	155	0.240	176	0.272
27	1	25.4	1.000	507	0.785	157	0.243	203	0.314	269	0.416	304	0.471
35	1-1/4	34.0	1.340	908	1.410	281	0.437	363	0.564	481	0.747	545	0.846
41	1-1/2	39.9	1.570	1250	1.936	388	0.600	500	0.774	663	1.026	750	1.162
53	2	51.3	2.020	2067	3.205	641	0.993	827	1.282	1095	1.699	1240	1.923
63	2-1/2	—	—	—	—	—	—	—	—	—	—	—	—
78	3	—	—	—	—	—	—	—	—	—	—	—	—
91	3-1/2	—	—	—	—	—	—	—	—	—	—	—	—

Article 348—Flexible Metal Conduit (FMC)

12	3/8	9.7	0.384	74	0.116	23	0.036	30	0.046	39	0.061	44	0.069
16	1/2	16.1	0.635	204	0.317	63	0.098	81	0.127	108	0.168	122	0.190
21	3/4	20.9	0.824	343	0.533	106	0.165	137	0.213	182	0.283	206	0.320
27	1	25.9	1.020	527	0.817	163	0.253	211	0.327	279	0.433	316	0.490
35	1-1/4	32.4	1.275	824	1.277	256	0.396	330	0.511	437	0.677	495	0.766
41	1-1/2	39.1	1.538	1201	1.858	372	0.576	480	0.743	636	0.985	720	1.115
53	2	51.8	2.040	2107	3.269	653	1.013	843	1.307	1117	1.732	1264	1.961

Continued

Table 4 Dimensions and Percent Area of Conduit and Tubing
(Areas of Conduit or Tubing for the Combinations of Wires Permitted in Table 1) *Continued*

Metric Designator	Trade Size	Nominal Internal Diameter mm	Nominal Internal Diameter in.	Total Area 100% mm²	Total Area 100% in.²	2 Wires 31% mm²	2 Wires 31% in.²	Over 2 Wires 40% mm²	Over 2 Wires 40% in.²	1 Wire 53% mm²	1 Wire 53% in.²	60% mm²	60% in.²
\multicolumn{14}{l}{Article 348—Flexible Metal Conduit (FMC) *Continued*}
63	2-1/2	63.5	2.500	3167	4.909	982	1.522	1267	1.963	1678	2.602	1900	2.945
78	3	76.2	3.000	4560	7.069	1414	2.191	1824	2.827	2417	3.746	2736	4.241
91	3-1/2	88.9	3.500	6207	9.621	1924	2.983	2483	3.848	3290	5.099	3724	5.773
103	4	101.6	4.000	8107	12.566	2513	3.896	3243	5.027	4297	6.660	4864	7.540
\multicolumn{14}{l}{Article 342—Intermediate Metal Conduit (IMC)}
12	3/8	—	—	—	—	—	—	—	—	—	—	—	—
16	1/2	16.8	0.660	222	0.342	69	0.106	89	0.137	117	0.181	133	0.205
21	3/4	21.9	0.864	377	0.586	117	0.182	151	0.235	200	0.311	226	0.352
27	1	28.1	1.105	620	0.959	192	0.297	248	0.384	329	0.508	372	0.575
35	1-1/4	36.8	1.448	1064	1.647	330	0.510	425	0.659	564	0.873	638	0.988
41	1-1/2	42.7	1.683	1432	2.225	444	0.690	573	0.890	759	1.179	859	1.335

Conduit and Tubing Fill Tables

53	2	54.6	2.150	2341	3.630	726	1.125	937	1.452	1241	1.924	1405	2.178
63	2-1/2	64.9	2.557	3308	5.135	1026	1.592	1323	2.054	1753	2.722	1985	3.081
78	3	80.7	3.176	5115	7.922	1586	2.456	2046	3.169	2711	4.199	3069	4.753
91	3-1/2	93.2	3.671	6822	10.584	2115	3.281	2729	4.234	3616	5.610	4093	6.351
103	4	105.4	4.166	8725	13.631	2705	4.226	3490	5.452	4624	7.224	5235	8.179

Article 356—Liquidtight Flexible Nonmetallic Conduit (LFNC-B*)

12	3/8	12.5	0.494	123	0.192	38	0.059	49	0.077	65	0.102	74	0.115
16	1/2	16.1	0.632	204	0.314	63	0.097	81	0.125	108	0.166	122	0.187
21	3/4	21.1	0.830	350	0.541	108	0.168	140	0.216	185	0.287	210	0.325
27	1	26.8	1.054	564	0.873	175	0.270	226	0.349	299	0.462	338	0.524
35	1-1/4	35.4	1.395	984	1.528	305	0.474	394	0.611	522	0.810	591	0.917
41	1-1/2	40.3	1.588	1276	1.981	395	0.614	510	0.792	676	1.050	765	1.188
53	2	51.6	2.033	2091	3.246	648	1.006	836	1.298	1108	1.720	1255	1.948

Article 356—Liquidtight Flexible Nonmetallic Conduit (LFNC-A*)

12	3/8	12.6	0.495	125	0.192	39	0.060	50	0.077	66	0.102	75	0.115
16	1/2	16.0	0.630	201	0.312	62	0.097	80	0.125	107	0.165	121	0.187
21	3/4	21.0	0.825	346	0.535	107	0.166	139	0.214	184	0.283	208	0.321

Continued

Table 4 Dimensions and Percent Area of Conduit and Tubing
(Areas of Conduit or Tubing for the Combinations of Wires Permitted in Table 1) *Continued*

Metric Designator	Trade Size	Nominal Internal Diameter		Total Area 100%		2 Wires 31%		Over 2 Wires 40%		1 Wire 53%		60%	
		mm	in.	mm²	in.²	mm²	in.²	mm²	in.²	mm²	in.²	mm²	in.²
Article 356—Liquidtight Flexible Nonmetallic Conduit (LFNC-A*) Continued													
27	1	26.5	1.043	552	0.854	171	0.265	221	0.342	292	0.453	331	0.513
35	1-1/4	35.1	1.383	968	1.502	300	0.466	387	0.601	513	0.796	581	0.901
41	1-1/2	40.7	1.603	1301	2.018	403	0.626	520	0.807	690	1.070	781	1.211
53	2	52.4	2.063	2157	3.343	669	1.036	863	1.337	1143	1.772	1294	2.006
Article 350—Liquidtight Flexible Metal Conduit (LFMC)													
12	1/8	12.5	0.494	123	0.192	38	0.059	49	0.077	65	0.102	74	0.115
16	1/2	16.1	0.632	204	0.314	63	0.097	81	0.125	108	0.166	122	0.188
21	1/4	21.1	0.830	350	0.541	108	0.168	140	0.216	185	0.287	210	0.325
27	1	26.8	1.054	564	0.873	175	0.270	226	0.349	299	0.462	338	0.524
35	1-1/4	35.4	1.395	984	1.528	305	0.474	394	0.611	522	0.810	591	0.917
41	1-1/2	40.3	1.588	1276	1.981	395	0.614	510	0.792	676	1.050	765	1.188

Conduit and Tubing Fill Tables

53	2	51.6	2.033	2091	3.246	648	1.006	836	1.298	1108	1.720	1255	1.948
63	2-1/2	63.3	2.493	3147	4.881	976	1.513	1259	1.953	1668	2.587	1888	2.929
78	3	78.4	3.085	4827	7.475	1497	2.317	1931	2.990	2559	3.962	2896	4.485
91	3-1/2	89.4	3.520	6277	9.731	1946	3.017	2511	3.893	3327	5.158	3766	5.839
103	4	102.1	4.020	8187	12.692	2538	3.935	3275	5.077	4339	6.727	4912	7.615
129	5	—	—	—	—	—	—	—	—	—	—	—	—
155	6	—	—	—	—	—	—	—	—	—	—	—	—
					Article 344—Rigid Metal Conduit								
12	3/8	—	—	—	—	—	—	—	—	—	—	—	—
16	1/2	16.1	0.632	203	0.314	63	0.097	81	0.125	108	0.166	122	0.188
21	3/4	21.0	0.836	353	0.549	109	0.170	141	0.220	187	0.291	212	0.329
27	1	27.0	1.063	573	0.887	177	0.275	229	0.355	303	0.470	344	0.532
35	1-1/4	35.4	1.394	984	1.526	305	0.473	394	0.610	522	0.809	591	0.916
41	1-1/2	41.2	1.624	1333	2.071	413	0.642	533	0.829	707	1.098	800	1.243
53	2	52.9	2.083	2198	3.408	681	1.056	879	1.363	1165	1.806	1319	2.045
63	2-1/2	63.2	2.489	3137	4.866	972	1.508	1255	1.946	1663	2.579	1882	2.919
78	3	78.5	3.090	4840	7.499	1500	2.325	1936	3.000	2565	3.974	2904	4.499
91	3-1/2	90.7	3.570	6461	10.010	2003	3.103	2584	4.004	3424	5.305	3877	6.006
103	4	102.9	4.050	8316	12.882	2578	3.994	3326	5.153	4408	6.828	4990	7.729

Continued

Table 4 Dimensions and Percent Area of Conduit and Tubing
(Areas of Conduit or Tubing for the Combinations of Wires Permitted in Table 1) *Continued*

Metric Designator	Trade Size	Nominal Internal Diameter		Total Area 100%		2 Wires 31%		Over 2 Wires 40%		1 Wire 53%		60%	
		mm	in.	mm²	in.²	mm²	in.²	mm²	in.²	mm²	in.²	mm²	in.²
Article 344—Rigid Metal Conduit *Continued*													
129	5	128.9	5.073	13050	20.212	4045	6.266	5220	8.085	6916	10.713	7830	12.127
155	6	154.8	6.093	18812	29.158	5834	9.039	7528	11.663	9975	15.454	11292	17.495
Article 352—Rigid PVC Conduit (RNC), Schedule 80													
12	1/8	—	—	—	—	—	—	—	—	—	—	—	—
16	1/2	13.4	0.526	141	0.217	44	0.067	56	0.087	75	0.115	85	0.130
21	3/4	18.3	0.722	263	0.409	82	0.127	105	0.164	139	0.217	158	0.246
27	1	23.8	0.936	445	0.688	138	0.213	178	0.275	236	0.365	267	0.413
35	1-1/4	31.9	1.255	799	1.237	248	0.383	320	0.495	424	0.656	480	0.742
41	1-1/2	37.5	1.476	1104	1.711	342	0.530	442	0.684	585	0.907	663	1.027
53	2	48.6	1.913	1855	2.874	575	0.891	742	1.150	983	1.52	1113	1.725
63	2-1/2	58.2	2.290	2660	4.119	825	1.277	1064	1.647	1410	2.183	1596	2.471

Conduit and Tubing Fill Tables

78	3	72.7	2.864	4151	6.442	1287	1.997	1660	2.577	2200	3.414	2491	3.865
91	3-1/2	84.5	3.326	5608	8.688	1738	2.693	2243	3.475	2972	4.605	3365	5.213
103	4	96.2	3.786	7268	11.258	2253	3.490	2907	4.503	3852	5.967	4361	6.755
129	5	121.1	4.768	11518	17.855	3571	5.535	4607	7.142	6105	9.463	6911	10.713
155	6	145.0	5.709	16513	25.598	5119	7.935	6605	10.239	8752	13.567	9908	15.359

Article 352 — Rigid PVC Conduit (RNC), Schedule 40, and HDPE Conduit

12	1/8	—	—	184	0.285	57	0.088	74	0.114	97	0.151	110	0.171
16	1/2	15.3	0.602	327	0.508	101	0.157	131	0.203	173	0.269	196	0.305
21	3/4	20.4	0.804	535	0.832	166	0.258	214	0.333	284	0.441	321	0.499
27	1	26.1	1.029	935	1.453	290	0.450	374	0.581	495	0.770	561	0.872
35	1-1/4	34.5	1.360	1282	1.986	397	0.616	513	0.794	679	1.052	769	1.191
41	1-1/2	40.4	1.590	2124	3.291	658	1.020	849	1.316	1126	1.744	1274	1.975
53	2	52.0	2.047	3029	4.695	939	1.455	1212	1.878	1605	2.488	1817	2.817
63	2-1/2	62.1	2.445	4693	7.268	1455	2.253	1877	2.907	2487	3.852	2816	4.361
78	3	77.3	3.042	6277	9.737	1946	3.018	2511	3.895	3327	5.161	3766	5.842
91	3-1/2	89.4	3.521	8091	12.554	2508	3.892	3237	5.022	4288	6.654	4855	7.532
103	4	101.5	3.998	12748	19.761	3952	6.126	5099	7.904	6756	10.473	7649	11.856
129	5	127.4	5.016	18433	28.567	5714	8.856	7373	11.427	9770	15.141	11060	17.140
155	6	153.2	6.031										

Table 5 Dimensions of Insulated Conductors and Fixture Wires

Type	Size (AWG or kcmil)	Approximate Diameter		Approximate Area	
		mm	in.	mm²	in.²
Type: FFH-2, RFH-1, RFH-2, RHH*, RHW*, RHW-2*, RHH, RHW, RHW-2, SF-1, SF-2, SFF-1, SFF-2, TF, TFF, THHW, THW, THW-2, TW, XF, XFF					
RFH-2, FFH-2	18	3.454	0.136	9.355	0.0145
	16	3.759	0.148	11.10	0.0172
RHW-2, RHH, RHW	14	4.902	0.193	18.90	0.0293
	12	5.385	0.212	22.77	0.0353
	10	5.994	0.236	28.19	0.0437
	8	8.280	0.326	53.87	0.0835
	6	9.246	0.364	67.16	0.1041
	4	10.46	0.412	86.00	0.1333
	3	11.18	0.440	98.13	0.1521
	2	11.99	0.472	112.9	0.1750
	1	14.78	0.582	171.6	0.2660
	1/0	15.80	0.622	196.1	0.3039
	2/0	16.97	0.668	226.1	0.3505
	3/0	18.29	0.720	262.7	0.4072
	4/0	19.76	0.778	306.7	0.4754

Conduit and Tubing Fill Tables

	250	22.73	0.895	405.9	0.6291
	300	24.13	0.950	457.3	0.7088
	350	25.43	1.001	507.7	0.7870
	400	26.62	1.048	556.5	0.8626
	500	28.78	1.133	650.5	1.0082
	600	31.57	1.243	782.9	1.2135
	700	33.38	1.314	874.9	1.3561
	750	34.24	1.348	920.8	1.4272
	800	35.05	1.380	965.0	1.4957
	900	36.68	1.444	1057	1.6377
	1000	38.15	1.502	1143	1.7719
	1250	43.92	1.729	1515	2.3479
	1500	47.04	1.852	1738	2.6938
	1750	49.94	1.966	1959	3.0357
	2000	52.63	2.072	2175	3.3719
SF-2, SFF-2	18	3.073	0.121	7.419	0.0115
	16	3.378	0.133	8.968	0.0139
	14	3.759	0.148	11.10	0.0172

Continued

Table 5 Dimensions of Insulated Conductors and Fixture Wires *Continued*

Type	Size (AWG or kcmil)	Approximate Diameter		Approximate Area	
		mm	in.	mm²	in.²
Type: FFH-2, RFH-1, RFH-2, RHH*, RHW*, RHW-2*, RHH, RHW, RHW-2, SF-1, SF-2, SFF-1, SFF-2, TF, TFF, THHW, THW, THW-2, TW, XF, XFF *Continued*					
SF-1, SFF-1	18	2.311	0.091	4.194	0.0065
RFH-1, XF, XFF	18	2.692	0.106	5.161	0.0080
TF, TFF, XF, XFF	16	2.997	0.118	7.032	0.0109
TW, XF, XFF, THHW, THW, THW-2	14	3.378	0.133	8.968	0.0139
TW, THHW, THW, THW-2	12	3.861	0.152	11.68	0.0181
	10	4.470	0.176	15.68	0.0243
	8	5.994	0.236	28.19	0.0437
RHH*, RHW*, RHW-2*	14	4.140	0.163	13.48	0.0209
RHH*, RHW*, RHW-2*, XF, XFF	12	4.623	0.182	16.77	0.0260

Conduit and Tubing Fill Tables

Type	Size			
RHH*, RHW*, RHW-2*, XF, XFF				
RHH*, RHW*, RHW-2*	10	5.232	0.206	21.48	0.0333
RHH*, RHW*, RHW-2*	8	6.756	0.266	35.87	0.0556
TW, THW, THHW, THW-2, RHH*, RHW*, RHW-2*	6	7.722	0.304	46.84	0.0726
	4	8.941	0.652	62.77	0.0973
	3	9.652	0.380	73.16	0.1134
	2	10.46	0.412	86.00	0.1333
	1	12.50	0.492	122.6	0.1901
	1/0	13.51	0.532	143.4	0.2223
	2/0	14.68	0.578	169.3	0.2624
	3/0	16.00	0.630	201.1	0.3117
	4/0	17.48	0.688	239.9	0.3718
	250	19.43	0.765	296.5	0.4596
	300	20.93	0.820	340.7	0.5281
	350	22.12	0.871	384.4	0.5958
	400	23.32	0.918	427.0	0.6619

Continued

Table 5 Dimensions of Insulated Conductors and Fixture Wires *Continued*

Type	Size (AWG or kcmil)	Approximate Diameter		Approximate Area	
		mm	in.	mm²	in.²
Type: RHH*, RHW*, RHW-2*, THHN, THHW, THW, THW-2, TFN, TFFN, THWN, THWN-2, XF, XFF *Continued*					
TW, THW, THHW, THW-2, RHH*, RHW*, RHW-2*	500	25.48	1.003	509.7	0.7901
	600	28.27	1.113	627.7	0.9729
	700	30.07	1.184	710.3	1.1010
	750	30.94	1.218	751.7	1.1652
	800	31.75	1.250	791.7	1.2272
	900	33.38	1.314	874.9	1.3561
	1000	34.85	1.372	953.8	1.4784
	1250	39.09	1.539	1200	1.8602
	1500	42.21	1.662	1400	2.1695
	1750	45.11	1.776	1598	2.4773
	2000	47.80	1.882	1795	2.7818

Conduit and Tubing Fill Tables

TFN, TFFN	18	2.134	0.084	3.548	0.0055
	16	2.438	0.096	4.645	0.0072
THHN, THWN, THWN-2	14	2.819	0.111	6.258	0.0097
	12	3.302	0.130	8.581	0.0133
	10	4.166	0.164	13.61	0.0211
	8	5.486	0.216	23.61	0.0366
	6	6.452	0.254	32.71	0.0507
	4	8.230	0.324	53.16	0.0824
	3	8.941	0.352	62.77	0.0973
	2	9.754	0.384	74.71	0.1158
	1	11.33	0.446	100.8	0.1562
	1/0	12.34	0.486	119.7	0.1855
	2/0	13.51	0.532	143.4	0.2223
	3/0	14.83	0.584	172.8	0.2679
	4/0	16.31	0.642	208.8	0.3237
	250	18.06	0.711	256.1	0.3970
	300	19.46	0.766	297.3	0.4608

Continued

Table 5 Dimensions of Insulated Conductors and Fixture Wires *Continued*

Type	Size (AWG or kcmil)	Approximate Diameter		Approximate Area	
		mm	in.	mm²	in.²
Type: FEP, FEPB, PAF, PAFF, PF, PFA, PFAH, PFF, PGF, PGFF, PTF, PTFF, TFE, THHN, THWN, THWN-2, Z, ZF, ZFF					
THHN, THWN, THWN-2	350	20.75	0.817	338.2	0.5242
	400	21.95	0.864	378.3	0.5863
	500	24.10	0.949	456.3	0.7073
	600	26.70	1.051	559.7	0.8676
	700	28.50	1.122	637.9	0.9887
	750	29.36	1.156	677.2	1.0496
	800	30.18	1.188	715.2	1.1085
	900	31.80	1.252	794.3	1.2311
	1000	33.27	1.310	869.5	1.3478
PF, PGFF, PGF, PFF, PTF, PAF, PTFF, PAFF	18	2.184	0.086	3.742	0.0058
	16	2.489	0.098	4.839	0.0075
PF, PGFF, PGF, PFF, PTF, PAF, PTFF, PAFF, TFE, FEP, PFA, FEPB, PFAH	14	2.870	0.113	6.452	0.0100

Conduit and Tubing Fill Tables

TFE, FEP, PFA, FEPB, PFAH	12	3.353	0.132	8.839	0.0137
	10	3.962	0.156	12.32	0.0191
	8	5.232	0.206	21.48	0.0333
	6	6.198	0.244	30.19	0.0468
	4	7.417	0.292	43.23	0.0670
	3	8.128	0.320	51.87	0.0804
	2	8.941	0.352	62.77	0.0973
TFE, PFAH	1	10.72	0.422	90.26	0.1399
TFE, PFA, PFAH, Z	1/0	11.73	0.462	108.1	0.1676
	2/0	12.90	0.508	130.8	0.2027
	3/0	14.22	0.560	158.9	0.2463
	4/0	15.70	0.618	193.5	0.3000
ZF, ZFF	18	1.930	0.076	2.903	0.0045
	16	2.235	0.088	3.935	0.0061
Z, ZF, ZFF	14	2.616	0.103	5.355	0.0083
Z	12	3.099	0.122	7.548	0.0117
	10	3.962	0.156	12.32	0.0191

Continued

Table 5 Dimensions of Insulated Conductors and Fixture Wires *Continued*

Type	Size (AWG or kcmil)	Approximate Diameter mm	Approximate Diameter in.	Approximate Area mm²	Approximate Area in.²
Type: FEP, FEPB, PAF, PAFF, PF, PFA, PFAH, PFF, PGF, PGFF, PTF, PTFE, THHN, THWN, THWN-2, Z, ZF, ZFF *Continued*					
Z	8	4.978	0.196	19.48	0.0302
	6	5.944	0.234	27.74	0.0430
	4	7.163	0.282	40.32	0.0625
	3	8.382	0.330	55.16	0.0855
	2	9.195	0.362	66.39	0.1029
	1	10.21	0.402	81.87	0.1269
Type: KF-1, KF-2, KFF-1, KFF-2, XHH, XHHW, XHHW-2, ZW					
XHHW, ZW, XHHW-2, XHH	14	3.378	0.133	8.968	0.0139
	12	3.861	0.152	11.68	0.0181
	10	4.470	0.176	15.68	0.0243
	8	5.994	0.236	28.19	0.0437
	6	6.960	0.274	38.06	0.0590
	4	8.179	0.322	52.52	0.0814
	3	8.890	0.350	62.06	0.0962
	2	9.703	0.382	73.94	0.1146

Conduit and Tubing Fill Tables

XHHW, XHHW-2, XHH	1	11.23		0.442		98.97		0.1534
	1/0	12.24		0.482		117.7		0.1825
	2/0	13.41		0.528		141.3		0.2190
	3/0	14.73		0.580		170.5		0.2642
	4/0	16.21		0.638		206.3		0.3197
	250	17.91		0.705		251.9		0.3904
	300	19.30		0.760		292.6		0.4536
	350	20.60		0.811		333.3		0.5166
	400	21.79		0.858		373.0		0.5782
	500	23.95		0.943		450.6		0.6984
	600	26.75		1.053		561.9		0.8709
	700	28.55		1.124		640.2		0.9923
	750	29.41		1.158		679.5		1.0532
	800	30.23		1.190		717.5		1.1122
	900	31.85		1.254		796.8		1.2351
	1000	33.32		1.312		872.2		1.3519
	1250	37.57		1.479		1108		1.7180
	1500	40.69		1.602		1300		2.0157

Continued

Table 5 Dimensions of Insulated Conductors and Fixture Wires *Continued*

Type	Size (AWG or kcmil)	Approximate Diameter		Approximate Area	
		mm	in.	mm²	in.²
Type: KF-1, KF-2, KFF-1, KFF-2, XHH, XHHW, XHHW-2, ZW *Continued*					
XHHW, XHHW-2, XHH	1750	43.59	1.716	1492	2.3127
	2000	46.28	1.822	1682	2.6073
KF-2, KFF-2	18	1.600	0.063	2.000	0.0031
	16	1.905	0.075	2.839	0.0044
	14	2.286	0.090	4.129	0.0064
	12	2.769	0.109	6.000	0.0093
	10	3.378	0.133	8.968	0.0139
KF-1, KFF-1	18	1.448	0.057	1.677	0.0026
	16	1.753	0.069	2.387	0.0037
	14	2.134	0.084	3.548	0.0055
	12	2.616	0.103	5.355	0.0083
	10	3.226	0.127	8.194	0.0127

*Types RHH, RHW, and RHW-2 without outer covering.

Conduit and Tubing Fill Tables

Table 8 Conductor Properties

Size (AWG or kcmil)	Area		Conductors						Direct-Current Resistance at 75°C (167°F)					
			Stranding		Overall				Copper				Aluminum	
					Diameter		Area		Uncoated		Coated			
	mm²	Circular mils	Qty	mm	in.	mm	in.	mm²	in.²	ohm/km	ohm/kFT	ohm/km	ohm/kFT	ohm/km	ohm/kFT
18	0.823	1620	1	—	—	1.02	0.040	0.823	0.001	25.5	7.77	26.5	8.08	42.0	12.8
18	0.823	1620	7	0.39	0.015	1.16	0.046	1.06	0.002	26.1	7.95	27.7	8.45	42.8	13.1
16	1.31	2580	1	—	—	1.29	0.051	1.31	0.002	16.0	4.89	16.7	5.08	26.4	8.05
16	1.31	2580	7	0.49	0.019	1.46	0.058	1.68	0.003	16.4	4.99	17.3	5.29	26.9	8.21
14	2.08	4110	1	—	—	1.63	0.064	2.08	0.003	10.1	3.07	10.4	3.19	16.6	5.06
14	2.08	4110	7	0.62	0.024	1.85	0.073	2.68	0.004	10.3	3.14	10.7	3.26	16.9	5.17
12	3.31	6530	1	—	—	2.05	0.081	3.31	0.005	6.34	1.93	6.57	2.01	10.45	3.18
12	3.31	6530	7	0.78	0.030	2.32	0.092	4.25	0.006	6.50	1.98	6.73	2.05	10.69	3.25
10	5.261	10380	1	—	—	2.588	0.102	5.26	0.008	3.984	1.21	4.148	1.26	6.561	2.00
10	5.261	10380	7	0.98	0.308	2.95	0.116	6.76	0.011	4.070	1.24	4.226	1.29	6.679	2.04

Continued

Table 8 Conductor Properties *Continued*

Size (AWG or kcmil)	Area mm²	Area Circular mils	Stranding Qty	Stranding Diameter mm	Stranding Diameter in.	Conductors Overall Diameter mm	Conductors Overall Diameter in.	Conductors Overall Area mm²	Conductors Overall Area in.²	DC Resistance at 75°C (167°F) Copper Uncoated ohm/km	Copper Uncoated ohm/kFT	Copper Coated ohm/km	Copper Coated ohm/kFT	Aluminum ohm/km	Aluminum ohm/kFT
8	8.367	16510	1	—	—	3.264	0.128	8.37	0.013	2.506	0.764	2.579	0.786	4.125	1.26
8	8.367	16510	7	1.23	0.049	3.71	0.146	10.76	0.017	2.551	0.778	2.653	0.809	4.204	1.28
6	13.30	26240	7	1.56	0.061	4.67	0.184	17.09	0.027	1.608	0.491	1.671	0.510	2.652	0.808
4	21.15	41740	7	1.96	0.077	5.89	0.232	27.19	0.042	1.010	0.308	1.053	0.321	1.666	0.508
3	26.67	52620	7	2.20	0.087	6.60	0.260	34.28	0.053	0.802	0.245	0.833	0.254	1.320	0.403
2	33.62	66360	7	2.47	0.097	7.42	0.292	43.23	0.067	0.634	0.194	0.661	0.201	1.045	0.319
1	42.41	83690	19	1.69	0.066	8.43	0.332	55.80	0.087	0.505	0.154	0.524	0.160	0.829	0.253
1/0	53.49	105600	19	1.89	0.074	9.45	0.372	70.41	0.109	0.399	0.122	0.415	0.127	0.660	0.201
2/0	67.43	133100	19	2.13	0.084	10.62	0.418	88.74	0.137	0.3170	0.0967	0.329	0.101	0.523	0.159
3/0	85.01	167800	19	2.39	0.094	11.94	0.470	111.9	0.173	0.2512	0.0766	0.2610	0.0797	0.413	0.126
4/0	107.2	211600	19	2.68	0.106	13.41	0.528	141.1	0.219	0.1996	0.0608	0.2050	0.0626	0.328	0.100
250	—	—	37	2.09	0.082	14.61	0.575	168	0.260	0.1687	0.0515	0.1753	0.0535	0.2778	0.0847

Conduit and Tubing Fill Tables

300	—	37	2.29	0.090	16.00	0.630	201	0.312	0.1409	0.0429	0.1463	0.0446	0.2318	0.0707
350	—	37	2.47	0.097	17.30	0.681	235	0.364	0.1205	0.0367	0.1252	0.0382	0.1984	0.0605
400	—	37	2.64	0.104	18.49	0.728	268	0.416	0.1053	0.0321	0.1084	0.0331	0.1737	0.0529
500	—	37	2.95	0.116	20.65	0.813	336	0.519	0.0845	0.0258	0.0869	0.0265	0.1391	0.0424
600	—	61	2.52	0.099	22.68	0.893	404	0.626	0.0704	0.0214	0.0732	0.0223	0.1159	0.0353
700	—	61	2.72	0.107	24.49	0.964	471	0.730	0.0603	0.0184	0.0622	0.0189	0.0994	0.0303
750	—	61	2.82	0.111	25.35	0.998	505	0.782	0.0563	0.0171	0.0579	0.0176	0.0927	0.0282
800	—	61	2.91	0.114	26.16	1.030	538	0.834	0.0528	0.0161	0.0544	0.0166	0.0868	0.0265
900	—	61	3.09	0.122	27.79	1.094	606	0.940	0.0470	0.0143	0.0481	0.0147	0.0770	0.0235
1000	—	61	3.25	0.128	29.26	1.152	673	1.042	0.0423	0.0129	0.0434	0.0132	0.0695	0.0212
1250	—	91	2.98	0.117	32.74	1.289	842	1.305	0.0338	0.0103	0.0347	0.0106	0.0554	0.0169
1500	—	91	3.26	0.128	35.86	1.412	1011	1.566	0.02814	0.00858	0.02814	0.00883	0.0464	0.0141
1750	—	127	2.98	0.117	38.76	1.526	1180	1.829	0.02410	0.00735	0.02410	0.00756	0.0397	0.0121
2000	—	127	3.19	0.126	41.45	1.632	1349	2.092	0.02109	0.00643	0.02109	0.00662	0.0348	0.0106

Notes:
1. These resistance values are valid **only** for the parameters as given. Using conductors having coated strands, different stranding type, and, especially, other temperatures changes the resistance.
2. Formula for temperature change: $R_2 = R_1 [1 + \alpha(T_2 - 75)]$ where $\alpha_{cu} = 0.00323$, $\alpha_{Al} = 0.00330$ at 75°C.
3. Conductors with compact and compressed stranding have about 9 percent and 3 percent, respectively, smaller bare conductor diameters than those shown. See Table 5A for actual compact cable dimensions.
4. The IACS conductivities used: bare copper = 100%, aluminum = 61%.
5. Class B stranding is listed as well as solid for some sizes. Its overall diameter and area is that of its circumscribing circle.

Table 9 Alternating-Current Resistance and Reactance for 600-Volt Cables, 3-Phase, 60 Hz, 75°C (167°F)—Three Single Conductors in Conduit

Ohms to Neutral per Kilometer
Ohms to Neutral per 1000 Feet

Size (AWG or kcmil)	X_L (Reactance) for All Wires		Alternating-Current Resistance for Uncoated Copper Wires			Alternating-Current Resistance for Aluminum Wires			Effective Z at 0.85 PF for Uncoated Copper Wires			Effective Z at 0.85 PF for Aluminum Wires			Size (AWG or kcmil)
	PVC, Aluminum Conduits	Steel Conduit	PVC Conduit	Aluminum Conduit	Steel Conduit	PVC Conduit	Aluminum Conduit	Steel Conduit	PVC Conduit	Aluminum Conduit	Steel Conduit	PVC Conduit	Aluminum Conduit	Steel Conduit	
14	0.190 / 0.058	0.240 / 0.073	10.2 / 3.1	10.2 / 3.1	10.2 / 3.1	—	—	—	8.9 / 2.7	8.9 / 2.7	8.9 / 2.7	—	—	—	14
12	0.177 / 0.054	0.223 / 0.068	6.6 / 2.0	6.6 / 2.0	6.6 / 2.0	10.5 / 3.2	10.5 / 3.2	10.5 / 3.2	5.6 / 1.7	5.6 / 1.7	5.6 / 1.7	9.2 / 2.8	9.2 / 2.8	9.2 / 2.8	12
10	0.164 / 0.050	0.207 / 0.063	3.9 / 1.2	3.9 / 1.2	3.9 / 1.2	6.6 / 2.0	6.6 / 2.0	6.6 / 2.0	3.6 / 1.1	3.6 / 1.1	3.6 / 1.1	5.9 / 1.8	5.9 / 1.8	5.9 / 1.8	10
8	0.171 / 0.052	0.213 / 0.065	2.56 / 0.78	2.56 / 0.78	2.56 / 0.78	4.3 / 1.3	4.3 / 1.3	4.3 / 1.3	2.26 / 0.69	2.26 / 0.69	2.30 / 0.70	3.6 / 1.1	3.6 / 1.1	3.6 / 1.1	8

6	0.167	0.210	1.61	1.61	1.61	2.66	2.66	2.66	1.44	1.18	1.18	2.33	2.36	2.36	6
	0.051	0.064	0.49	0.49	0.49	0.81	0.81	0.81	0.44	0.45	0.45	0.71	0.72	0.72	
4	0.157	0.197	1.02	1.02	1.02	1.67	1.67	1.67	0.95	0.95	0.98	1.51	1.51	1.51	4
	0.048	0.060	0.31	0.31	0.31	0.51	0.51	0.51	0.29	0.29	0.30	0.46	0.46	0.46	
3	0.154	0.194	0.82	0.82	0.82	1.31	1.35	1.31	0.75	0.79	0.79	1.21	1.21	1.21	3.
	0.047	0.059	0.25	0.25	0.25	0.40	0.41	0.40	0.23	0.24	0.24	0.37	0.037	0.37	
2	0.148	0.187	0.62	0.66	0.66	1.05	1.05	1.05	0.62	0.62	0.66	0.98	0.98	0.98	2
	0.045	0.057	0.19	0.20	0.20	0.32	0.32	0.32	0.19	0.19	0.20	0.30	0.30	0.30	
1	0.151	0.187	0.49	0.52	0.52	0.82	0.85	0.82	0.52	0.52	0.52	0.79	0.79	0.82	1
	0.046	0.057	0.15	0.16	0.16	0.25	0.26	0.25	0.16	0.16	0.16	0.24	0.24	0.25	
1/0	0.144	0.180	0.39	0.43	0.39	0.66	0.69	0.66	0.43	0.43	0.43	0.62	0.66	0.66	1/0
	0.044	0.055	0.12	0.13	0.12	0.20	0.21	0.20	0.13	0.13	0.13	0.19	0.20	0.20	
2/0	0.141	0.177	0.33	0.33	0.33	0.52	0.52	0.52	0.36	0.36	0.36	0.52	0.52	0.52	2/0
	0.043	0.054	0.10	0.10	0.10	0.16	0.16	0.16	0.11	0.11	0.11	0.16	0.16	0.16	
3/0	0.138	0.171	0.253	0.269	0.259	0.43	0.43	0.43	0.289	0.302	0.308	0.43	0.43	0.46	3/0
	0.042	0.052	0.077	0.082	0.079	0.13	0.13	0.13	0.088	0.092	0.094	0.13	0.13	0.14	

Continued

Table 9 Alternating-Current Resistance and Reactance for 600-Volt Cables, 3-Phase, 60 Hz, 75°C (167°F)—Three Single Conductors in Conduit *Continued*

			Ohms to Neutral per Kilometer / Ohms to Neutral per 1000 Feet										
Size (AWG or kcmil)	X_L (Reactance) for All Wires		Alternating-Current Resistance for Uncoated Copper Wires			Alternating-Current Resistance for Aluminum Wires			Effective Z at 0.85 PF for Uncoated Copper Wires			Effective Z at 0.85 PF for Aluminum Wires			Size (AWG or kcmil)
	PVC, Aluminum Conduits	Steel Conduit	PVC Conduit	Aluminum Conduit	Steel Conduit	PVC Conduit	Aluminum Conduit	Steel Conduit	PVC Conduit	Aluminum Conduit	Steel Conduit	PVC Conduit	Aluminum Conduit	Steel Conduit	
4/0	0.135 / 0.041	0.167 / 0.051	0.203 / 0.062	0.220 / 0.067	0.207 / 0.063	0.33 / 0.10	0.36 / 0.11	0.33 / 0.10	0.243 / 0.074	0.256 / 0.078	0.262 / 0.080	0.36 / 0.11	0.36 / 0.11	0.36 / 0.11	4/0
250	0.135 / 0.041	0.171 / 0.052	0.171 / 0.052	0.187 / 0.057	0.177 / 0.054	0.279 / 0.085	0.295 / 0.090	0.282 / 0.086	0.217 / 0.066	0.230 / 0.070	0.240 / 0.073	0.308 / 0.094	0.322 / 0.098	0.33 / 0.10	250
300	0.135 / 0.041	0.167 / 0.051	0.144 / 0.044	0.161 / 0.049	0.148 / 0.045	0.233 / 0.071	0.249 / 0.076	0.236 / 0.072	0.194 / 0.059	0.207 / 0.063	0.213 / 0.065	0.269 / 0.082	0.282 / 0.086	0.289 / 0.088	300
350	0.131 / 0.040	0.164 / 0.050	0.125 / 0.038	0.141 / 0.043	0.128 / 0.039	0.200 / 0.061	0.217 / 0.066	0.207 / 0.063	0.174 / 0.053	0.190 / 0.058	0.197 / 0.060	0.240 / 0.073	0.253 / 0.077	0.262 / 0.080	350

400	0.131 0.040	0.161 0.049	0.108 0.033	0.125 0.038	0.115 0.035	0.177 0.054	0.180 0.055	0.161 0.049	0.174 0.053	0.184 0.056	0.217 0.066	0.233 0.071	0.240 0.073
500	0.128 0.039	0.157 0.048	0.089 0.027	0.105 0.032	0.095 0.029	0.141 0.043	0.148 0.045	0.141 0.043	0.157 0.048	0.164 0.050	0.187 0.057	0.200 0.061	0.210 0.064
600	0.128 0.039	0.157 0.048	0.075 0.023	0.092 0.028	0.082 0.025	0.118 0.036	0.125 0.038	0.131 0.040	0.144 0.044	0.154 0.047	0.167 0.051	0.180 0.055	0.190 0.058
750	0.125 0.038	0.157 0.048	0.062 0.019	0.079 0.024	0.069 0.021	0.095 0.029	0.102 0.031	0.118 0.036	0.131 0.040	0.141 0.043	0.148 0.045	0.161 0.049	0.171 0.052
1000	0.121 0.037	0.151 0.046	0.049 0.015	0.062 0.019	0.059 0.018	0.075 0.023	0.082 0.025	0.105 0.032	0.118 0.036	0.131 0.040	0.128 0.039	0.138 0.042	0.151 0.046

Notes:

1. These values are based on the following constants: UL-Type RHH wires with Class B stranding, in cradled configuration. Wire conductivities are 100 percent IACS copper and 61 percent IACS aluminum, and aluminum conduit is 45 percent IACS. Capacitive reactance is ignored, since it is negligible at these voltages. These resistance values are valid only at 75°C (167°F) and for the parameters as given, but are representative for 600-volt wire types operating at 60 Hz.

2. *Effective Z* is defined as $R \cos(\theta) + X \sin(\theta)$, where θ is the power factor angle of the circuit. Multiplying current by effective impedance gives a good approximation for line-to-neutral voltage drop. Effective impedance values shown in this table are valid only at 0.85 power factor. For another circuit power factor (*PF*), effective impedance (*Ze*) can be calculated from *R* and X_L values given in this table as follows: $Ze = R \times PF + X_L \sin[\arccos(PF)]$.

ANNEX C

Conduit and Tubing Fill Tables for Conductors and Fixture Wires of the Same Size

Table C1 Maximum Number of Conductors or Fixture Wires in Electrical Metallic Tubing (EMT) (Based on Table 1)

Type	Conductor Size (AWG/kcmil)	\multicolumn{10}{c}{CONDUCTORS — Metric Designator (Trade Size)}									
		16 (1/2)	21 (3/4)	27 (1)	35 (1-1/4)	41 (1-1/2)	53 (2)	63 (2-1/2)	78 (3)	91 (3-1/2)	103 (4)
RHH, RHW, RHW-2	14	4	7	11	20	27	46	80	120	157	201
	12	3	6	9	17	23	38	66	100	131	167
	10	2	5	8	13	18	30	53	81	105	135
	8	1	2	4	7	9	16	28	42	55	70
	6	1	1	3	5	8	13	22	34	44	56
	4	1	1	2	4	6	10	17	26	34	44
	3	1	1	1	3	5	9	15	23	30	38
	2	1	1	1	3	4	7	13	20	26	33
	1	0	1	1	1	3	5	9	13	17	22
	1/0	0	1	1	1	2	4	7	11	15	19
	2/0	0	1	1	1	2	4	6	10	13	17
	3/0	0	0	1	1	1	3	5	8	11	14
	4/0	0	0	1	1	1	3	5	7	9	12

Conduit and Tubing Fill Tables

Size										
250	0	0	0	0	1	1	1	3	5	7	9
300	0	0	0	0	1	1	1	3	5	6	8
350	0	0	0	0	1	1	1	3	4	6	7
400	0	0	0	0	1	1	1	2	4	5	7
500	0	0	0	0	0	1	1	2	3	4	6
600	0	0	0	0	0	1	1	1	3	4	5
700	0	0	0	0	0	0	1	1	2	3	4
750	0	0	0	0	0	0	1	1	2	3	4
800	0	0	0	0	0	0	1	1	2	3	4
900	0	0	0	0	0	0	1	1	1	3	3
1000	0	0	0	0	0	0	1	1	1	2	3
1250	0	0	0	0	0	0	0	1	1	1	2
1500	0	0	0	0	0	0	0	1	1	1	1
1750	0	0	0	0	0	0	0	1	1	1	1
2000	0	0	0	0	0	0	0	1	1	1	1
TW, THHW	14	8	15	25	43	58	96	168	254	332	424
THW, THW-2	12	6	11	19	33	45	74	129	195	255	326
	10	5	8	14	24	33	55	96	145	190	243
	8	2	5	8	13	18	30	53	81	105	135

Continued

Table C1 Maximum Number of Conductors or Fixture Wires in Electrical Metallic Tubing (EMT) (Based on Table 1) *Continued*

CONDUCTORS

Type	Conductor Size (AWG/kcmil)	Metric Designator (Trade Size)								
		16 (1/2)	21 (3/4)	27 (1)	35 (1-1/4)	41 (1-1/2)	53 (2)	63 (2-1/2)	78 (3)	91 (3-1/2)	103 (4)
RHH*, RHW*, RHW-2*	14	6	10	16	28	39	64	112	169	221	282
	12	4	8	13	23	31	51	90	136	177	227
	10	3	6	10	18	24	40	70	106	138	177
	8	1	4	6	10	14	24	42	63	83	106
RHH*, RHW*, RHW-2*, TW, THW, THHW, THW-2	6	1	3	4	8	11	18	32	48	63	81
	4	1	1	3	6	8	13	24	36	47	60
	3	1	1	3	5	7	12	20	31	40	52
	2	1	1	2	4	6	10	17	26	34	44
	1	1	1	1	3	4	7	12	18	24	31
	1/0	0	1	1	2	3	6	10	16	20	26
	2/0	0	1	1	1	3	5	9	13	17	22
	3/0	0	1	1	1	2	4	7	11	15	19
	4/0	0	0	1	1	1	3	6	9	12	16

Conduit and Tubing Fill Tables

250	0	0	0	0	1	1	1	3	5	7	10	13
300	0	0	0	0	1	1	1	2	4	6	8	11
350	0	0	0	0	0	1	1	1	4	6	7	10
400	0	0	0	0	0	1	1	1	3	5	7	9
500	0	0	0	0	0	1	1	1	3	4	6	7
600	0	0	0	0	0	1	1	1	2	3	4	6
700	0	0	0	0	0	0	1	1	1	3	4	5
750	0	0	0	0	0	0	1	1	1	3	4	5
800	0	0	0	0	0	0	1	1	1	3	3	5
900	0	0	0	0	0	0	0	1	1	2	3	4
1000	0	0	0	0	0	0	0	1	1	2	3	4
1250	0	0	0	0	0	0	0	1	1	1	2	3
1500	0	0	0	0	0	0	0	1	1	1	1	2
1750	0	0	0	0	0	0	0	0	1	1	1	1
2000	0	0	0	0	0	0	0	0	1	1	1	1

THHN, THWN, THWN-2												
14	12	22	35	61	84	138	241	364	476	608
12	9	16	26	45	61	101	176	266	347	443
10	5	10	16	28	38	63	111	167	219	279

Continued

Table C1 Maximum Number of Conductors or Fixture Wires in Electrical Metallic Tubing (EMT) (Based on Table 1) *Continued*

CONDUCTORS

Type	Conductor Size (AWG/kcmil)	16 (1/2)	21 (3/4)	27 (1)	35 (1-1/4)	41 (1-1/2)	53 (2)	63 (2-1/2)	78 (3)	91 (3-1/2)	103 (4)
THHN, THWN, THWN-2	8	3	6	9	16	22	36	64	96	126	161
	6	2	4	7	12	16	26	46	69	91	116
	4	1	2	4	7	10	16	28	43	56	71
	3	1	1	3	6	8	13	24	36	47	60
	2	1	1	3	5	7	11	20	30	40	51
	1	1	1	1	4	5	8	15	22	29	37
	1/0	1	1	1	3	4	7	12	19	25	32
	2/0	0	1	1	2	3	6	10	16	20	26
	3/0	0	1	1	1	3	5	8	13	17	22
	4/0	0	1	1	1	2	4	7	11	14	18
	250	0	0	1	1	1	3	6	9	11	15
	300	0	0	1	1	1	3	5	7	10	13
	350	0	0	1	1	1	2	4	6	9	11

Conduit and Tubing Fill Tables

Type	Size										
FEP, FEPB, PFA, PFAH, TFE	400	0	0	0	0	1	1	1	4	6	8	10
	500	0	0	0	0	1	1	1	3	5	6	8
	600	12	21	34	60	81	134	234	354	462	590
	700	9	15	25	43	59	98	171	258	337	430
	750	6	11	18	31	42	70	122	185	241	309
	800	6	11	18	31	42	70	122	185	241	309
	900	3	6	10	18	24	40	70	106	138	177
	1000	2	4	7	12	17	28	50	75	98	126
	4	1	3	5	9	12	20	35	53	69	88
	3	1	2	4	7	10	16	29	44	57	73
	2	1	1	3	6	8	13	24	36	47	60
PFA, PFAH, TFE	1	1	1	2	4	6	9	16	25	33	42

Continued

Table C1 Maximum Number of Conductors or Fixture Wires in Electrical Metallic Tubing (EMT) (Based on Table 1) *Continued*

CONDUCTORS

Type	Conductor Size (AWG/kcmil)	16 (1/2)	21 (3/4)	27 (1)	35 (1-1/4)	41 (1-1/2)	53 (2)	63 (2-1/2)	78 (3)	91 (3-1/2)	103 (4)
PFAH, TFE, PFA, PFAH, TFE, Z	1/0	1	1	1	3	5	8	14	21	27	35
	2/0	0	1	1	3	4	6	11	17	22	29
	3/0	0	1	1	2	3	5	9	14	18	24
	4/0	0	1	1	1	2	4	8	11	15	19
Z	14	14	25	41	72	98	161	282	426	556	711
	12	10	18	29	51	69	114	200	302	394	504
	10	6	11	18	31	42	70	122	185	241	309
	8	4	7	11	20	27	44	77	117	153	195
	6	3	5	8	14	19	31	54	82	107	137
	4	1	3	5	9	13	21	37	56	74	94
	3	1	2	4	7	9	15	27	41	54	69
	2	1	1	3	6	8	13	22	34	45	57
	1	1	1	2	4	6	10	18	28	36	46

Conduit and Tubing Fill Tables

Type	Size											
XHH, XHHW, XHHW-2, ZW	14	8	15	25	43	58	96	168	254	332	424
	12	6	11	19	33	45	74	129	195	255	326
	10	5	8	14	24	33	55	96	145	190	243
	8	2	5	8	13	18	30	53	81	105	135
	6	1	3	6	10	14	22	39	60	78	100
	4	1	2	4	7	10	16	28	43	56	72
	3	1	1	3	6	8	14	24	36	48	61
	2	1	1	3	5	7	11	20	31	40	51
XHH, XHHW, XHHW-2	1	1	1	1	4	5	8	15	23	30	38
	1/0	1	1	1	3	4	7	13	19	25	32
	2/0	1	1	1	2	3	6	10	16	21	27
	3/0	0	1	1	1	3	5	9	13	17	22
	4/0	0	1	1	1	2	4	7	11	14	18
	250	0	0	1	1	1	3	6	9	12	15
	300	0	0	1	1	1	3	5	8	10	13
	350	0	0	1	1	1	2	4	7	9	11
	400	0	0	1	1	1	1	4	6	8	10
	500	0	0	0	1	1	1	3	5	6	8

Continued

Table C1 Maximum Number of Conductors or Fixture Wires in Electrical Metallic Tubing (EMT) (Based on Table 1) *Continued*

CONDUCTORS

Type	Conductor Size (AWG/kcmil)	Metric Designator (Trade Size)									
		16 (1/2)	21 (3/4)	27 (1)	35 (1-1/4)	41 (1-1/2)	53 (2)	63 (2-1/2)	78 (3)	91 (3-1/2)	103 (4)
XHH, XHHW, XHHW-2	600	0	0	0	1	1	1	2	4	5	6
	700	0	0	0	0	1	1	2	3	4	6
	750	0	0	0	0	1	1	1	3	4	5
	800	0	0	0	0	1	1	1	3	4	5
	900	0	0	0	0	1	1	1	3	3	4
	1000	0	0	0	0	0	1	1	2	3	4
	1250	0	0	0	0	0	1	1	1	2	3
	1500	0	0	0	0	0	1	1	1	1	3
	1750	0	0	0	0	0	0	1	1	1	2
	2000	0	0	0	0	0	0	1	1	1	1

Conduit and Tubing Fill Tables

Table C2 Maximum Number of Conductors or Fixture Wires in Electrical Nonmetallic Tubing (ENT) (Based on Table 1)

	CONDUCTORS						
	Conductor Size (AWG/kcmil)	Metric Designator (Trade Size)					
Type		16 (1/2)	21 (3/4)	27 (1)	35 (1-1/4)	41 (1-1/2)	53 (2)
RHH, RHW, RHW-2	14	3	6	10	19	26	43
	12	2	5	9	16	22	36
RHH, RHW, RHW-2	10	1	4	7	13	17	29
	8	1	1	3	6	9	15
	6	1	1	3	5	7	12
	4	1	1	2	4	6	9
	3	1	1	1	3	5	8
	2	0	1	1	3	4	7
	1	0	1	1	1	3	5
	1/0	0	0	1	1	2	4
	2/0	0	0	1	1	1	3

Continued

Table C2 Maximum Number of Conductors or Fixture Wires in Electrical Nonmetallic Tubing (ENT) (Based on Table 1) *Continued*

CONDUCTORS

Type	Conductor Size (AWG/kcmil)	Metric Designator (Trade Size)						
		16 (1/2)	21 (3/4)	27 (1)	35 (1-1/4)	41 (1-1/2)	53 (2)
RHH, RHW, RHW-2	3/0	0	0	1	1	1	3
	4/0	0	0	1	1	1	2
	250	0	0	0	1	1	1
	300	0	0	0	1	1	1
	350	0	0	0	1	1	1
	400	0	0	0	1	1	1
	500	0	0	0	0	1	1
	600	0	0	0	0	1	1
	700	0	0	0	0	0	1
	750	0	0	0	0	0	1
	800	0	0	0	0	0	1
	900	0	0	0	0	0	1

Conduit and Tubing Fill Tables

	1000	0	0	0	0	0	0	1
	1250	0	0	0	0	0	0	0
	1500	0	0	0	0	0	0	0
	1750	0	0	0	0	0	0	0
	2000	0	0	0	0	0	0	0
TW, THHW, THW, THW-2	14	7	13	22	40	55	92
	12	5	10	17	31	42	71
	10	4	7	13	23	32	52
	8	1	4	7	13	17	29
RHH*, RHW*, RHW-2*	14	4	8	15	27	37	61
RHH*, RHW*, RHW-2*	12	3	7	12	21	29	49
	10	3	5	9	17	23	38
RHH*, RHW*, RHW-2*	8	1	3	5	10	14	23
RHH*, RHW*, RHW-2*, TW, THW, THHW, THW-2	6	1	2	4	7	10	17
	4	1	1	3	5	8	13
	3	1	1	2	5	7	11

Continued

Table C2 Maximum Number of Conductors or Fixture Wires in Electrical Nonmetallic Tubing (ENT) (Based on Table 1) *Continued*

CONDUCTORS

Type	Conductor Size (AWG/kcmil)	Metric Designator (Trade Size)					
		16 (1/2)	21 (3/4)	27 (1)	35 (1-1/4)	41 (1-1/2)	53 (2)
RHH*, RHW*, RHW-2*, TW, THW, THHW, THW-2	2	1	1	2	4	6	9
	1	0	1	1	3	4	6
	1/0	0	1	1	2	3	5
	2/0	0	1	1	1	3	5
	3/0	0	0	1	1	2	4
	4/0	0	0	1	1	1	3
	250	0	0	1	1	1	2
	300	0	0	0	1	1	2
	350	0	0	0	1	1	1
	400	0	0	0	1	1	1
	500	0	0	0	1	1	1
	600	0	0	0	0	1	1
	700	0	0	0	0	1	1

Conduit and Tubing Fill Tables

Size							
750	0	0	0	0	0	1	1
800	0	0	0	0	0	1	1
900	0	0	0	0	0	0	1

THHN, THWN, THWN-2

Size							
1000	0	0	0	0	0	0	1
1250	0	0	0	0	0	0	1
1500	0	0	0	0	0	0	0
1750	0	0	0	0	0	0	0
2000	0	0	0	0	0	0	0
14	10	18	32	58	80	132	
12	7	13	23	42	58	96	
10	4	8	15	26	36	60	
8	2	5	8	15	21	35	
6	1	3	6	11	15	25	
4	1	1	4	7	9	15	
3	1	1	3	5	8	13	
2	1	1	2	5	6	11	
1	1	1	1	3	5	8	
1/0	0	1	1	3	4	7	
2/0	0	1	1	2	3	5	

Continued

Table C2 Maximum Number of Conductors or Fixture Wires in Electrical Nonmetallic Tubing (ENT) (Based on Table 1) *Continued*

CONDUCTORS

Type	Conductor Size (AWG/kcmil)	Metric Designator (Trade size)					
		16 (1/2)	21 (3/4)	27 (1)	35 (1-1/4)	41 (1-1/2)	53 (2)
THHN, THWN, THWN-2	3/0	0	1	1	1	3	4
	4/0	0	0	1	1	2	4
	250	0	0	1	1	1	3
	300	0	0	1	1	1	2
	350	0	0	0	1	1	2
	400	0	0	0	1	1	1
	500	0	0	0	1	1	1
	600	0	0	0	1	1	1
	700	0	0	0	0	1	1
	750	0	0	0	0	1	1
	800	0	0	0	0	1	1
	900	0	0	0	0	1	1
	1000	0	0	0	0	0	1

Conduit and Tubing Fill Tables

Type	Size							
FEP, FEPB, PFA, PFAH, TFE	14	10	18	31	56	77	128
	12	7	13	23	41	56	93
	10	5	9	16	29	40	67
	8	3	5	9	17	23	38
	6	1	4	6	12	16	27
	4	1	2	4	8	11	19
	3	1	1	4	7	9	16
	2	1	1	3	5	8	13
PFA, PFAH, TFE	1	1	1	1	4	5	9
PFA, PFAH, TFE, Z	1/0	0	1	1	3	4	7
	2/0	0	1	1	2	4	6
	3/0	0	1	1	1	3	5
	4/0	0	1	1	1	2	4
Z	14	12	22	38	68	93	154
	12	8	15	27	48	66	109

Continued

Table C2 Maximum Number of Conductors or Fixture Wires in Electrical Nonmetallic Tubing (ENT) (Based on Table 1) *Continued*

CONDUCTORS

Type	Conductor Size (AWG/kcmil)	Metric Designator (Trade size)					
		16 (1/2)	21 (3/4)	27 (1)	35 (1-1/4)	41 (1-1/2)	53 (2)
Z	10	5	9	16	29	40	67
	8	3	6	10	18	25	42
	6	1	4	7	13	18	30
	4	1	3	5	9	12	20
	3	1	1	3	6	9	15
	2	1	1	3	5	7	12
	1	1	1	2	4	6	10
XHH, XHHW, XHHW-2, ZW	14	7	13	22	40	55	92
	12	5	10	17	31	42	71
	10	4	7	13	23	32	52
	8	1	4	7	13	17	29
	6	1	3	5	9	13	21

Conduit and Tubing Fill Tables

XHH, XHHW, XHHW-2, ZW	4	1	1	1	4	7		15
	3	1	1	1	3	6	9	13
	2	1	1	1	2	5	8	11
XHH, XHHW, XHHW-2	1	1	1	1	1	3	6	8
	1/0	0	1	1	1	3	4	7
	2/0	0	1	1	1	2	3	6
	3/0	0	1	1	1	1	3	5
	4/0	0	0	1	1	1	2	4
	250	0	0	0	1	1	1	3
	300	0	0	0	1	1	1	3
	350	0	0	0	1	1	1	2
	400	0	0	0	0	1	1	1
	500	0	0	0	0	1	1	1
	600	0	0	0	0	1	1	1
	700	0	0	0	0	0	1	1
	750	0	0	0	0	0	1	1
	800	0	0	0	0	0	1	1
	900	0	0	0	0	0	1	1

Continued

Table C2 Maximum Number of Conductors or Fixture Wires in Electrical Nonmetallic Tubing (ENT) (Based on Table 1) *Continued*

CONDUCTORS

Type	Conductor Size (AWG/kcmil)	Metric Designator (Trade Size)					
		16 (1/2)	21 (3/4)	27 (1)	35 (1-1/4)	41 (1-1/2)	53 (2)
XHH, XHHW, XHHW-2	1000	0	0	0	0	0	1
	1250	0	0	0	0	0	1
	1500	0	0	0	0	0	1
	1750	0	0	0	0	0	0
	2000	0	0	0	0	0	0

Conduit and Tubing Fill Tables

Table C3 Maximum Number of Conductors or Fixture Wires in Flexible Metal Conduit (FMC) (Based on Table 1).

CONDUCTORS

Type	Conductor Size (AWG/kcmil)	16 (1/2)	21 (3/4)	27 (1)	35 (1-1/4)	41 (1-1/2)	53 (2)	63 (2-1/2)	78 (3)	91 (3-1/2)	103 (4)
RHH, RHW, RHW-2	14	4	7	11	17	25	44	67	96	131	171
	12	3	6	9	14	21	37	55	80	109	142
RHH, RHW, RHW-2	10	3	5	7	11	17	30	45	64	88	115
	8	1	2	4	6	9	15	23	34	46	60
	6	1	1	3	5	7	12	19	27	37	48
	4	1	1	2	4	5	10	14	21	29	37
	3	1	1	1	3	5	8	13	18	25	33
	2	1	1	1	3	4	7	11	16	22	28
	1	0	1	1	1	2	5	7	10	14	19
	1/0	0	1	1	2	2	4	6	9	12	16
	2/0	0	1	1	1	1	3	5	8	11	14

Continued

Table C3 Maximum Number of Conductors or Fixture Wires in Flexible Metal Conduit (FMC) (Based on Table 1) *Continued*

CONDUCTORS

Type	Conductor Size (AWG/kcmil)	16 (1/2)	21 (3/4)	27 (1)	35 (1-1/4)	41 (1-1/2)	53 (2)	63 (2-1/2)	78 (3)	91 (3-1/2)	103 (4)
RHH, RHW, RHW-2	3/0	0	0	1	1	1	1	3	5	7	9	12
	4/0	0	0	1	1	1	1	2	4	6	8	10
	250	0	0	0	1	1	1	1	3	4	6	8
	300	0	0	0	1	1	1	1	2	4	5	7
	350	0	0	0	1	1	1	1	2	3	5	6
	400	0	0	0	0	1	1	1	1	3	4	6
	500	0	0	0	0	1	1	1	1	3	4	5
	600	0	0	0	0	0	1	1	1	2	3	4
	700	0	0	0	0	0	1	1	1	1	3	3
	750	0	0	0	0	0	1	1	1	1	2	3
	800	0	0	0	0	0	1	1	1	1	2	3
	900	0	0	0	0	0	1	1	1	1	2	3

Conduit and Tubing Fill Tables

Type	Size										
	1000	0	0	0	0	0	0	1	1	1	3
	1250	0	0	0	0	0	0	0	1	1	1
	1500	0	0	0	0	0	0	0	1	1	1
	1750	0	0	0	0	0	0	0	1	1	1
	2000	0	0	0	0	0	0	0	0	1	1
TW, THHW, THW, THW-2	14	9	15	23	36	53	94	141	203	277	361
	12	7	11	18	28	41	72	108	156	212	277
	10	5	8	13	21	30	54	81	116	158	207
	8	3	5	7	11	17	30	45	64	88	115
RHH*, RHW*, RHW-2	14	6	10	15	24	35	62	94	135	184	240
RHH*, RHW*, RHW-2	12	5	8	12	19	28	50	75	108	148	193
	10	4	6	10	15	22	39	59	85	115	151
RHH*, RHW*, RHW-2	8	1	4	6	9	13	23	35	51	69	90

Continued

Table C3 Maximum Number of Conductors or Fixture Wires in Flexible Metal Conduit (FMC) (Based on Table 1) *Continued*

CONDUCTORS

Type	Conductor Size (AWG/kcmil)	16 (1/2)	21 (3/4)	27 (1)	35 (1-1/4)	41 (1-1/2)	53 (2)	63 (2-1/2)	78 (3)	91 (3-1/2)	103 (4)
RHH*, RHW*, RHW-2*, TW, THW, THHW, THW-2	6	1	3	4	7	10	18	27	39	53	69
	4	1	1	3	5	7	13	20	29	39	51
	3	1	1	3	4	6	11	17	25	34	44
	2	1	1	2	4	5	10	14	21	29	37
	1	1	1	1	2	4	7	10	15	20	26
	1/0	0	1	1	1	3	6	9	12	17	22
	2/0	0	1	1	1	3	5	7	10	14	19
	3/0	0	0	1	1	2	4	6	9	12	16
	4/0	0	0	1	1	1	3	5	7	10	13
	250	0	0	1	1	1	3	4	6	8	11
	300	0	0	0	1	1	2	3	5	7	9
	350	0	0	0	1	1	1	3	4	6	8

Conduit and Tubing Fill Tables

	400	0	0	0	0	1	1	1	3	4	6	7
	500	0	0	0	0	1	1	1	2	3	5	6
	600	0	0	0	0	0	1	1	1	3	4	5
	700	0	0	0	0	0	1	1	1	2	3	4
	750	0	0	0	0	0	1	1	1	2	3	4
	800	0	0	0	0	0	1	1	1	1	3	4
	900	0	0	0	0	0	1	1	1	1	3	3
	1000	0	0	0	0	0	0	1	1	1	2	3
	1250	0	0	0	0	0	0	0	1	1	1	2
	1500	0	0	0	0	0	0	0	1	1	1	1
	1750	0	0	0	0	0	0	0	1	1	1	1
	2000	0	0	0	0	0	0	0	1	1	1	1
THHN, THWN, THWN-2	14	13	22	33	52	76	134	202	291	396	518	
	12	9	16	24	38	56	98	147	212	289	378	
	10	6	10	15	24	35	62	93	134	182	238	
	8	3	6	9	14	20	35	53	77	105	137	
	6	2	4	6	10	14	25	38	55	76	99	

Continued

Table C3 Maximum Number of Conductors or Fixture Wires in Flexible Metal Conduit (FMC) (Based on Table 1) *Continued*

CONDUCTORS

Type	Conductor Size (AWG/Kcmil)	16 (1/2)	21 (3/4)	27 (1)	35 (1-1/4)	41 (1-1/2)	53 (2)	63 (2-/2)	78 (3)	91 (3-1/2)	103 (4)
THHN, THWN, THWN-2	4	1	2	4	6	9	16	24	34	46	61
	3	1	1	3	5	7	13	20	29	39	51
	2	1	1	3	4	6	11	17	24	33	43
	1	1	1	1	3	4	8	12	18	24	32
	1/0	1	1	1	2	4	7	10	15	20	27
	2/0	0	1	1	1	3	6	9	12	17	22
	3/0	0	1	1	1	2	5	7	10	14	18
	4/0	0	1	1	1	1	4	6	8	12	15
	250	0	0	1	1	1	3	5	7	9	12
	300	0	0	1	1	1	3	4	6	8	11
	350	0	0	1	1	1	2	3	5	7	9
	400	0	0	0	1	1	1	3	5	6	8
	500	0	0	0	1	1	1	2	4	5	7

Conduit and Tubing Fill Tables

Type	Size										
FEP, FEPB, PFA, PFAH, TFE	600	0	0	0	0	1	1	1	3	4	5
	700	0	0	0	0	1	1	1	3	4	5
	750	0	0	0	0	1	1	1	2	3	4
	800	0	0	0	0	1	1	1	2	3	4
	900	0	0	0	0	0	1	1	1	3	4
	1000	0	0	0	0	0	1	1	1	3	3
	14	12	21	32	51	74	130	196	282	385	502
	12	9	15	24	37	54	95	143	206	281	367
	10	6	11	17	26	39	68	103	148	201	263
	8	4	6	10	15	22	39	59	85	115	151
	6	2	4	7	11	16	28	42	60	82	107
PFA, PFAH, TFE	4	1	3	5	7	11	19	29	42	57	75
	3	1	2	4	6	9	16	24	35	48	62
	2	1	1	3	5	7	13	20	29	39	51
PFA, PFAH, TFE	1	1	1	2	3	5	9	14	20	27	36
PFA, PFAH, TFE, Z	1/0	1	1	1	3	4	8	11	17	23	30
	2/0	1	1	1	2	3	6	9	14	19	24

Continued

Table C3 Maximum Number of Conductors or Fixture Wires in Flexible Metal Conduit (FMC) (Based on Table 1) *Continued*

CONDUCTORS

Type	Conductor Size (AWG/kcmil)	Metric Designator (Trade Size)									
		16 (1/2)	21 (3/4)	27 (1)	35 (1-1/4)	41 (1-1/2)	53 (2)	63 (2-1/2)	78 (3)	91 (3-1/2)	103 (4)
PFA, PFAH, TFE	3/0	0	1	1	1	3	5	8	11	15	20
	4/0	0	1	1	1	2	4	6	9	13	16
Z	14	15	25	39	61	89	157	236	340	463	605
	12	11	18	28	43	63	111	168	241	329	429
	10	6	11	17	26	39	68	103	148	201	263
	8	4	7	11	17	24	43	65	93	127	166
	6	3	5	7	12	17	30	45	65	89	117
	4	1	3	5	8	12	21	31	45	61	80
	3	1	2	4	6	8	15	23	33	45	58
	2	1	1	3	5	7	12	19	27	37	49
	1	1	1	2	4	6	10	15	22	30	39

Conduit and Tubing Fill Tables

XHH, XHHW, XHHW-2, ZW	14	9	15	23	36	53	94	141	203	277	361	
	12	7	11	18	28	41	72	108	156	212	277	
	10	5	8	13	21	30	54	81	116	158	207	
	8	3	5	7	11	17	30	45	64	88	115	
	6	1	3	5	8	12	22	33	48	65	85	
	4	1	2	4	6	9	16	24	34	47	61	
	3	1	1	3	5	7	13	20	29	40	52	
	2	1	1	3	4	6	11	17	24	33	44	
XHH, XHHW, XHHW-2	1	1	1	1	3	5	8	13	18	25	32	
	1/0	1	1	1	2	4	7	10	15	21	27	
	2/0	0	1	1	2	3	6	9	13	17	23	
	3/0	0	1	1	1	3	5	7	10	14	19	
	4/0	0	1	1	1	2	4	6	9	12	15	
	250	0	0	1	1	1	3	5	7	10	13	
	300	0	0	1	1	1	3	4	6	8	11	
	350	0	0	1	1	1	2	4	5	7	9	
	400	0	0	0	1	1	1	3	5	6	8	
	500	0	0	0	1	1	1	3	4	5	7	

Continued

Table C3 Maximum Number of Conductors or Fixture Wires in Flexible Metal Conduit (FMC) (Based on Table 1) *Continued*

		CONDUCTORS									
		Metric Designator (Trade Size)									
Type	Conductor Size (AWG/kcmil)	16 (1/2)	21 (3/4)	27 (1)	35 (1-1/4)	41 (1-1/2)	53 (2)	63 (2-1/2)	78 (3)	91 (3-1/2)	103 (4)
XHH, XHHW, XHHW-2	600	0	0	0	0	1	1	1	3	4	5
	700	0	0	0	0	1	1	1	3	4	5
	750	0	0	0	0	1	1	1	2	3	4
	800	0	0	0	0	1	1	1	2	3	4
	900	0	0	0	0	0	1	1	1	3	4
	1000	0	0	0	0	0	0	1	1	3	3
	1250	0	0	0	0	0	0	1	1	1	3
	1500	0	0	0	0	0	0	1	1	1	2
	1750	0	0	0	0	0	0	0	1	1	1
	2000	0	0	0	0	0	0	0	1	1	1

*Types RHH, RHW, and RHW-2 without outer covering.

Conduit and Tubing Fill Tables

Table C4 Maximum Number of Conductors or Fixture Wires in Intermediate Metal Conduit (IMC) (Based on Table 1)

CONDUCTORS

Type	Conductor Size (AWG/kcmil)	Metric Designator (Trade Size)									
		16 (1/2)	21 (3/4)	27 (1)	35 (1-1/4)	41 (1-1/2)	53 (2)	63 (2-1/2)	78 (3)	91 (3-1/2)	103 (4)
RHH, RHW, RHW-2	14	4	8	13	22	30	49	70	108	144	186
	12	4	6	11	18	25	41	58	89	120	154
RHH, RHW, RHW-2	10	3	5	8	15	20	33	47	72	97	124
	8	1	3	4	8	10	17	24	38	50	65
	6	1	1	3	6	8	14	19	30	40	52
	4	1	1	3	5	6	11	15	23	31	41
	3	1	1	2	4	6	9	13	21	28	36
	2	1	1	1	3	5	8	11	18	24	31
	1	0	1	1	2	3	5	7	12	16	20
	1/0	0	1	1	1	3	4	6	10	14	18
	2/0	0	1	1	1	2	4	6	9	12	15

Continued

Table C4 Maximum Number of Conductors or Fixture Wires in Intermediate Metal Conduit (IMC) (Based on Table 1) *Continued*

CONDUCTORS

Type	Conductor Size (AWG/kcmil)	Metric Designator (Trade Size)									
		16 (1/2)	21 (3/4)	27 (1)	35 (1-1/4)	41 (1-1/2)	53 (2)	63 (2-1/2)	78 (3)	91 (3-1/2)	103 (4)
RHH, RHW, RHW-2	3/0	0	0	1	1	1	3	5	7	10	13
	4/0	0	0	1	1	1	3	4	6	9	11
	250	0	0	0	1	1	1	3	5	6	8
	300	0	0	0	1	1	1	3	4	6	7
	350	0	0	0	1	1	1	2	4	5	7
	400	0	0	0	1	1	1	2	3	5	6
	500	0	0	0	1	1	1	1	3	4	5
	600	0	0	0	0	1	1	1	2	3	4
	700	0	0	0	0	1	1	1	2	3	4
	750	0	0	0	0	1	1	1	1	3	4
	800	0	0	0	0	0	1	1	1	3	3
	900	0	0	0	0	0	1	1	1	2	3

Conduit and Tubing Fill Tables

Type	Size														
	1000	0	0	0	0	0	0	0	0	0	0	1	1	2	3
	1250	0	0	0	0	0	0	0	0	0	1	1	1	1	2
	1500	0	0	0	0	0	0	0	0	0	1	1	1	1	1
	1750	0	0	0	0	0	0	0	0	0	0	1	1	1	1
	2000	0	0	0	0	0	0	0	0	0	0	1	1	1	1
TW, THHW, THW, THW-2	14	0	0	0	0	10	17	27	47	64	104	147	228	304	392
	12	0	0	0	0	7	13	21	36	49	80	113	175	234	301
	10	0	0	0	0	5	9	15	27	36	59	84	130	174	224
	8	0	0	0	0	3	5	8	15	20	33	47	72	97	124
RHH*, RHW*, RHW-2*	14	0	0	0	0	6	11	18	31	42	69	98	151	202	261
RHH*, RHW*, RHW-2*	12	0	0	0	0	5	9	14	25	34	56	79	122	163	209
	10	0	0	0	0	4	7	11	19	26	43	61	95	127	163
RHH*, RHW*, RHW-2*	8	0	0	0	0	2	4	7	12	16	26	37	57	76	98
RHH*, RHW*, RHW-2*	6	0	0	0	0	1	3	5	9	12	20	28	43	58	75
	4	0	0	0	0	1	2	4	6	9	15	21	32	43	56

Continued

Table C4 Maximum Number of Conductors or Fixture Wires in Intermediate Metal Conduit (IMC) (Based on Table 1) *Continued*

CONDUCTORS

Type	Conductor Size (AWG/kcmil)	Metric Designator (Trade Size)									
		16 (1/2)	21 (3/4)	27 (1)	35 (1-1/4)	41 (1-1/2)	53 (2)	63 (2-1/2)	78 (3)	91 (3-1/2)	103 (4)
TW, THW, THHW, THW-2	3	1	1	3	6	8	13	18	28	37	48
	2	1	1	3	5	6	11	15	23	31	41
	1	1	1	1	3	4	7	11	16	22	28
	1/0	1	1	1	3	4	6	9	14	19	24
	2/0	0	1	1	2	3	5	8	12	16	20
	3/0	0	1	1	1	3	4	6	10	13	17
	4/0	0	1	1	1	2	4	5	8	11	14
	250	0	0	1	1	1	3	4	7	9	12
	300	0	0	1	1	1	2	4	6	8	10
	350	0	0	1	1	1	2	3	5	7	9
	400	0	0	0	1	1	1	3	4	6	8
	500	0	0	0	1	1	1	2	4	5	7

Conduit and Tubing Fill Tables

Size											
600	0	0	0	0	0	1	1	1	3	4	5
700	0	0	0	0	0	0	1	1	3	4	5
750	0	0	0	0	0	0	1	1	2	3	4
800	0	0	0	0	0	0	1	1	2	3	4
900	0	0	0	0	0	0	1	1	2	3	4
1000	0	0	0	0	0	0	1	1	1	3	3
1250	0	0	0	0	0	0	1	1	1	1	3
1500	0	0	0	0	0	0	1	1	1	1	2
1750	0	0	0	0	0	0	0	1	1	1	1
2000	0	0	0	0	0	0	0	1	1	1	1

THHN, THWN, THWN-2

Size											
14	14	24	39	68	91	149	211	326	436	562	
12	10	17	29	49	67	109	154	238	318	410	
10	6	11	18	31	42	68	97	150	200	258	
8	3	6	10	18	24	39	56	86	115	149	
6	2	4	7	13	17	28	40	62	83	107	
4	1	3	4	8	10	17	25	38	51	66	
3	1	2	4	6	9	15	21	32	43	56	

Continued

Table C4 Maximum Number of Conductors or Fixture Wires in Intermediate Metal Conduit (IMC) (Based on Table 1) *Continued*

CONDUCTORS

Type	Conductor Size (AWG/kcmil)	Metric Designator (Trade Size)									
		16 (1/2)	21 (3/4)	27 (1)	35 (1-1/4)	41 (1-1/2)	53 (2)	63 (2-1/2)	78 (3)	91 (3-1/2)	103 (4)
THHN, THWN, THWN-2	2	1	1	3	5	7	12	17	27	36	47
	1	1	1	2	4	5	9	13	20	27	35
	1/0	1	1	1	3	4	8	11	17	23	29
	2/0	1	1	1	3	4	6	9	14	19	24
	3/0	0	1	1	2	3	5	7	12	16	20
	4/0	0	1	1	1	2	4	6	9	13	17
	250	0	0	1	1	1	3	5	8	10	13
	300	0	0	1	1	1	3	4	7	9	12
	350	0	0	1	1	1	2	4	6	8	10
	400	0	0	1	1	1	2	3	5	7	9
	500	0	0	0	1	1	1	3	4	6	7

Conduit and Tubing Fill Tables

	600	0	0	0	0	1	1	1	2	3	5	6
	700	0	0	0	0	1	1	1	1	3	4	5
	750	0	0	0	0	1	1	1	1	3	4	5
	800	0	0	0	0	1	1	1	1	3	4	5
	900	0	0	0	0	0	1	1	1	2	3	4
	1000	0	0	0	0	0	1	1	1	2	3	4
FEP, FEPB, PFA, PFAH, TFE	14	13	23	38	66	89	145	205	317	423	545
	12	10	17	28	48	65	106	150	231	309	398
	10	7	12	20	34	46	76	107	166	221	285
	8	4	7	11	19	26	43	61	95	127	163
	6	3	5	8	14	19	31	44	67	90	116
	4	1	3	5	10	13	21	30	47	63	81
	3	1	3	4	8	11	18	25	39	52	68
	2	1	2	4	6	9	15	21	32	43	56
PFA, PFAH, TFE	1	1	1	2	4	6	10	14	22	30	39
PFA, PFAH, TFE, Z	1/0	1	1	1	4	5	8	12	19	25	32
	2/0	1	1	1	3	4	7	10	15	21	27

Continued

Table C4 Maximum Number of Conductors or Fixture Wires in Intermediate Metal Conduit (IMC) (Based on Table 1) *Continued*

		CONDUCTORS									
	Conductor Size (AWG/kcmil)	Metric Designator (Trade Size)									
Type		16 (1/2)	21 (3/4)	27 (1)	35 (1-1/4)	41 (1-1/2)	53 (2)	63 (2-1/2)	78 (3)	91 (3-1/2)	103 (4)
PFA, PFAH, TFE, Z	3/0	0	1	1	2	3	6	8	13	17	22
	4/0	0	1	1	1	3	5	7	10	14	18
Z	14	16	28	46	79	107	175	247	381	510	657
	12	11	20	32	56	76	124	175	271	362	466
	10	7	12	20	34	46	76	107	166	221	285
	8	4	7	12	21	29	48	68	105	140	180
	6	3	5	9	15	20	33	47	73	98	127
	4	1	3	6	10	14	23	33	50	67	87
	3	1	2	4	7	10	17	24	37	49	63
	2	1	1	3	6	8	14	20	30	41	53
	1	1	1	3	5	7	11	16	25	33	43

Conduit and Tubing Fill Tables

XHH, XHHW, XHHW-2, ZW	14	10	17	27	47	64	104	147	228	304	392
	12	7	13	21	36	49	80	113	175	234	301
	10	5	9	15	27	36	59	84	130	174	224
	8	3	5	8	15	20	33	47	72	97	124
	6	1	4	6	11	15	24	35	53	71	92
XHH, XHHW, XHHW-2	4	1	3	4	8	11	18	25	39	52	67
	3	1	2	4	7	9	15	21	33	44	56
	2	1	1	3	5	7	12	18	27	37	47
	1	1	1	2	4	5	9	13	20	27	35
	1/0	1	1	1	3	5	8	11	17	23	30
	2/0	1	1	1	3	4	6	9	14	19	25
	3/0	1	1	1	2	3	5	7	12	16	20
	4/0	0	1	1	1	2	4	6	10	13	17
	250	0	0	1	1	1	3	5	8	11	14
	300	0	0	1	1	1	3	4	7	9	12
	350	0	0	1	1	1	3	4	6	8	10
	400	0	0	1	1	1	2	3	5	7	10
	500	0	0	0	1	1	1	3	4	6	8

Continued

Table C4 Maximum Number of Conductors or Fixture Wires in Intermediate Metal Conduit (IMC) (Based on Table 1) *Continued*

CONDUCTORS

Type	Conductor Size (AWG/kcmil)	Metric Designator (Trade Size)								
		16 (1/2)	21 (3/4)	27 (1)	35 (1-1/4)	41 (1-1/2)	53 (2)	63 (2-1/2)	78 (3)	91 (3-1/2)	103 (4)
XHH, XHHW, XHHW-2	600	0	0	0	1	1	1	2	3	5	6
	700	0	0	0	1	1	1	1	3	4	5
	750	0	0	0	1	1	1	1	3	4	5
	800	0	0	0	0	1	1	1	3	4	5
	900	0	0	0	0	1	1	1	2	3	4
	1000	0	0	0	0	1	1	1	2	3	4
	1250	0	0	0	0	0	1	1	1	2	3
	1500	0	0	0	0	0	1	1	1	1	2
	1750	0	0	0	0	0	1	1	1	1	2
	2000	0	0	0	0	0	0	1	1	1	1

Conduit and Tubing Fill Tables

Table C5 Maximum Number of Conductors or Fixture Wires in Liquidtight Flexible Nonmetallic Conduit (Type LFNC-B*) (Based on Table 1)

CONDUCTORS

Type	Conductor Size (AWG/kcmil)	Metric Designator (Trade Size)						
		12 (3/8)	16 (1/2)	21 (3/4)	27 (1)	35 (1-1/4)	41 (1-1/2)	53 (2)
RHH, RHW, RHW-2	14	2	4	7	12	21	27	44
	12	1	3	6	10	17	22	36
RHH, RHW, RHW-2	10	1	3	5	8	14	18	29
	8	1	1	2	4	7	9	15
	6	1	1	1	3	6	7	12
	4	0	1	1	2	4	6	9
	3	0	1	1	1	4	5	8
	2	0	1	1	1	3	4	7
	1	0	0	1	1	1	3	5
	1/0	0	0	1	1	1	2	4
	2/0	0	0	1	1	1	1	3

Continued

Table C5 Maximum Number of Conductors or Fixture Wires in Liquidtight Flexible Nonmetallic Conduit (Type LFNC-B*) (Based on Table 1) *Continued*

Type	Conductor Size (AWG/kcmil)	CONDUCTORS						
		Metric Designator (Trade Size)						
		12 (3/8)	16 (1/2)	21 (3/4)	27 (1)	35 (1-1/4)	41 (1-1/2)	53 (2)
RHH, RHW, RHW-2	3/0	0	0	0	0	1	1	3
	4/0	0	0	0	0	1	1	2
	250	0	0	0	0	1	1	1
	300	0	0	0	0	1	1	1
	350	0	0	0	0	1	1	1
	400	0	0	0	0	1	1	1
	500	0	0	0	0	1	1	1
	600	0	0	0	0	0	1	1
	700	0	0	0	0	0	0	1
	750	0	0	0	0	0	0	1
	800	0	0	0	0	0	0	1
	900	0	0	0	0	0	0	1

Conduit and Tubing Fill Tables

TW, THHW, THW, THW-2	1000	0	0	0	0	0	0	0	1
	1250	0	0	0	0	0	0	0	0
	1500	0	0	0	0	0	0	0	0
	1750	0	0	0	0	0	0	0	0
	2000	0	0	0	0	0	0	0	0
	14	5	9	15	25	44	57	93
	12	4	7	12	19	33	43	71
	10	3	5	9	14	25	32	53
	8	1	3	5	8	14	18	29
RHH†, RHW†, RHW-2†	14	3	6	10	16	29	38	62
RHH†, RHW†, RHW-2†	12	3	5	8	13	23	30	50
	10	1	3	6	10	18	23	39
RHH†, RHW†, RHW-2†	8	1	1	4	6	11	14	23

Continued

Table C5 Maximum Number of Conductors or Fixture Wires in Liquidtight Flexible Nonmetallic Conduit (Type LFNC-B**) (Based on Table 1) *Continued*

CONDUCTORS

Type	Conductor Size (AWG/kcmil)	Metric Designator (Trade Size)							
		12 (3/8)	16 (1/2)	21 (3/4)	27 (1)	35 (1-1/4)	41 (1-1/2)	53 (2)
RHH†, RHW†, RHW-2†, TW, THW	6	1	1	3	5	8	11	18
	4	1	1	1	3	6	8	13
	3	1	1	1	3	5	7	11
THHW, THW-2	2	0	1	1	2	4	6	9
	1	0	1	1	1	3	4	7
	1/0	0	0	1	1	2	3	6
	2/0	0	0	1	1	2	3	5
	3/0	0	0	1	1	1	2	4
	4/0	0	0	0	1	1	1	3
	250	0	0	0	1	1	1	3
	300	0	0	0	1	1	1	2
	350	0	0	0	0	1	1	1

Conduit and Tubing Fill Tables

400	0	0	0	0	0	1	1
500	0	0	0	0	0	1	1
600	0	0	0	0	1	1	1
700	0	0	0	0	0	1	1
750	0	0	0	0	0	1	1
800	0	0	0	0	0	1	1
900	0	0	0	0	0	1	1
1000	0	0	0	0	0	0	1
1250	0	0	0	0	0	0	0
1500	0	0	0	0	0	0	0
1750	0	0	0	0	0	0	0
2000	0	0	0	0	0	0	0
THHN, THWN, THWN-2							
14	8	13	22	36	63	81	133
12	5	9	16	26	46	59	97
10	3	6	10	16	29	37	61
8	1	3	6	9	16	21	35
6	1	2	4	7	12	15	25

Continued

Table C5 Maximum Number of Conductors or Fixture Wires in Liquidtight Flexible Nonmetallic Conduit (Type LFNC-B*) (Based on Table 1) *Continued*

CONDUCTORS

Type	Conductor Size (AWG/kcmil)	12 (3/8)	16 (1/2)	21 (3/4)	27 (1)	35 (1-1/4)	41 (1-1/2)	53 (2)
THHN, THWN, THWN-2	4	1	1	2	4	7	9	15
	3	1	1	1	3	6	8	13
	2	1	1	1	3	5	7	11
	1	0	1	1	1	4	5	8
	1/0	0	1	1	1	3	4	7
	2/0	0	0	1	1	2	3	6
	3/0	0	0	1	1	1	3	5
	4/0	0	0	1	1	1	2	4
	250	0	0	0	1	1	1	3
	300	0	0	0	1	1	1	3
	350	0	0	0	1	1	1	2
	400	0	0	0	0	1	1	1
	500	0	0	0	0	1	1	1

Metric Designator (Trade Size)

Conduit and Tubing Fill Tables

FEP, FEPB, PFA, PFAH, TFE	600	0	0	0	0	0	1	1
	700	0	0	0	0	0	1	1
	750	0	0	0	0	0	1	1
	800	0	0	0	0	0	1	1
	900	0	0	0	0	0	1	1
	1000	0	0	0	0	0	0	1
	14	7	12	21	35	61	79	129
	12	5	9	15	25	44	57	94
	10	4	6	11	18	32	41	68
	8	1	3	6	10	18	23	39
	6	1	2	4	7	13	17	27
	4	1	1	3	5	9	12	19
	3	1	1	2	4	7	10	16
	2	1	1	1	3	6	8	13
PFA, PFAH, TFE	1	0	1	1	2	4	5	9
PFA, PFAH	1/0	0	1	1	1	3	4	7

Continued

Table C5 Maximum Number of Conductors or Fixture Wires in Liquidtight Flexible Nonmetallic Conduit (Type LFNC-B*) (Based on Table 1) *Continued*

CONDUCTORS

Type	Conductor Size (AWG/kcmil)	Metric Designator (Trade Size)							
		12 (3/8)	16 (1/2)	21 (3/4)	27 (1)	35 (1-1/4)	41 (1-1/2)	53 (2)
TFE, Z	2/0	0	1	1	1	3	4	6
	3/0	0	0	1	1	2	3	5
	4/0	0	0	1	1	1	2	4
Z	14	9	15	26	42	73	95	156
	12	6	10	18	30	52	67	111
	10	4	6	11	18	32	41	68
	8	2	4	7	11	20	26	43
	6	1	3	5	8	14	18	30
	4	1	1	3	5	9	12	20
	3	1	1	2	4	7	9	15
	2	0	1	1	3	6	7	12
	1	0	1	1	2	5	6	10

Conduit and Tubing Fill Tables

XHH, XHHW, XHHW-2, ZW	14	5	9	15	25	44	57	93	
	12	4	7	12	19	33	43	71	
	10	3	5	9	14	25	32	53	
	8	1	3	5	8	14	18	29	
	6	1	1	3	6	10	13	22	
	4	1	1	2	4	7	9	16	
	3	1	1	1	3	6	8	13	
	2	1	1	1	3	5	7	11	
XHH, XHHW, XHHW-2	1	0	1	1	1	4	5	8	
	1/0	0	1	1	1	3	4	7	
	2/0	0	0	1	1	2	3	6	
	3/0	0	0	1	1	1	3	5	
	4/0	0	0	1	1	1	2	4	
	250	0	0	0	1	1	1	3	
	300	0	0	0	1	1	1	3	
	350	0	0	0	1	1	1	2	

Continued

Table C5 Maximum Number of Conductors or Fixture Wires in Liquidtight Flexible Nonmetallic Conduit (Type LFNC-B*) (Based on Table 1) *Continued*

CONDUCTORS

Type	Conductor Size (AWG/kcmil)	Metric Designator (Trade Size)							
		12 (3/8)	16 (1/2)	21 (3/4)	27 (1)	35 (1-1/4)	41 (1-1/2)	53 (2)
XHH, XHHW, XHHW-2	400	0	0	0	0	1	1	1
	500	0	0	0	0	1	1	1
	600	0	0	0	0	1	1	1
	700	0	0	0	0	1	1	1
	750	0	0	0	0	0	1	1
	800	0	0	0	0	0	1	1
	900	0	0	0	0	0	1	1
	1000	0	0	0	0	0	0	1
	1250	0	0	0	0	0	0	1
	1500	0	0	0	0	0	0	1
	1750	0	0	0	0	0	0	0
	2000	0	0	0	0	0	0	0

Conduit and Tubing Fill Tables

Table C6 Maximum Number of Conductors or Fixture Wires in Liquidtight Flexible Nonmetallic Conduit (Type LFNC-A*) (Based on Table 1)

CONDUCTORS

Type	Conductor Size (AWG/kcmil)	Metric Designator (Trade Size)						
		12 (3/8)	16 (1/2)	21 (3/4)	27 (1)	35 (1-1/4)	41 (1-1/2)	53 (2)
RHH, RHW, RHW-2	14	2	4	7	11	20	27	45
	12	1	3	6	9	17	23	38
	10	1	3	5	8	13	18	30
	8	1	1	2	4	7	9	16
	6	1	1	1	3	5	7	13
	4	0	1	1	2	4	6	10
	3	0	1	1	1	4	5	8
	2	0	1	1	1	3	4	7
	1	0	0	1	1	1	3	5
	1/0	0	0	1	1	1	2	4
	2/0	0	0	1	1	1	1	4

Continued

Table C6 Maximum Number of Conductors or Fixture Wires in Liquidtight Flexible Nonmetallic Conduit (Type LFNC-A*) (Based on Table 1) *Continued*

		CONDUCTORS							
		Metric Designator (Trade Size)							
Type	Conductor Size (AWG/kcmil)	12 (3/8)	16 (1/2)	21 (3/4)	27 (1)	35 (1-1/4)	41 (1-1/2)	53 (2)
RHH, RHW, RHW-2	3/0	0	0	0	0	1	1	3
	4/0	0	0	0	1	1	1	3
	250	0	0	0	0	1	1	1
	300	0	0	0	0	1	1	1
	350	0	0	0	0	1	1	1
	400	0	0	0	0	1	1	1
	500	0	0	0	0	0	1	1
	600	0	0	0	0	0	1	1
	700	0	0	0	0	0	0	1
	750	0	0	0	0	0	0	1
	800	0	0	0	0	0	0	1
	900	0	0	0	0	0	0	1

Conduit and Tubing Fill Tables

	1000	0	0	0	0	0	0	0	1
	1250	0	0	0	0	0	0	0	0
	1500	0	0	0	0	0	0	0	0
	1750	0	0	0	0	0	0	0	0
	2000	0	0	0	0	0	0	0	0
TW, THHW, THW, THW-2	14	5	9	15	24	43	58	96
	12	4	7	12	19	33	44	74
	10	3	5	9	14	24	33	55
	8	1	3	5	8	13	18	30
RHH†, RHW†, RHW-2†	14	3	6	10	16	28	38	64
	12	3	4	8	13	23	31	51
	10	1	3	6	10	18	24	40
	8	1	1	4	6	10	14	24
RHH†, RHW†, RHW-2†, TW, THW, THHW, THW-2	6	1	1	3	4	8	11	18
	4	1	1	1	3	6	8	13
	3	1	1	1	2	5	7	11
	2	0	1	1	2	4	6	10
	1	0	1	1	1	3	4	7

Continued

Table C6 Maximum Number of Conductors or Fixture Wires in Liquidtight Flexible Nonmetallic Conduit (Type LFNC-A*) (Based on Table 1) *Continued*

Type	Conductor Size (AWG/kcmil)	CONDUCTORS Metric Designator (Trade Size)							
		12 (3/8)	16 (1/2)	21 (3/4)	27 (1)	35 (1-1/4)	41 (1-1/2)	53 (2)	
RHH†, RHW†, RHW-2†, TW, THW, THHW, THW-2	1/0	0	0	1	1	2	3	6	
	2/0	0	0	1	1	1	3	5	
	3/0	0	0	1	1	1	2	4	
	4/0	0	0	0	1	1	1	3	
	250	0	0	0	1	1	1	3	
	300	0	0	0	0	1	1	2	
	350	0	0	0	0	1	1	1	
	400	0	0	0	0	1	1	1	
	500	0	0	0	0	1	1	1	
	600	0	0	0	0	0	1	1	
	700	0	0	0	0	0	1	1	
	750	0	0	0	0	0	1	1	
	800	0	0	0	0	0	1	1	
	900	0	0	0	0	0	0	1	

Conduit and Tubing Fill Tables

	1000	0	0	0	0	0	0	0	0	1
	1250	0	0	0	0	0	0	0	0	1
	1500	0	0	0	0	0	0	0	0	1
	1750	0	0	0	0	0	0	0	0	0
	2000	0	0	0	0	0	0	0	0	0
THHN, THWN, THWN-2	14	8	13	22	35	62	83	137
	12	5	9	16	25	45	60	100
	10	3	6	10	16	28	38	63
	8	1	3	6	9	16	22	36
	6	1	2	4	6	12	16	26
	4	1	1	2	4	7	9	16
	3	1	1	1	3	6	8	13
	2	1	1	1	3	5	7	11
	1	0	1	1	1	4	5	8
	1/0	0	1	1	1	3	4	7
	2/0	0	0	1	1	2	3	6
	3/0	0	0	1	1	1	3	5
	4/0	0	0	1	1	1	2	4

Continued

Table C6 Maximum Number of Conductors or Fixture Wires in Liquidtight Flexible Nonmetallic Conduit (Type LFNC-A*) (Based on Table 1) *Continued*

Type	Conductor Size (AWG/kcmil)	CONDUCTORS Metric Designator (Trade Size)							
		12 (3/8)	16 (1/2)	21 (3/4)	27 (1)	35 (1-1/4)	41 (1-1/2)	53 (2)
THHN, THWN, THWN-2	250	0	0	0	1	1	1	3
	300	0	0	0	1	1	1	3
	350	0	0	0	1	1	1	2
	400	0	0	0	0	1	1	1
	500	0	0	0	0	1	1	1
	600	0	0	0	0	1	1	1
	700	0	0	0	0	1	1	1
	750	0	0	0	0	0	1	1
	800	0	0	0	0	0	1	1
	900	0	0	0	0	0	1	1
	1000	0	0	0	0	0	0	1
FEP, FEPB, PFA, PFAH, TFE	14	7	12	21	34	60	80	133
	12	5	9	15	25	44	59	97

Conduit and Tubing Fill Tables

	10	4		6	11	18	31	42	70
	8	1		3	6	10	18	24	40
	6	1		2	4	7	13	17	28
	4	1		1	3	5	9	12	20
	3	1		1	2	4	7	10	16
	2	1		1	1	3	6	8	13
PFA, PFAH, TFE	1	0		1	1	2	4	5	9
PFA, PFAH, TFE, Z	1/0	0		1	1	1	3	5	8
	2/0	0		1	1	1	3	4	6
	3/0	0		0	1	1	2	3	5
	4/0	0		0	1	1	1	2	4
Z	14	9	15	25	41	72	97	161
	12	6	10	18	29	51	69	114
	10	4	6	11	18	31	42	70
	8	2	4	7	11	20	26	44
	6	1	3	5	8	14	18	31

Continued

Table C6 Maximum Number of Conductors or Fixture Wires in Liquidtight Flexible Nonmetallic Conduit (Type LFNC-A*) (Based on Table 1) *Continued*

CONDUCTORS

Type	Conductor Size (AWG/kcmil)	Metric Designator (Trade Size)							
		12 (3/8)	16 (1/2)	21 (3/4)	27 (1)	35 (1-1/4)	41 (1-1/2)	53 (2)
Z	4	1	1	3	5	9	13	21
	3	1	1	2	4	7	9	15
	2	1	1	1	3	6	8	13
	1	1	1	1	2	4	6	10
XHH, XHHW, XHHW-2, ZW	14	5	9	15	24	43	58	96
	12	4	7	12	19	33	44	74
	10	3	5	9	14	24	33	55
	8	1	3	5	8	13	18	30
	6	1	1	3	5	10	13	22
	4	1	1	2	4	7	10	16
	3	1	1	1	3	6	8	14
	2	1	1	1	3	5	7	11

Conduit and Tubing Fill Tables

XHH, XHHW, XHHW-2

Size									
	1		0	1	1	1	1	4	5	8
1		0	1	1	1	1	3	4	7
1/0		0	0	1	1	1	2	3	6
2/0		0	0	1	1	1	1	3	5
3/0		0	0	0	1	1	1	2	4
4/0		0	0	0	1	1	1	2	4
250		0	0	0	0	1	1	1	3
300		0	0	0	0	1	1	1	3
350		0	0	0	0	1	1	1	2
400		0	0	0	0	0	1	1	1
500		0	0	0	0	0	1	1	1
600		0	0	0	0	0	1	1	1
700		0	0	0	0	0	0	1	1
750		0	0	0	0	0	0	1	1
800		0	0	0	0	0	0	1	1
900		0	0	0	0	0	0	1	1
1000		0	0	0	0	0	0	0	1
1250		0	0	0	0	0	0	0	1
1500		0	0	0	0	0	0	0	1

Continued

Table C6 Maximum Number of Conductors or Fixture Wires in Liquidtight Flexible Nonmetallic Conduit (Type LFNC-A*) (Based on Table 1) *Continued*

Type	Conductor Size (AWG/kcmil)	Metric Designator (Trade Size)						
		12 (3/8)	16 (1/2)	21 (3/4)	27 (1)	35 (1-1/4)	41 (1-1/2)	53 (2)
XHH, XHHW, XHHW-2	1750	0	0	0	0	0	0	0
	2000	0	0	0	0	0	0	0

Conduit and Tubing Fill Tables

Table C7 Maximum Number of Conductors or Fixture Wires in Liquidtight Flexible Metal Conduit (LFMC) (Based on Table 1)

CONDUCTORS

Type	Conductor Size (AWG/kcmil)	16 (1/2)	21 (3/4)	27 (1)	35 (1-1/4)	41 (1-1/2)	53 (2)	63 (2-1/2)	78 (3)	91 (3-1/2)	103 (4)
RHH, RHW, RHW-2	14	4	7	12	21	27	44	66	102	133	173
	12	3	6	10	17	22	36	55	84	110	144
	10	3	5	8	14	18	29	44	68	89	116
	8	1	2	4	7	9	15	23	36	46	61
	6	1	1	3	6	7	12	18	28	37	48
	4	1	1	2	4	6	9	14	22	29	38
	3	1	1	1	3	5	8	13	19	25	33
	2	1	1	1	3	4	7	11	17	22	29
	1	0	1	1	1	3	5	7	11	14	19
	1/0	0	1	1	1	2	4	6	10	13	16
	2/0	0	1	1	1	1	3	5	8	11	14

Continued

Table C7 Maximum Number of Conductors or Fixture Wires in Liquidtight Flexible Metal Conduit (LFMC) (Based on Table 1) *Continued*

CONDUCTORS

Type	Conductor Size (AWG/kcmil)	Metric Designator (Trade Size)									
		16 (1/2)	21 (3/4)	27 (1)	35 (1-1/4)	41 (1-1/2)	53 (2)	63 (2-1/2)	78 (3)	91 (3-1/2)	103 (4)
RHH, RHW, RHW-2	3/0	0	0	1	1	1	3	4	7	9	12
	4/0	0	0	1	1	1	2	4	6	8	10
	250	0	0	0	1	1	1	3	4	6	8
	300	0	0	0	1	1	1	2	4	5	7
	350	0	0	0	1	1	1	2	3	5	6
	400	0	0	0	1	1	1	1	3	4	6
	500	0	0	0	1	1	1	1	3	4	5
	600	0	0	0	0	1	1	1	2	3	4
	700	0	0	0	0	0	1	1	1	3	3
	750	0	0	0	0	0	1	1	1	2	3
	800	0	0	0	0	0	1	1	1	2	3
	900	0	0	0	0	0	1	1	1	2	3

Conduit and Tubing Fill Tables

Type	Size										
	1000	0	0	0	0	0	1	1	1	1	3
	1250	0	0	0	0	0	0	1	1	1	1
	1500	0	0	0	0	0	0	1	1	1	1
	1750	0	0	0	0	0	0	0	1	1	1
	2000	0	0	0	0	0	0	0	1	1	1
TW, THHW, THW, THW-2	14	9	15	25	44	57	93	140	215	280	365
	12	7	12	19	33	43	71	108	165	215	280
	10	5	9	14	25	32	53	80	123	160	209
	8	3	5	8	14	18	29	44	68	89	116
RHH*, RHW*, RHW-2*	14	6	10	16	29	38	62	93	143	186	243
	12	5	8	13	23	30	50	75	115	149	195
	10	3	6	10	18	23	39	58	89	117	152
	8	1	4	6	11	14	23	35	53	70	91
RHH*, RHW*, RHW-2*, TW, THW, THHW, THW-2	6	1	3	5	8	11	18	27	41	53	70
	4	1	1	3	6	8	13	20	30	40	52
	3	1	1	3	5	7	11	17	26	34	44
	2	1	1	2	4	6	9	14	22	29	38
	1	1	1	1	3	4	7	10	15	20	26

Continued

Table C7 Maximum Number of Conductors or Fixture Wires in Liquidtight Flexible Metal Conduit (LFMC) (Based on Table 1) *Continued*

CONDUCTORS

Type	Conductor Size (AWG/kcmil)	16 (1/2)	21 (3/4)	27 (1)	35 (1-1/4)	41 (1-1/2)	53 (2)	63 (2-1/2)	78 (3)	91 (3-1/2)	103 (4)
RHH*, RHW*, RHW-2*, TW, THW, THHW, THW-2	1/0	0	1	1	2	3	6	8	13	17	23
	2/0	0	1	1	2	3	5	7	11	15	19
	3/0	0	1	1	1	2	4	6	9	12	16
	4/0	0	0	1	1	1	3	5	8	10	13
	250	0	0	1	1	1	3	4	6	8	11
	300	0	0	0	1	1	2	3	5	7	9
	350	0	0	0	1	1	1	3	5	6	8
	400	0	0	0	1	1	1	3	4	6	7
	500	0	0	0	1	1	1	2	3	5	6
	600	0	0	0	0	1	1	1	3	4	5
	700	0	0	0	0	1	1	1	2	3	4
	750	0	0	0	0	1	1	1	2	3	4
	800	0	0	0	0	1	1	1	2	3	4

Conduit and Tubing Fill Tables

	900	0	0	0	0	0	0	0	1	1	3	3
	1000	0	0	0	0	0	0	0	1	1	2	3
THHN, THWN, THWN-2	14	13	22	36	63	81	133	201	308	401	523
	12	9	16	26	46	59	97	146	225	292	381
	10	6	10	16	29	37	61	92	141	184	240
	8	3	6	9	16	21	35	53	81	106	138
	6	2	4	7	12	15	25	38	59	76	100
	4	1	2	4	7	9	15	23	36	47	61
	3	1	1	3	6	8	13	20	30	40	52
	2	1	1	3	5	7	11	17	26	33	44
	1	1	1	1	4	5	8	12	19	25	32
	1/0	1	1	1	3	4	7	10	16	21	27
	2/0	0	1	1	2	3	6	8	13	17	23

Continued

Table C7 Maximum Number of Conductors or Fixture Wires in Liquidtight Flexible Metal Conduit (LFMC) (Based on Table 1) *Continued*

CONDUCTORS

Type	Conductor Size (AWG/kcmil)	Metric Designator (Trade Size)									
		16 (1/2)	21 (3/4)	27 (1)	35 (1-1/4)	41 (1-1/2)	53 (2)	63 (2-1/2)	78 (3)	91 (3-1/2)	103 (4)
THHN, THWN, THWN-2	3/0	0	1	1	1	3	5	7	11	14	19
	4/0	0	1	1	1	2	4	6	9	12	15
	250	0	0	1	1	1	3	5	7	10	12
	300	0	0	1	1	1	3	4	6	8	11
	350	0	0	1	1	1	2	3	5	7	9
	400	0	0	0	1	1	1	3	5	6	8
	500	0	0	0	1	1	1	2	4	5	7
	600	0	0	0	1	1	1	1	3	4	6
	700	0	0	0	1	1	1	1	3	4	5
	750	0	0	0	0	1	1	1	3	3	5
	800	0	0	0	0	1	1	1	2	3	4
	900	0	0	0	0	1	1	1	2	3	4
	1000	0	0	0	0	0	1	1	1	3	3

Conduit and Tubing Fill Tables

FEP, FEPB, PFA, PFAH, TFE	14	12	21	35	61	79	129	195	299	389	507
	12	9	15	25	44	57	94	142	218	284	370
	10	6	11	18	32	41	68	102	156	203	266
	8	3	6	10	18	23	39	58	89	117	152
	6	2	4	7	13	17	27	41	64	83	108
PFA, PFAH, TFE	4	1	3	5	9	12	19	29	44	58	75
	3	1	2	4	7	10	16	24	37	48	63
	2	1	1	3	6	8	13	20	30	40	52
PFA, PFAH, TFE	1	1	1	2	4	5	9	14	21	28	36
PFA, PFAH, TFE, Z	1/0	1	1	1	3	4	7	11	18	23	30
	2/0	1	1	1	3	4	6	9	14	19	25
	3/0	0	1	1	2	3	5	8	12	16	20
	4/0	0	1	1	1	2	4	6	10	13	17
Z	14	20	26	42	73	95	156	235	360	469	611
	12	14	18	30	52	67	111	167	255	332	434
	10	8	11	18	32	41	68	102	156	203	266

Continued

Table C7 Maximum Number of Conductors or Fixture Wires in Liquidtight Flexible Metal Conduit (LFMC) (Based on Table 1) *Continued*

CONDUCTORS

Type	Conductor Size (AWG/kcmil)	Metric Designator (Trade Size)									
		16 (1/2)	21 (3/4)	27 (1)	35 (1-1/4)	41 (1-1/2)	53 (2)	63 (2-1/2)	78 (3)	91 (3-1/2)	103 (4)
Z	8	5	7	11	20	26	43	64	99	129	168
	6	4	5	8	14	18	30	45	69	90	118
	4	2	3	5	9	12	20	31	48	62	81
	3	2	2	4	7	9	15	23	35	45	59
	2	1	1	3	6	7	12	19	29	38	49
	1	1	1	2	5	6	10	15	23	30	40
XHH, XHHW, XHHW-2, ZW	14	9	15	25	44	57	93	140	215	280	365
	12	7	12	19	33	43	71	108	165	215	280
	10	5	9	14	25	32	53	80	123	160	209
	8	3	5	8	14	18	29	44	68	89	116
	6	1	3	6	10	13	22	33	50	66	86
	4	1	2	4	7	9	16	24	36	48	62

Conduit and Tubing Fill Tables

Type	Size										
XHH, XHHW, XHHW-2	3	52	40	31	20	13	8	6	3	1	1
	2	44	34	26	17	11	7	5	3	1	1
	1	33	25	19	12	8	5	4	1	1	1
	1/0	28	21	16	10	7	4	3	1	1	1
	2/0	23	17	13	9	6	3	2	1	1	1
	3/0	19	14	11	7	5	3	1	1	1	1
	4/0	16	12	9	6	4	2	1	1	0	0
	250	13	10	7	5	3	1	1	1	0	0
	300	11	8	6	4	3	1	1	1	0	0
	350	10	7	5	3	2	1	1	0	0	0
	400	8	6	5	3	1	1	1	0	0	0
	500	7	5	4	2	1	1	1	0	0	0
	600	6	4	3	1	1	1	1	0	0	0
	700	5	4	3	1	1	1	0	0	0	0
	750	5	3	3	1	1	1	0	0	0	0
	800	4	3	2	1	1	1	0	0	0	0
	900	4	3	2	1	1	1	0	0	0	0

Continued

Table C7 Maximum Number of Conductors or Fixture Wires in Liquidtight Flexible Metal Conduit (LFMC) (Based on Table 1) *Continued*

CONDUCTORS

Type	Conductor Size (AWG/kcmil)	Metric Designator (Trade Size)								
		16 (1/2)	21 (3/4)	27 (1)	35 (1-1/4)	41 (1-1/2)	53 (2)	63 (2-1/2)	78 (3)	91 (3-1/2)	103 (4)
XHH, XHHW, XHHW-2	1000	0	0	0	0	0	1	1	1	1	3
	1250	0	0	0	0	0	1	1	1	3	3
	1500	0	0	0	0	0	1	1	1	1	2
	1750	0	0	0	0	0	0	1	1	1	2
	2000	0	0	0	0	0	0	1	1	1	2

*Types RHH, RHW, and RHW-2 without outer covering.

Conduit and Tubing Fill Tables

Table C8 Maximum Number of Conductors or Fixture Wires in Rigid Metal Conduit (RMC) (Based on Table 1)

		CONDUCTORS										
		Metric Designator (Trade Size)										
Type	Conductor Size (AWG/kcmil)	16 (1/2)	21 (3/4)	27 (1)	35 (1-1/4)	41 (1-1/2)	53 (2)	63 (2-1/2)	78 (3)	91 (3-1/2)	103 (4)	129 (5)	155 (6)
RHH, RHW, RHW-2	14	4	7	12	21	28	46	66	102	136	176	276	398
	12	3	6	10	17	23	38	55	85	113	146	229	330
	10	3	5	8	14	19	31	44	68	91	118	185	267
	8	1	2	4	7	10	16	23	36	48	61	97	139
	6	1	1	3	6	8	13	18	29	38	49	77	112
	4	1	1	2	4	6	10	14	22	30	38	60	87
	3	1	1	2	4	5	9	12	19	26	34	53	76
	2	1	1	1	3	4	7	11	17	23	29	46	66
	1	0	1	1	1	3	5	7	11	15	19	30	44
	1/0	0	1	1	1	2	4	6	10	13	17	26	38
	2/0	0	1	1	1	2	4	5	8	11	14	23	33

Continued

Table C8 Maximum Number of Conductors or Fixture Wires in Rigid Metal Conduit (RMC) (Based on Table 1) *Continued*

CONDUCTORS

Type	Conductor Size (AWG/kcmil)	Metric Designator (Trade Size)									
		16 (1/2)	21 (3/4)	27 (1)	35 (1-1/4)	41 (1-1/2)	53 (2)	63 (2-1/2)	78 (3)	91 (3-1/2)	103 (4)	129 (5)	155 (6)
RHH, RHW, RHW-2	3/0	0	0	1	1	1	3	4	7	10	12	20	28
	4/0	0	0	1	1	1	3	4	6	8	11	17	24
	250	0	0	0	1	1	1	3	4	6	8	13	18
	300	0	0	0	1	1	1	2	4	5	7	11	16
	350	0	0	0	1	1	1	2	4	5	6	10	15
	400	0	0	0	1	1	1	1	3	4	6	9	13
	500	0	0	0	1	1	1	1	3	4	5	8	11
	600	0	0	0	0	1	1	1	2	3	4	6	9
	700	0	0	0	0	1	1	1	2	3	4	6	8
	750	0	0	0	0	0	1	1	1	3	3	5	8
	800	0	0	0	0	0	1	1	1	2	3	5	7
	900	0	0	0	0	0	1	1	1	2	3	5	7

Conduit and Tubing Fill Tables

	1000	0	0	0	0	0	1	1	1	1	3	4	6
	1250	0	0	0	0	0	0	1	1	1	1	3	5
	1500	0	0	0	0	0	0	1	1	1	1	3	4
	1750	0	0	0	0	0	0	1	1	1	1	2	4
	2000	0	0	0	0	0	0	0	1	1	1	2	3
TW, THHW, THW, THW-2	14	9	15	25	44	59	98	140	216	288	370	581	839
	12	7	12	19	33	45	75	107	165	221	284	446	644
	10	5	9	14	25	34	56	80	123	164	212	332	480
	8	3	5	8	14	19	31	44	68	91	118	185	267
RHH*, RHW*, RHW-2*	14	6	10	17	29	39	65	93	143	191	246	387	558
RHH*, RHW*, RHW-2*	12	5	8	13	23	32	52	75	115	154	198	311	448
	10	3	6	10	18	25	41	58	90	120	154	242	350
RHH*, RHW*, RHW-2*	8	1	4	6	11	15	24	35	54	72	92	145	209

Continued

Table C8 Maximum Number of Conductors or Fixture Wires in Rigid Metal Conduit (RMC) (Based on Table 1) *Continued*

Type	Conductor Size (AWG/kcmil)	CONDUCTORS / Metric Designator (Trade Size)											
		16 (1/2)	21 (3/4)	27 (1)	35 (1-1/4)	41 (1-1/2)	53 (2)	63 (2-1/2)	78 (3)	91 (3-1/2)	103 (4)	129 (5)	155 (6)
RHH*, RHW*, RHW-2*, TW, THW, THHW, THW-2	6	1	3	5	8	11	18	27	41	55	71	111	160
	4	1	1	3	6	8	14	20	31	41	53	83	120
	3	1	1	2	5	7	12	17	26	35	45	71	103
	2	1	1	2	4	6	10	14	22	30	38	60	87
	1	1	1	1	3	4	7	10	15	21	27	42	61
	1/0	0	1	1	2	3	6	8	13	18	23	36	52
	2/0	0	1	1	2	3	5	7	11	15	19	31	44
	3/0	0	1	1	1	2	4	6	9	13	16	26	37
	4/0	0	0	1	1	1	3	5	8	10	14	21	31
	250	0	0	1	1	1	3	4	6	8	11	17	25
	300	0	0	0	1	1	2	3	5	7	9	15	22
	350	0	0	0	1	1	1	3	5	6	8	13	19

Conduit and Tubing Fill Tables

400	0	0	0	0	1	1	3	4	6	7	12	17
500	0	0	0	0	1	1	2	3	5	6	10	14
600	0	0	0	0	1	1	1	3	4	5	8	12
700	0	0	0	0	1	1	1	2	3	4	7	10
750	0	0	0	0	1	1	1	2	3	4	7	10
800	0	0	0	0	0	1	1	2	3	4	6	9
900	0	0	0	0	0	1	1	1	3	4	6	8
1000	0	0	0	0	0	1	1	1	2	3	5	8
1250	0	0	0	0	0	0	1	1	1	2	4	6
1500	0	0	0	0	0	0	1	1	1	2	3	5
1750	0	0	0	0	0	0	1	1	1	2	3	4
2000	0	0	0	0	0	0	1	1	1	1	3	4

THHN, THWN, THWN-2

14	13	22	36	63	85	140	200	309	412	531	833	1202
12	9	16	26	46	62	102	146	225	301	387	608	877
10	6	10	17	29	39	64	92	142	189	244	383	552
8	3	6	9	16	22	37	53	82	109	140	221	318
6	2	4	7	12	16	27	38	59	79	101	159	230

Continued

Table C8 Maximum Number of Conductors or Fixture Wires in Rigid Metal Conduit (RMC) (Based on Table 1) *Continued*

Type	Conductor Size (AWG/kcmil)	CONDUCTORS Metric Designator (Trade Size)									
		16 (1/2)	21 (3/4)	27 (1)	35 (1-1/4)	41 (1-1/2)	53 (2)	63 (2-1/2)	78 (3)	91 (3-1/2)	103 (4)	129 (5)	155 (6)
THHN, THWN, THWN-2	4	1	2	4	7	10	16	23	36	48	62	98	141
	3	1	1	3	6	8	14	20	31	41	53	83	120
	2	1	1	3	5	7	11	17	26	34	44	70	100
	1	1	1	1	4	5	8	12	19	25	33	51	74
	1/0	1	1	1	3	4	7	10	16	21	27	43	63
	2/0	0	1	1	2	3	6	8	13	18	23	36	52
	3/0	0	1	1	1	3	5	7	11	15	19	30	43
	4/0	0	1	1	1	2	4	6	9	12	16	25	36
	250	0	0	1	1	1	3	5	7	10	13	20	29
	300	0	0	1	1	1	3	4	6	8	11	17	25
	350	0	0	1	1	1	2	3	5	7	10	15	22
	400	0	0	1	1	1	2	3	5	7	8	13	20
	500	0	0	0	1	1	1	2	4	5	7	11	16

Conduit and Tubing Fill Tables

FEP, FEPB, PFA, PFAH, TFE	14	12	9	6	3	2	12	22	35	61	83	136	194	300	400	515	808	1166
	12	9	6	3	2	1	9	16	26	44	60	99	142	219	292	376	590	851
	10	6	4	2	1	1	6	11	18	32	43	71	102	157	209	269	423	610
	8	3	2	1	1	1	3	6	10	18	25	41	58	90	120	154	242	350
	6	2	1	1	1	1	2	4	7	13	17	29	41	64	85	110	172	249
	4	1	1	0	0	0	1	3	5	9	12	20	29	44	59	77	120	174
	3	1	1	0	0	0	1	2	4	7	10	17	24	37	50	64	100	145
	2	1	1	0	0	0	1	1	3	6	8	14	20	31	41	53	83	120
PFA, PFAH, TFE	1	1	0	0	0	0	1	1	1	4	6	9	14	21	28	37	57	83
PFA, PFAH, TFE, Z	1/0	1	0	0	0	0	1	1	1	3	5	8	11	18	24	30	48	69
	2/0	1	0	0	0	0	1	1	1	3	4	6	9	14	19	25	40	57

Continued

Table C8 Maximum Number of Conductors or Fixture Wires in Rigid Metal Conduit (RMC) (Based on Table 1) *Continued*

CONDUCTORS

Type	Conductor Size (AWG/kcmil)	Metric Designator (Trade Size)										
		16 (1/2)	21 (3/4)	27 (1)	35 (1-1/4)	41 (1-1/2)	53 (2)	63 (2-1/2)	78 (3)	91 (3-1/2)	103 (4)	129 (5)	155 (6)
PFA, PFAH, TFE, Z	3/0	0	1	1	2	3	5	8	12	16	21	33	47
	4/0	0	1	1	1	2	4	6	10	13	17	27	39
Z	14	15	26	42	73	100	164	234	361	482	621	974	1405
	12	10	18	30	52	71	116	166	256	342	440	691	997
	10	6	11	18	32	43	71	102	157	209	269	423	610
	8	4	7	11	20	27	45	64	99	132	170	267	386
	6	3	5	8	14	19	31	45	69	93	120	188	271
	4	1	3	5	9	13	22	31	48	64	82	129	186
	3	1	2	4	7	9	16	22	35	47	60	94	136
	2	1	1	3	6	8	13	19	29	39	50	78	113
	1	1	1	2	5	6	10	15	23	31	40	63	92

Conduit and Tubing Fill Tables

XHH, XHHW, XHHW-2, ZW	14	9	15	25	44	59	98	140	216	288	370	581	839
	12	7	12	19	33	45	75	107	165	221	284	446	644
	10	5	9	14	25	34	56	80	123	164	212	332	480
	8	3	5	8	14	19	31	44	68	91	118	185	267
	6	1	3	6	10	14	23	33	51	68	87	137	197
	4	1	2	4	7	10	16	24	37	49	63	99	143
	3	1	1	3	6	8	14	20	31	41	53	84	121
	2	1	1	3	5	7	12	17	26	35	45	70	101
XHH, XHHW, XHHW-2	1	1	1	1	4	5	9	12	19	26	33	52	76
	1/0	1	1	1	3	4	7	10	16	22	28	44	64
	2/0	0	1	1	2	3	6	9	13	18	23	37	53
	3/0	0	1	1	1	3	5	7	11	15	19	30	44
	4/0	0	1	1	1	2	4	6	9	12	16	25	36
	250	0	0	1	1	1	3	5	7	10	13	20	30
	300	0	0	1	1	1	3	4	6	9	11	18	25
	350	0	0	1	1	1	2	3	6	7	10	15	22
	400	0	0	1	1	1	2	3	5	7	9	14	20
	500	0	0	0	1	1	1	2	4	5	7	11	16

Continued

Table C8 Maximum Number of Conductors or Fixture Wires in Rigid Metal Conduit (RMC) (Based on Table 1) *Continued*

Type	Conductor Size (AWG/kcmil)	CONDUCTORS Metric Designator (Trade Size)										
		16 (1/2)	21 (3/4)	27 (1)	35 (1-1/4)	41 (1-1/2)	53 (2)	63 (2-1/2)	78 (3)	91 (3-1/2)	103 (4)	129 (5)	155 (6)
XHH, XHHW, XHHW-2	600	0	0	0	1	1	1	1	3	4	6	9	13
	700	0	0	0	1	1	1	1	3	4	5	8	11
	750	0	0	0	0	1	1	1	3	4	5	7	11
	800	0	0	0	0	1	1	1	2	3	4	7	10
	900	0	0	0	0	1	1	1	2	3	4	6	9
	1000	0	0	0	0	0	1	1	1	3	4	6	8
	1250	0	0	0	0	0	1	1	1	2	3	4	6
	1500	0	0	0	0	0	1	1	1	1	2	4	5
	1750	0	0	0	0	0	0	1	1	1	1	3	5
	2000	0	0	0	0	0	0	1	1	1	1	3	4

Conduit and Tubing Fill Tables

Table C9 Maximum Number of Conductors or Fixture Wires in Rigid PVC Conduit, Schedule 80 (Based on Table 1)

		CONDUCTORS										
		Metric Designator (Trade Size)										
Type	Conductor Size (AWG/kcmil)	16 (1/2)	21 (3/4)	27 (1)	35 (1-1/4)	41 (1-1/2)	53 (2)	63 (2-1/2)	78 (3)	91 (3-1/2)	103 (4)	129 (5)	155 (6)
RHH, RHW, RHW-2	14	3	5	9	17	23	39	56	88	118	153	243	349
	12	2	4	7	14	19	32	46	73	98	127	202	290
	10	1	3	6	11	15	26	37	59	79	103	163	234
	8	1	1	3	6	8	13	19	31	41	54	85	122
	6	1	1	2	4	6	11	16	24	33	43	68	98
	4	1	1	1	3	5	8	12	19	26	33	53	77
	3	0	1	1	3	4	7	11	17	23	29	47	67
	2	0	1	1	3	4	6	9	14	20	25	41	58
	1	0	1	1	1	2	4	6	9	13	17	27	38
	1/0	0	0	1	1	1	3	5	8	11	15	23	33
	2/0	0	0	1	1	1	3	4	7	10	13	20	29

Continued

Table C9 Maximum Number of Conductors or Fixture Wires in Rigid PVC Conduit, Schedule 80 (Based on Table 1) *Continued*

CONDUCTORS

Type	Conductor Size (AWG/kcmil)	Metric Designator (Trade Size)										
		16 (1/2)	21 (3/4)	27 (1)	35 (1-1/4)	41 (1-1/2)	53 (2)	63 (2-1/2)	78 (3)	91 (3-1/2)	103 (4)	129 (5)	155 (6)
RHH, RHW, RHW-2	3/0	0	0	1	1	1	3	4	6	8	11	17	25
	4/0	0	0	0	1	1	2	3	5	7	9	15	21
	250	0	0	0	1	1	1	2	4	5	7	11	16
	300	0	0	0	1	1	1	2	3	5	6	10	14
	350	0	0	0	1	1	1	1	3	4	5	9	13
	400	0	0	0	0	1	1	1	3	4	5	8	12
	500	0	0	0	0	1	1	1	2	3	4	7	10
	600	0	0	0	0	0	1	1	1	3	3	6	8
	700	0	0	0	0	0	1	1	1	2	3	5	7
	750	0	0	0	0	0	1	1	1	2	3	5	7
	800	0	0	0	0	0	1	1	1	2	3	4	7

Conduit and Tubing Fill Tables

TW, THHW, THW, THW-2	1000	0	0	0	0	0	0	0	1	1	1	1	2	4	5
	1250	0	0	0	0	0	0	0	0	1	1	1	1	3	4
	1500	0	0	0	0	0	0	0	0	1	1	1	1	2	4
	1750	0	0	0	0	0	0	0	0	0	1	1	1	2	3
	2000	0	0	0	0	0	0	0	0	0	1	1	1	1	3
RHH*, RHW*, RHW-2*	14	6	11	20	35	49	82	118	185	250	324	514	736		
	12	5	9	15	27	38	63	91	142	192	248	394	565		
	10	4	7	12	22	31	51	74	115	156	201	319	457		
	10	3	6	11	20	28	47	67	106	143	185	294	421		
	8	1	3	6	11	15	26	37	59	79	103	163	234		
RHH*, RHW*, RHW-2*, TW, THW, THHW, THW-2	6	4	8	13	23	32	55	79	123	166	215	341	490		
	4	3	6	10	19	26	44	63	99	133	173	274	394		
	3	2	5	8	15	20	34	49	77	104	135	214	307		
	2	1	3	5	9	12	20	29	46	62	81	128	184		
	6	1	1	3	7	9	16	22	35	48	62	98	141		
	4	1	1	3	5	7	12	17	26	35	46	73	105		
	3	1	1	2	4	6	10	14	22	30	39	63	90		
	2	1	1	1	3	5	8	12	19	26	33	53	77		
	1	0	1	1	2	3	6	8	13	18	23	37	54		

Continued

Table C9 Maximum Number of Conductors or Fixture Wires in Rigid PVC Conduit, Schedule 80 (Based on Table 1) *Continued*

CONDUCTORS

Type	Conductor Size (AWG/kcmil)	16 (1/2)	21 (3/4)	27 (1)	35 (1-1/4)	41 (1-1/2)	53 (2)	63 (2-1/2)	78 (3)	91 (3-1/2)	103 (4)	129 (5)	155 (6)
RHH*, RHW*, RHW-2*, TW, THW, THHW, THW-2	1/0	0	1	1	1	3	5	7	11	15	20	32	46
	2/0	0	1	1	1	2	4	6	10	13	17	27	39
	3/0	0	0	1	1	1	3	5	8	11	14	23	33
	4/0	0	0	1	1	1	3	4	7	9	12	19	27
	250	0	0	0	1	1	2	3	5	7	9	15	22
	300	0	0	0	1	1	1	3	5	6	8	13	19
	350	0	0	0	1	1	1	2	4	6	7	12	17
	400	0	0	0	1	1	1	2	4	5	7	10	15
	500	0	0	0	1	0	1	1	3	4	5	9	13
	600	0	0	0	0	0	1	1	2	3	4	7	10
	700	0	0	0	0	0	1	1	2	3	4	6	9
	750	0	0	0	0	0	1	1	1	3	4	6	8

Conduit and Tubing Fill Tables

	Size (AWG/kcmil)												
	800	0	0	0	0	0	0	0	0	1	1	3	6	7
	900	0	0	0	0	0	0	0	0	1	1	3	5	7
	1000	0	0	0	0	0	0	0	0	1	1	3	5	7
	1250	0	0	0	0	0	0	0	0	1	1	2	4	5
	1500	0	0	0	0	0	0	0	0	1	1	1	3	4
	1750	0	0	0	0	0	0	0	0	0	1	1	3	4
	2000	0	0	0	0	0	0	0	0	0	1	1	2	3
THHN, THWN, THWN-2	14	9	17	28	51	70	118	170	265	358	464	736	1055
	12	6	12	20	37	51	86	124	193	261	338	537	770
	10	4	7	13	23	32	54	78	122	164	213	338	485
	8	2	4	7	13	18	31	45	70	95	123	195	279
	6	1	3	5	9	13	22	32	51	68	89	141	202
	4	1	1	3	6	8	14	20	31	42	54	86	124
	3	1	1	3	5	7	12	17	26	35	46	73	105
	2	1	1	2	4	6	10	14	22	30	39	61	88
	1	0	1	1	3	4	7	10	16	22	29	45	65
	1/0	0	1	1	2	3	6	9	14	18	24	38	55
	2/0	0	1	1	1	3	5	7	11	15	20	32	46

Continued

Table C9 Maximum Number of Conductors or Fixture Wires in Rigid PVC Conduit, Schedule 80 (Based on Table 1) *Continued*

CONDUCTORS

Type	Conductor Size (AWG/kcmil)	Metric Designator (Trade Size)										
		16 (1/2)	21 (3/4)	27 (1)	35 (1-1/4)	41 (1-1/2)	53 (2)	63 (2-1/2)	78 (3)	91 (3-1/2)	103 (4)	129 (5)	155 (6)
THHN, THWN, THWN-2	3/0	0	1	1	1	2	4	6	9	13	17	26	38
	4/0	0	0	1	1	1	3	5	8	10	14	22	31
	250	0	0	0	1	1	3	4	6	8	11	18	25
	300	0	0	0	1	1	2	3	5	7	9	15	22
	350	0	0	0	1	1	1	3	5	6	8	13	19
	400	0	0	0	1	1	1	3	4	6	7	12	17
	500	0	0	0	1	1	1	2	3	5	6	10	14
	600	0	0	0	0	1	1	1	3	4	5	8	12
	700	0	0	0	0	1	1	1	2	3	4	7	10
	750	0	0	0	0	1	1	1	2	3	4	7	9
	800	0	0	0	0	1	1	1	2	3	4	6	9
	900	0	0	0	0	0	1	1	1	3	3	6	8
	1000	0	0	0	0	0	1	1	1	2	3	5	7

Conduit and Tubing Fill Tables

Type	Size												
FEP, FEPB, PFA, PFAH, TFE	14	8	16	27	49	68	115	164	257	347	450	714	1024
	12	6	12	20	36	50	84	120	188	253	328	521	747
	10	4	8	14	26	36	60	86	135	182	235	374	536
	8	2	5	8	15	20	34	49	77	104	135	214	307
	6	1	3	6	10	14	24	35	55	74	96	152	218
	4	1	2	4	7	10	17	24	38	52	67	106	153
	3	1	1	3	6	8	14	20	32	43	56	89	127
	2	1	1	3	5	7	12	17	26	35	46	73	105
PFA, PFAH, TFE	1	1	1	1	3	5	8	11	18	25	32	51	73
PFA, PFAH, TFE, Z	1/0	0	1	1	3	4	7	10	15	20	27	42	61
	2/0	0	1	1	2	3	5	8	12	17	22	35	50
	3/0	0	1	1	1	2	4	6	10	14	18	29	41
	4/0	0	0	1	1	1	4	5	8	11	15	24	34
Z	14	10	19	33	59	82	138	198	310	418	542	860	1233
	12	7	14	23	42	58	98	141	220	297	385	610	875
	10	4	8	14	26	36	60	86	135	182	235	374	536

Continued

Table C9 Maximum Number of Conductors or Fixture Wires in Rigid PVC Conduit, Schedule 80 (Based on Table 1) *Continued*

CONDUCTORS

Type	Conductor Size (AWG/kcmil)	16 (1/2)	21 (3/4)	27 (1)	35 (1-1/4)	41 (1-1/2)	53 (2)	63 (2-1/2)	78 (3)	91 (3-1/2)	103 (4)	129 (5)	155 (6)
Z	8	3	5	9	16	22	38	54	85	115	149	236	339
	6	2	4	6	11	16	26	38	60	81	104	166	238
	4	1	2	4	8	11	18	26	41	55	72	114	164
	3	1	1	3	5	8	13	19	30	40	52	83	119
	2	1	1	2	5	6	11	16	25	33	43	69	99
	1	0	1	2	4	5	9	13	20	27	35	56	80
XHH, XHHW, XHHW-2, ZW	14	6	11	20	35	49	82	118	185	250	324	514	736
	12	5	9	15	27	38	63	91	142	192	248	394	565
	10	3	6	11	20	28	47	67	106	143	185	294	421
	8	1	3	6	11	15	26	37	59	79	103	163	234
	6	1	2	4	8	11	19	28	43	59	76	121	173

Conduit and Tubing Fill Tables

Size													
4	1	1	1	3	6	8	14	20	31	42	55	87	125
3	1	1	1	3	5	7	12	17	26	36	47	74	106
2	1	1	1	2	4	6	10	14	22	30	39	62	89
1	0	1	1	1	3	4	7	10	16	22	29	46	66
1/0	0	0	1	1	2	3	6	9	14	19	24	39	56
2/0	0	0	1	1	1	3	5	7	11	16	20	32	46
3/0	0	0	1	1	1	2	4	6	9	13	17	27	38
4/0	0	0	0	1	1	1	3	5	8	11	14	22	32
250	0	0	1	1	1	1	3	4	6	9	11	18	26
300	0	0	0	1	1	1	2	3	5	7	10	15	22
350	0	0	0	1	1	1	1	3	5	6	8	14	20
400	0	0	0	1	1	1	1	3	4	6	7	12	17
500	0	0	0	1	1	1	1	2	3	5	6	10	14
600	0	0	0	0	0	1	1	1	3	4	5	8	11
700	0	0	0	0	0	1	1	1	2	3	4	7	10
750	0	0	0	0	0	1	1	1	2	3	4	6	9

XHH, XHHW, XHHW-2

Continued

Table C9 Maximum Number of Conductors or Fixture Wires in Rigid PVC Conduit, Schedule 80 (Based on Table 1) *Continued*

CONDUCTORS

Type	Conductor Size (AWG/kcmil)	16 (1/2)	21 (3/4)	27 (1)	35 (1-1/4)	41 (1-1/2)	53 (2)	63 (2-1/2)	78 (3)	91 (3-1/2)	103 (4)	129 (5)	155 (6)
XHH, XHHW, XHHW-2	800	0	0	0	0	1	1	1	1	3	4	6	9
	900	0	0	0	0	0	1	1	—	3	3	5	8
	1000	0	0	0	0	0	1	1	1	2	3	5	7
	1250	0	0	0	0	0	1	1	1	1	2	4	6
	1500	0	0	0	0	0	0	1	1	1	1	3	5
	1750	0	0	0	0	0	0	1	1	1	1	3	4
	2000	0	0	0	0	0	0	1	1	1	1	2	4

Conduit and Tubing Fill Tables

Table C10 Maximum Number of Conductors or Fixture Wires in Rigid PVC Conduit, Schedule 40 and HDPE Conduit (Based on Table 1)

CONDUCTORS

Type	Conductor Size (AWG/kcmil)	Metric Designator (Trade Size)										
		16 (1/2)	21 (3/4)	27 (1)	35 (1-1/4)	41 (1-1/2)	53 (2)	63 (2-1/2)	78 (3)	91 (3-1/2)	103 (4)	129 (5)	155 (6)
RHH, RHW, RHW-2	14	4	7	11	20	27	45	64	99	133	171	269	390
	12	3	5	9	16	22	37	53	82	110	142	224	323
	10	2	4	7	13	18	30	43	66	89	115	181	261
	8	1	2	4	7	9	15	22	35	46	60	94	137
	6	1	1	3	5	7	12	18	28	37	48	76	109
	4	1	1	2	4	6	10	14	22	29	37	59	85
	3	1	1	1	4	5	8	12	19	25	33	52	75
	2	1	1	1	3	4	7	10	16	22	28	45	65
	1	0	1	1	1	3	5	7	11	14	19	29	43
	1/0	0	0	1	1	2	4	6	9	13	16	26	37
	2/0	0	0	1	1	1	3	5	8	11	14	22	32

Continued

Table C10 Maximum Number of Conductors or Fixture Wires in Rigid PVC Conduit, Schedule 40 and HDPE Conduit (Based on Table 1) *Continued*

CONDUCTORS

Type	Conductor Size (AWG/kcmil)	16 (1/2)	21 (3/4)	27 (1)	35 (1-1/4)	41 (1-1/2)	53 (2)	63 (2-1/2)	78 (3)	91 (3-1/2)	103 (4)	129 (5)	155 (6)
RHH, RHW, RHW-2	3/0	0	0	1	1	1	3	4	7	9	12	19	28
	4/0	0	0	1	1	1	2	4	6	8	10	16	24
	250	0	0	0	1	1	1	3	4	6	8	12	18
	300	0	0	0	1	1	1	2	4	5	7	11	16
	350	0	0	0	1	1	1	2	3	5	6	10	14
	400	0	0	0	1	1	1	1	3	4	6	9	13
	500	0	0	0	0	1	1	1	3	4	5	8	11
	600	0	0	0	0	1	1	1	2	3	4	6	9
	700	0	0	0	0	0	1	1	1	3	3	6	8
	750	0	0	0	0	0	1	1	1	3	3	5	8
	800	0	0	0	0	0	1	1	1	2	3	5	7
	900	0	0	0	0	0	1	1	1	2	3	5	7

Conduit and Tubing Fill Tables

Type	Size												
TW, THHW, THW, THW-2	1000	0	0	0	0	0	0	1	1	1	1	3	4	6
	1250	0	0	0	0	0	0	0	1	1	1	1	3	5
	1500	0	0	0	0	0	0	0	1	1	1	1	3	4
	1750	0	0	0	0	0	0	0	0	1	1	1	2	3
	2000	0	0	0	0	0	0	0	0	1	1	1	2	3
RHH*, RHW*, RHW-2*	14	8	14	24	42	57	94	135	209	280	361	568	822
	12	6	11	18	32	44	72	103	160	215	277	436	631
	10	4	8	13	24	32	54	77	119	160	206	325	470
	8	2	4	7	13	18	30	43	66	89	115	181	261
TW, THW, THHW, THW-2	14	5	9	16	28	38	63	90	139	186	240	378	546
	12	4	8	12	22	30	50	72	112	150	193	304	439
	10	3	6	10	17	24	39	56	87	117	150	237	343
	8	1	3	6	10	14	23	33	52	70	90	142	205
TW, THW, THHW, THW-2	6	1	2	4	8	11	18	26	40	53	69	109	157
	4	1	1	3	6	8	13	19	30	40	51	81	117
	3	1	1	3	5	7	11	16	25	34	44	69	100
	2	1	1	2	4	6	10	14	22	29	37	59	85
	1	0	1	1	3	4	7	10	15	20	26	41	60

Continued

Table C10 Maximum Number of Conductors or Fixture Wires in Rigid PVC Conduit, Schedule 40 and HDPE Conduit (Based on Table 1) *Continued*

CONDUCTORS

Type	Conductor Size (AWG/kcmil)	16 (1/2)	21 (3/4)	27 (1)	35 (1-1/4)	41 (1-1/2)	53 (2)	63 (2-1/2)	78 (3)	91 (3-1/2)	103 (4)	129 (5)	155 (6)
TW, THW, THHW, THW-2	1/0	0	1	1	2	3	6	8	13	17	22	35	51
	2/0	0	1	1	1	3	5	7	11	15	19	30	43
	3/0	0	0	1	1	2	4	6	9	12	16	25	36
	4/0	0	0	1	1	1	3	5	8	10	13	21	30
	250	0	0	0	1	1	3	4	6	8	11	17	25
	300	0	0	0	1	1	2	3	5	7	9	15	21
	350	0	0	0	1	1	1	3	5	6	8	13	19
	400	0	0	0	1	1	1	3	4	6	7	12	17
	500	0	0	0	1	1	1	2	3	5	6	10	14
	600	0	0	0	0	1	1	1	3	4	5	8	11
	700	0	0	0	0	1	1	1	2	3	4	7	10
	750	0	0	0	0	1	1	1	2	3	4	6	10

Conduit and Tubing Fill Tables

Type	Size											
	800	0	0	0	0	0	0	1	1	2	3	6	9
	900	0	0	0	0	0	0	0	1	1	3	6	8
THHN, THWN, THWN-2	1000	0	0	0	0	0	0	1	1	1	2	5	7
	1250	0	0	0	0	0	0	1	1	1	1	4	6
	1500	0	0	0	0	0	0	1	1	1	1	3	5
	1750	0	0	0	0	0	0	0	1	1	1	3	4
	2000	0	0	0	0	0	0	0	1	1	1	3	4
	14	11	21	34	60	82	135	193	299	401	517	815	1178
	12	8	15	25	43	59	99	141	218	293	377	594	859
	10	5	9	15	27	37	62	89	137	184	238	374	541
	8	3	5	9	16	21	36	51	79	106	137	216	312
	6	1	4	6	11	15	26	37	57	77	99	156	225
	4	1	2	4	7	9	16	22	35	47	61	96	138
	3	1	1	3	6	8	13	19	30	40	51	81	117
	2	1	1	3	5	7	11	16	25	33	43	68	98
	1	1	1	1	3	5	8	12	18	25	32	50	73
	1/0	1	1	1	3	4	7	10	15	21	27	42	61
	2/0	0	1	1	2	3	6	8	13	17	22	35	51

Continued

Table C10 Maximum Number of Conductors or Fixture Wires in Rigid PVC Conduit, Schedule 40 and HDPE Conduit (Based on Table 1) *Continued*

CONDUCTORS

Type	Conductor Size (AWG/kcmil)	16 (1/2)	21 (3/4)	27 (1)	35 (1-1/4)	41 (1-1/2)	53 (2)	63 (2-1/2)	78 (3)	91 (3-1/2)	103 (4)	129 (5)	155 (6)
THHN, THWN, THWN-2	3/0	0	1	1	1	3	5	7	11	14	18	29	42
	4/0	0	1	1	1	2	4	6	9	12	15	24	35
	250	0	0	1	1	1	3	4	7	10	12	20	28
	300	0	0	1	1	1	3	4	6	8	11	17	24
	350	0	0	1	1	1	2	3	5	7	9	15	21
	400	0	0	0	1	1	2	3	5	6	8	13	19
	500	0	0	0	1	1	1	2	4	5	7	11	16
	600	0	0	0	0	1	1	1	3	4	5	9	13
	700	0	0	0	0	1	1	1	3	4	5	8	11
	750	0	0	0	0	1	1	1	2	3	4	7	11
	800	0	0	0	0	1	1	1	2	3	4	7	10
	900	0	0	0	0	0	1	1	2	3	4	6	9
	1000	0	0	0	0	0	1	1	1	3	3	6	8

Conduit and Tubing Fill Tables

FEP, FEPB, PFA, PFAH, TFE	14	11	20	33	58	79	131	188	290	389	502	790	1142
	12	8	15	24	42	58	96	137	212	284	366	577	834
	10	6	10	17	30	41	69	98	152	204	263	414	598
	8	3	6	10	17	24	39	56	87	117	150	237	343
	6	2	4	7	12	17	28	40	62	83	107	169	244
	4	1	3	5	8	12	19	28	43	58	75	118	170
	3	1	2	4	7	10	16	23	36	48	62	98	142
	2	1	1	3	6	8	13	19	30	40	51	81	117
PFA, PFAH, TFE	1	1	1	2	4	5	9	13	20	28	36	56	81
PFA, PFAH, TFE, Z	1/0	1	1	1	3	4	8	11	17	23	30	47	68
	2/0	0	1	1	3	4	6	9	14	19	24	39	56
	3/0	0	1	1	2	3	5	7	12	16	20	32	46
	4/0	0	1	1	1	2	4	6	9	13	16	26	38
Z	14	13	24	40	70	95	158	226	350	469	605	952	1376
	12	9	17	28	49	68	112	160	248	333	429	675	976
	10	6	10	17	30	41	69	98	152	204	263	414	598
	8	3	6	11	19	26	43	62	96	129	166	261	378

Continued

Table C10 Maximum Number of Conductors or Fixture Wires in Rigid PVC Conduit, Schedule 40 and HDPE Conduit (Based on Table 1) *Continued*

CONDUCTORS

Type	Conductor Size (AWG/kcmil)	16 (1/2)	21 (3/4)	27 (1)	35 (1-1/4)	41 (1-1/2)	53 (2)	63 (2-1/2)	78 (3)	91 (3-1/2)	103 (4)	129 (5)	155 (6)
Z	6	2	4	7	13	18	30	43	67	90	116	184	265
	4	1	3	5	9	12	21	30	46	62	80	126	183
	3	1	2	4	6	9	15	22	34	45	58	92	133
	2	1	1	3	5	7	12	18	28	38	49	77	111
	1	1	1	2	4	6	10	14	23	30	39	62	90
XHH, XHHW, XHHW-2, ZW	14	8	14	24	42	57	94	135	209	280	361	568	822
	12	6	11	18	32	44	72	103	160	215	277	436	631
	10	4	8	13	24	32	54	77	119	160	206	325	470
	8	2	4	7	13	18	30	43	66	89	115	181	261
	6	1	3	5	10	13	22	32	49	66	85	134	193
	4	1	2	4	7	9	16	23	35	48	61	97	140
	3	1	1	3	6	8	13	19	30	40	52	82	118
	2	1	1	3	5	7	11	16	25	34	44	69	99

Conduit and Tubing Fill Tables

XHH, XHHW, XHHW-2	1	1	1	3	5	8	12	19	25	32	51	74
1/0	1	1	1	3	4	7	10	16	21	27	43	62
2/0	0	1	1	2	3	6	8	13	17	23	36	52
3/0	0	1	1	1	3	5	7	11	14	19	30	43
4/0	0	1	1	1	2	4	6	9	12	15	24	35
250	0	0	1	1	1	3	5	7	10	13	20	29
300	0	0	1	1	1	3	4	6	8	11	17	25
350	0	0	1	1	1	2	3	5	7	9	15	22
400	0	0	0	1	1	1	3	5	6	8	13	19
500	0	0	0	1	1	1	2	4	5	7	11	16
600	0	0	0	1	1	1	1	3	4	5	9	13
700	0	0	0	0	1	1	1	3	4	5	8	11
750	0	0	0	0	1	1	1	2	3	4	7	11
800	0	0	0	0	1	1	1	2	3	4	7	10
900	0	0	0	0	1	1	1	2	3	4	6	9
1000	0	0	0	0	0	1	1	1	3	3	6	8
1250	0	0	0	0	0	1	1	1	1	3	4	6

Continued

Table C10 Maximum Number of Conductors or Fixture Wires in Rigid PVC Conduit, Schedule 40 and HDPE Conduit (Based on Table 1) *Continued*

CONDUCTORS

Type	Conductor Size (AWG/kcmil)	Metric Designator (Trade Size)										
		16 (1/2)	21 (3/4)	27 (1)	35 (1-1/4)	41 (1-1/2)	53 (2)	63 (2-1/2)	78 (3)	91 (3-1/2)	103 (4)	129 (5)	155 (6)
XHH, XHHW, XHHW-2	1500	0	0	0	0	0	1	1	1	1	2	4	5
	1750	0	0	0	0	0	0	1	1	1	1	3	5
	2000	0	0	0	0	0	0	1	1	1	1	3	4

INDEX

AC and DC conductors in same enclosure 189
AC resistance and reactance conversion. 384–387
AC systems
   Conductor to be grounded 92–93
   Grounding connections 90–92
   Grounding electrode conductor 104–105
   Grounding of. 89–90, 93–97
   In same metallic enclosures 49, 202–204
Access and working space. *See also* Working space
   Not over 600 volts 11–16
   Switchboards 56
Accessible
   Air-conditioning and refrigeration disconnects 296
   Conduit bodies, junction, pull, and outlet boxes 141
   Grounding electrode connection 106
   Motor disconnects 275
   Overcurrent devices 74
   Services 41–42
   Splices and taps in wireways 180, 183
   Transformers and vaults 317–318
AC-DC general-use snap switches
   Motors 271
   Panelboards, use in 58
   Ratings, type loads 57
Air conditioning and refrigerating equipment 291–303
   Branch circuit Conductors 292, 298–299
   Outlets 224–225
   Selection current 292–293
   Short-circuit and ground-fault protection 296–298
   Equipment for 297–298
Air plenums 205
Alternators. *See* Generators
Ampacities
   Conductors 207–213
   Grounding 119–121
   Tables, 0–2000 volts 210–213
Attachment plugs (caps).
   Construction of 228–229

Index

Back fed devices 58
Backfill 194–195
Ballasts, electric discharge lamps 247
   Protection in recessed HID fixtures 247
Bare conductors, ampacities 209
Bathroom, receptacles in 217
Bathtubs, luminaires (lighting fixtures) ...237–238
Bodies, conduit. *See* Conduit bodies
Bonding 107–113
   Grounding-type receptacles 123–124
   Loosely jointed raceways 110
   Other enclosures 109–110
   Outside raceway 111
   Over 250 volts 109
   Piping systems and exposed structural steel ..112–113
   Receptacles 123–124
   Service equipment 108–109
   Swimming pools 336–339
Bonding jumpers. *See* Jumpers, bonding
Bored holes through studs, joists 189
Boxes (outlet, device, pull, and junction) ..128–141
   Accessibility 141
   Concealed work 133
   Conductors, number in box 128–133
   Covers 138, 140
   Cutout. *See* Cabinets, cutout boxes, and meter socket enclosures
   Damp locations 128
   Depth 138
   Exposed surface extensions 134
   Fill calculations 129, 132–133
   Floor, for receptacles 139
   Grounding 114–121, 124–125
   Junction, pull. *See* Junction boxes
   Lighting (luminaire) outlets 138, 238–239, 240
   Metal 124–141
   Repairing plaster around 134
   Required location 199–201
   Round 128
   Secured supports 134–138, 197–199
   Unused openings, closed 7
   Vertical raceway runs 201–202
   Volume calculations 129–131

Wall or ceiling	133
Wet locations	128
Branch circuits	23–30, 215–222
Air conditioners	298–299
Busways as branch circuits	177–178
Calculation of loads	24–26
Conductors, minimum ampacity and size	218
General	216–218
Individual	218–222, 262–264
Motor, on individual branch circuit	254–256
Multiwire	216–217
Overcurrent protection	219
Permissible loads	220–222
Ratings	216, 218–222
Requirements for	222
Taps from	65, 218, 221–222
Through luminaires (fixtures)	238, 245
Two or more outlets on	220–222
Bushings	
Generators	307–308
Insulated	142, 191, 243–244
Rigid metal conduit	157
Rigid nonmetallic conduit	163
Underground installations	195
Use in lieu of box or terminal fitting	201
Busways	175–179
Branches from	177
Dead ends closed	176
Extension through walls and floors	176
Feeder or branch circuits	178
Overcurrent protection	178
Reduction in size	178
Support	175
Through walls and floors	176
Use	175–176
Cabinets, cutout boxes, and meter socket enclosures	141–143
Damp, wet, or hazardous (classified) locations	141
Deflection of conductors	142
Insulation at bushings	191
Position in walls	141

Cabinets *(continued)*
  Switch enclosures, splices, taps 142
  Wire bending space at terminals 142
Cables
  Continuity 199
  Flat conductor (Type FCC). *See* Flat conductor
    cable (Type FCC)
  Installed in shallow grooves 191
  Metal-clad cable (Type MC). *See* Metal-clad
    cable (Type MC)
  Other types of. *See* names of systems
  Protection against physical damage 189–191
  Secured 197–199
  Splices in boxes 199–201
  Through studs, joists, rafters 189–190
  Underground 300.5, 300.50, 36–37, 191–196
Canopies
  Boxes and fittings 138
  Luminaires (lighting) fixtures 238
Ceiling fans 327, 340–341
  Support of 139
Circuit breakers 63–80
  Accessibility and grouping 60–61
  Disconnection of grounded circuits 232
  General 7, 64–67
  Overcurrent protection 64–67, 77–78, 262,
                                        277–279, 306–307
  Panelboards 56–59
  Rating, nonadjustable trip 66
  Service overcurrent protection 45–46
  Services, disconnecting means 41
  Switches, use as 61, 77
  Wet locations, in 60
Circuit directory, panelboards 10–11
Circuit-interrupters, ground-fault. *See* Ground-fault
  circuit-interrupters
Circuits
  Branch. *See* Branch circuits
  Grounding 85–125
  Impedance 7
  Motor 254–256
  Motor control 267–270
  Number of, in enclosures 4
Clamp fill, boxes 129–132

Index

Clamps, ground 88, 106–107
Clean surfaces, grounding conductor connections 89
Clearances
   Conductors, service drop 34–36
   Lighting luminaires (fixtures) 246, 247–248
   Swimming pools 323
   Switchboards 55–56
CO/ALR
   Receptacles 225
   Switches 234
Color code
   Branch circuits 216–217
   Grounded conductor 82–83
   Grounding conductor 117–118, 199
   Higher voltage to ground 40–41, 49–50, 55
Common grounding electrode. *See* Electrodes,
   grounding, common
Common neutral
   Feeders 49
Conductor fill
   Boxes 129, 132
   Outlet boxes, etc. 128–133
   Surface raceways 183–185
   Wireways 182
Conductors
   Aluminum, properties of 254–256, 382–383
   Ampacities of 207–213
   Bare. *See* Bare conductors
   Boxes and fittings, junction 128–133
   Branch circuits 215–222
   Busways. *See* Busways
   Cabinets and cutout boxes 142
   Combinations 360
   Conduit or tubing, number in 359–486
   Cooling of electric equipment 8
   Copper 210–213, 381–383
   Corrosive conditions 196, 207
   Deflection of. *See* Deflection of conductors
   Different systems 189
   Dimensions of 362–369
   Electrical metallic tubing. *See* Electrical metallic tubing
   Electrical nonmetallic tubing. *See* Electrical
     nonmetallic tubing
   Enclosure, grounding 86, 107

Conductors *(continued)*
   Equipment grounding. *See* Equipment grounding conductors
   Feeder..47–51
   General wiring................................206–213
   Generators, size....................................307
   Grounded 41–42, 43, 45–46, 74, 81–84, 91–93, 232, 260
   Grounding. *See* Grounding conductors
   Grounding electrode. *See* Grounding electrode conductors
   Insulating materials................................199
   Insulation...............................206–213, 242
   Insulation at bushings, No. 4 and larger..........191
   Length in boxes...................................199
   Lightning rods, spacing from.....................113
   Liquidtight flexible metal conduit. *See* Liquidtight flexible metal conduit
   Liquidtight flexible nonmetallic conduit. *See* Liquidtight flexible nonmetallic conduit
   Metal enclosures, spacing from lightning rods.....113
   Metal-clad cable. *See* Metal-clad cable
   Minimum size......................................38
   Motor circuits................................254–256
   Multiple..............110–111, 202–204, 206–207
   Number of, in. *See* Conductor fill
   Outlet boxes, temperature limits..................238
   Overcurrent protection..........................64–66
   Paralleled...........110–111, 202–204, 206–207
   Properties of.................................381–383
   Raceways, number of conductors in. *See* Conductor fill
   Rigid metal conduit. *See* Rigid metal conduit
   Same circuit......................................195
   Service. *See* Service-entrance conductors
   Service-entrance. *See* Service-entrance conductors
   Sizes. *See also* Conductors, minimum size...370–380
   Stranded...206
   Support of, in vertical raceways.............201–202
   Surface raceway. *See* Surface metal raceways; Surface nonmetallic raceways
   Switchboards and panelboards.....................54
   Ungrounded...........................45, 216–217
   Wet locations.....................................207

Index

Wireways. *See* Wireways, metal; Wireways, nonmetallic
Conduit bodies. *See also* Boxes, outlet, device, pull, and junction
    Number of conductors 133
    Pull and junction box 139–141
Conduit fill. *See also* Conductor fill
    Equipment grounding conductors............ 360
    Grounding conductor 209
Conduit nipples 360
Conduits
    Boxes supporting 137–138
    Conductors, number in 360, 388–486
    Dimensions........................... 362–369
    Electrical metallic tubing. *See* Electrical metallic tubing
    Electrical nonmetallic tubing. *See* Electrical nonmetallic tubing
    Flexible metal. *See* Flexible metal conduit
    Liquidtight flexible metal. *See* Liquidtight flexible metal conduit
    Liquidtight flexible nonmetallic. *See* Liquidtight flexible nonmetallic conduit
    Rigid metal. *See* Rigid metal conduit
    Rigid nonmetallic. *See* Rigid nonmetallic conduit
Connectors
    Boxes 133
    Electrical metallic tubing.................... 172
    Flexible metal conduit...................... 158
    Liquidtight flexible metal conduit............. 159
    Liquidtight flexible nonmetallic conduit 169
    Rigid metal conduit........................ 157
Construction sites
    Assured equipment grounding conductor program ... 21
    Extension cord sets 20–21
    Ground-fault circuit-interrupter, protection for.... 20–21
Continuity
    Electrical, metal raceways and cables 107–109
    Grounding, metal boxes, grounding-type receptacles............................ 123–125
    Mechanical, raceways and cables 199
Continuous load
    Applications 48–49, 219

Control centers
   Guarding live parts 15
   Headroom. 14
   Illumination at 14
   Motor. 271, 274
   Working space 11–16
Control panels, working space 11–16
Controllers
   Cases 114
   Enclosures, grounding 114
   Motor. 270–271
   Ratings 270–271
Conversion table, AC conductor resistances
   and reactances 384–387
Cooling of equipment 8
Correction factors, ambient temperature 210–213
Corrosive conditions
   Conductor insulation 207
   Deteriorating agents 7
Couplings
   Electrical metallic tubing 172
   Rigid metal conduit 157
   Rigid nonmetallic conduit, expansion fittings 163
   Running threads at 157
   Threaded and threadless 108
Cove lighting, space for 238
Covers
   Boxes and fittings 138, 140
   Faceplates. *See* Faceplates
Cross sectional areas
   Conductors 370–383
   Conduits 362–369

Damp or wet locations. *See also* Wet locations 7
   Boxes and fittings 128
   Cabinets and cutout boxes 141
   Luminaires (lighting fixtures) 236
   Receptacles 229
Dead ends
   Busways 176
   Wireways. 181, 182
Deflection of conductors, cabinets and
   cutout boxes 142

Index

Delta-connected, identifying high leg,
    3-phase supply, 4-wire 40–41, 49–50, 55
Different systems, conductors in same enclosure 189
Dimensions
    Conductors.................................. 370–383
    Conduits and tubing 362–369
Direct burial
    Liquidtight flexible metal conduit.................. 159
    Liquidtight flexible nonmetal conduit............... 161
    Rigid nonmetallic conduit 161, 192–193
    Underground service cable 39, 192–193
Disconnecting means
    Air-conditioning and refrigerating
        equipment............................. 293–296
    Fuses and thermal cutouts..................... 75–76
    Identification 11
    Motors and controllers 274–279
    Pools, spas, and hot tubs 325
    Separate building on same premises............... 97
    Services................................... 41–44
Drip loops, service heads......................... 40
Driveways
    Clearance of service drop 35–36
    Protection of service cables 39
Drop. *See* Service drops; Voltage and volts, drop
Dry location. *See also* Damp or wet locations

Eccentric knockouts. *See* Knockouts, bonding service
    equipment
Electric discharge lighting, connection to luminaires
    (lighting fixtures) 238–239, 243–245
Electrical metallic tubing (Type EMT).......... 169–172
    Bends, how made............................. 170
    Bends, number in one run 171
    Connectors and couplings 171
    Grounding................................... 172
    Installation 169–172
    Maximum number of conductors and fixture
        wires in 388–396
    Reaming and threading 171
    Securing and supporting 171
    Uses not permitted........................... 170
    Uses permitted............................... 170

Index

Electrical nonmetallic tubing (Type ENT)	172–175
Bends	174
Bushings	175
Installation	172–175
Joints	175
Maximum number of conductors and fixture wires in	397–406
Securing and supporting	174
Through metal framing members	189–191
Trimming	173
Uses not permitted	173–174
Uses permitted	172–173

Electrodes, grounding
- Aluminum 100
- Common 102
- Concrete encased 99
- Gas piping as 100
- Made 98–100
- Metal frame of building as 99
- Metal water piping system 98
- Resistance to ground of 102
- Separately derived systems 94–95

Equipment. *See also* specific types of
- Approval 2–3, 6
- Cooling of 8
- Examination of 3–4, 6
- Grounding 114–121
- Installation, general provisions 5–15
- Mounting. *See* Mounting of equipment

Equipment grounding conductor fill and boxes 132

Equipment grounding conductors
- Connections at outlets 123–125
- Installation 49, 118–119, 121–122
- Sizing 119–121
- Types recognized 116–119, 122–123

Exits, emergency lighting for 348
Expansion joints (fittings) 109–110, 163, 197
Explanatory material FPN 3

Exposed
- Extensions, boxes, and fittings 134
- Live parts 11–16
- Structural steel, grounding 99

Index

Extensions
  Boxes and fittings, exposed 134
  Wireways 181, 182

Faceplates
  Grounding. 107, 114, 232–233
  Mounting surfaces, against. 227, 232–233
Fans, ceiling. *See* Ceiling fans
Feeders 23–30, 47–51
  Busways 177–178
  Calculation of loads 26–27, 48–49
  Ground-fault circuit-interrupters, with 50
  Grounding means. 49
  Kitchen equipment, commercial 28–29
  Loads. *See* Loads, feeder
  Motors 254–256
  Overcurrent protection, motor. 266–267
  Services 26–30
  Taps 69, 255
Fine Print Notes FPN, mandatory rules,
  permissive rules, and explanatory material 3
Fire spread, prevention of, wiring methods 204
Fire-stopped partitions. *See also* Firewalls 204
Firewalls, wiring through 204
Fittings 128–141
  Conduit bodies 128–133
  Expansion. 163
  Insulation 191
Fixed equipment, grounding. 114–115, 122–123
Fixtures. *See* Luminaires (lighting fixtures)
Flat conductor cable (Type FCC) 148–151
  Branch-circuit rating 148–149
  Cable connections and ends 150
  Construction. 149, 151
  Installation 148–151
  Polarization. 150
  Systems alterations 151
  Uses not permitted 149
  Uses permitted 148–149
  Bends. 158
  Couplings and connectors 158
  Grounding and bonding 158
  Installation 157–159

Flexible metal conduit *(continued)*
  Liquidtight. *See* Liquidtight flexible metal conduit
  Maximum number of conductors and fixture
    wires in 407–416
  Securing and supporting 158
  Trimming 158
  Uses not permitted 157
  Uses permitted 157
Floors, receptacles 124, 139
Fluorescent luminaires (lighting fixtures) 246–248
  Circuit breakers used to switch 77
  Connection of 238–239, 244–245
  Load calculations, branch circuits 219–222
  Raceways 245
  Snap switches for 223, 224

Generators 305–308
  Bushings 307–308
  Conductor, ampacity of 307
  Emergency systems 353–355
  Grounding 114–115
  Guards for attendants 307
  Location 306
  Overcurrent protection 306–307
  Protection of live parts 307
Ground clamps 88, 106–107
Grounded, solidly, definition 46
Grounded conductor. *See* Conductors,
  Grounded; Neutral
Ground-fault circuit-interrupters
  Fountains 343
  Permitted uses 50
  Receptacles 20–21, 217, 226–227, 322,
                                326–333, 339–343
Ground-fault protection
  Emergency systems, not required 357
  Equipment 50
  Service disconnecting means 45–46
Grounding 85–125
  AC systems 89–91
  Air-conditioning units 115
  Appliances 116
  Bonding 107–113
  Circuits 86–98

Index

    Clothes washers............................... 115
    Continuity 108–109
    Dishwashers................................... 115
    Electrode system 98–107
    Enclosures 107
    Equipment, cord- and plug-connected 115
    Fire alarm systems............................ 115
    Fixed equipment............... 114–115, 122–123
    Fixtures, lampholders, etc................. 242–245
    Fountains 345
    Freezers 115
    Generators 306
    Metal enclosures for conductors............... 107
    Metal faceplates 232–233
    Methods 121–122
    Motors and controllers 280–282
    Panelboards................................. 58–59
    Portable equipment 116
    Receptacles 123–125, 217, 225–227
    Refrigerators.................................. 115
    Separate buildings 97–98
    Separately derived systems 90, 93–96
    Spas and tubs 322, 340–342
    Swimming pools 322, 335–336
    Systems 86–98
    Tools, motor operated......................... 116
    Transformers.................................. 317
Grounding conductors 98–107, 114–121
    Earth as 87, 101
    Enclosures 107
    Identification, multiconductor cable 118
    Installation 102–104, 119
    Material....................................... 102
    Objectionable current over 87–88
    Sizes 119–121
Grounding electrode conductors
    Connection to electrodes 86–89, 98–107
    Installation 102–104
    Material............................... 102, 116–117
    Sizing 104–105
Group installation, motors 264–266
Grouping, switches, circuit breakers.
    See also Accessible 60–61
Grouping of disconnects 42–43

Index

Guarding, guards
    Circuit breaker handles...........................77
    Construction sites................................22
    Generators......................................307
    Live parts, general...............................15
    Motors and motor controllers..............280, 281
    Over 600 volts...................................22
    Transformers..............................316–317

Heavy-duty lampholders
    Branch circuits............................219–222
    Unit loads.......................................25
Hermetic refrigerant motor compressors.
    See also Air-conditioning and
    refrigerating equipment
    Ampacity and rating.......................293–294
    Rating and interrupting capacity...........293–294

Identification
    Disconnecting means.............................11
    Grounded conductors......................81–84
    High leg......................................40–41
    Panelboard circuits..............................56
    Service disconnecting means....................42
    Ungrounded conductors..................216–217
    Wiring device terminals........................121
In sight from, air-conditioning
    or refrigerating equipment....................296
Incandescent lamps...........................236–249
    Snap switches for..............................234
Insulation
    Conductors...............................206–213
    Equipment.......................................6
    Service conductors....................34, 36, 38
    Splices and joints.............................9, 19

Joints. *See also* Splices and taps
    Expansion. *See* Expansion joints
    Grounding electrode conductor.................103
    Insulating, fixtures.............................241
    Insulation of......................................9
Jumpers, bonding.........................93, 110–111
    Equipment................................110–111

Index

    Expansion joints, telescoping sections
      of raceways.............................. 109–110
    Grounding-type receptacles................. 123–124
    Hazardous (classified) locations................. 110
    Piping systems............................. 111–113
    Service equipment................ 93, 108, 110–111
Junction boxes. *See also* Boxes; Pull boxes
    Accessibility..................................... 141
    Covers... 140
    Motor controllers and disconnects............... 254
    Separation from motors......................... 281
    Size, conductors No. 4 and larger .. 128–133, 139–140
    Supports................................. 134–138
    Switch enclosures..................... 59–60, 142

Knockouts
    Bonding service equipment...................... 108
    Opening to be closed........................ 7, 133

Lampholders
    Branch circuits supplying................... 219–222
    Damp or wet locations.......................... 236
    Heavy-duty. *See* Heavy-duty lampholders
    Unswitched over combustible material........... 237
Lamps. *See also* Lighting...................... 236–249
    Electric discharge............... 244–245, 246–248
    Fluorescent. *See* Fluorescent lighting fixtures
    Guards. *See* Guarding, guards
    Incandescent. *See* Incandescent lamps
Lengths
    Free conductors at outlets and switches......... 199
    Pull and junction boxes..................... 139–141
    Taps......................... 69–79, 255, 265–266
Lighting
    Branch circuits, calculation of load................ 24
    Cove... 238
    Electric discharge. *See* Electric discharge lighting
    Emergency............................... 347–357
    Feeders, calculation of load...................... 27
    Outlets.. 236
Liquidtight flexible metal conduit (Type LFMC)... 159–160
    Bends... 159
    Couplings and connectors...................... 160

Liquidtight flexible metal conduit *(continued)*
   Grounding and bonding 160
   Installation............................159–160
   Maximum number of conductors and fixture
     wires in.............................447–456
   Securing and supporting......................160
   Uses not permitted 159
   Uses permitted 159
Liquidtight flexible nonmetallic conduit
  (Type LFNC) 163–169
   Bends 169
   Grounding and bonding 160
   Installation............................163–169
   Maximum number of conductors and
     fixture wires in.......................427–436
   Securing and supporting......................160
   Trimming 169
   Uses not permitted 168–169
   Uses permitted 168
Loads
   Branch circuits..........23–30, 216–217, 219–222
   Continuous. *See* Continuous load
   Feeder..................................23–30
   Motors, conductors 254–256
Lugs
   Connection to terminals 9
   Solderless type at electrodes 106–107
Luminaires (lighting fixtures)..................236–249
   Autotransformers, supply circuits.............50–51
   Boxes, canopies, pans...................238–239
   Branch circuits........24–26, 218, 219–222, 246–247
   Combustible material, near............237, 247–248
   Connection, fluorescent 238–239, 243–245
   Construction 241–245
   Corrosive..................................237
   Damp, wet or corrosive locations 236–237
   Dry-niche..................................331
   Ducts or hoods, in 237
   Electric discharge. *See* Electric discharge lighting
   Fluorescent. *See* Fluorescent lighting
     luminaires (fixtures)
   Flush..................................245–246
   Fountains 343–344

Index

Grounding	241–242
Location	236–237
Mounting	247–248
No-niche	331
Outlets required	236
Overcurrent protection, wires and cords	64–66
Polarization	242
Raceways	245
Recessed. *See* Recessed lighting fixtures	
Show windows	237
Spas and hot tubs	340–341
Supports	239–241
Swimming pools	326–333
Wet	236
Wet-niche	329–331
Wiring	242–245
Made electrodes	98–100
Mandatory rules	3
Metal frame of building, grounding electrode	99
Metal-clad cable (Type MC)	146–148
Bends	147
Boxes and fittings	148
Installation	146–148
Supports	147–148
Through or parallel to framing members	147, 189–190
Uses not permitted	147
Uses permitted	146
Metals, dissimilar	8–10, 106–107
Motors	251–289
Air conditioning units	292–303
Branch circuits	254–256, 264–266, 267
Circuit conductors	254–256
Conductors	254–256
Control circuits	267–270
Controllers. *See* Controllers, motor	
Disconnecting means	274–279
Feeders, calculation of load	26–27
General	252–254
Ground-fault protection	262–267
Grounding	114–121, 280–289
Grouped	255, 260–261, 264–266, 279
Guards for attendants	280

Index

Motors *(continued)*
   Location 254
   Maintenance 254
   Motor control centers 271, 274
   Overheating, dust accumulations 254
   Overload protection 257–262
   Protection of live parts 280
   Restarting, automatic 262
   Short circuit protection 262–267
   Tables 282–289
   Taps 255
   Three overload units 261
   Ultimate trip current 257–258
   Ventilation 254
   Wiring diagram 253
   Wiring space in enclosures 254
Mounting of equipment 8, 134–138, 233–234
Multiple conductors (conductors in parallel).
    See Conductors, multiple

Neat and workmanlike installation 7–8
Neutral. *See also* Conductors, grounded
   Bonding to service equipment 108
   Common. *See* Common neutral
   Continuity of 199
   Equipment, grounding to 123
   Feeder load 29–30
   Grounding of 86–87, 90–93
   Identification 81–84
   Ranges and dryers, grounding 123
Number of services 32–33

Occupancy, fighting loads 24–26
Openings in equipment to be closed 133
Outdoor receptacles 229–230, 326–327
Outlet boxes. *See* Boxes
Outlets
   Devices, branch circuits 217
   Heating, air-conditioning, and
     refrigeration equipment 224–225
   Laundry 24–26
   Lighting 236
   Required 236

Index

Outside of buildings, when services considered	33
Overcurrent devices	63–80
Standard	66
Overcurrent protection	63–80
Air-conditioning and refrigerating equipment	296–298
Branch circuits	218–219
Busways	177–178
Circuit breakers	77–78
Disconnecting and guarding	75–76
Emergency systems	357
Enclosures	75
Feeder taps	69–75, 77–80, 255
Fixture wires	64–66
Generators	306–307
Lighting track	249
Location	45, 67–75, 78–80
Motors, motor circuits, controllers	262–266, 267–270
Multiple fuses and circuit breakers (in parallel)	66–67
Occupant access to	74
Panelboards	57–58
Paralleled fuses and circuit breakers	66–67
Services, equipment	45–46
Single appliance	218–219
Supervised industrial installations	78–80
Supplementary	67
Transformers	310–314
Vertical position, enclosures	75
Panelboards	53–59
Circuit directory	55
General	56
Grounding	58–59
Installation	14–15
Lighting and appliance branch-circuit	57
Overcurrent protection	57–58
Service equipment	41–44, 54–55, 57
Support for busbars and conductors	54–55
Use as enclosure	142
Pans, fixture	238
Paralleled conductors. *See* Conductors, paralleled	
Pendants, connector, cord	224
Permanent plaque or directory	33

Plugs, attachment. *See* Attachment Plugs
Protection
   Corrosion. 154, 170, 196–197, 207
   Ground fault. *See* Ground fault protection
   Ground fault circuit-interrupter. *See* Ground
      fault circuit-interrupter
   Live parts 15, 307, 317
   Motor overload 257–262
   Overcurrent. *See* Overcurrent protection
   Physical damage. 39–40, 75, 146, 159, 161, 168,
               170, 173, 175, 178, 181, 184, 186,
               189–191, 194, 195, 248, 316
Protective devices. *See* Circuit breakers; Ground-fault
   circuit-interrupters; Overcurrent protection
Pull boxes. *See also* Boxes; Junction boxes
   Accessibility. 141
   Construction specifications 128–141
   Sizes 128–133, 139–140

Raceways
   Bonding. 107–113
   Busways. *See* Busways
   Conductors in service 33
   Continuity 197, 199
   Drainage 40
   Electrical metallic tubing (Type EMT).
      See Electrical metallic tubing (Type EMT)
   Electrical nonmetallic tubing (Type ENT).
      See Electrical nonmetallic tubing (Type ENT)
   Emergency circuits, independent 351
   Expansion joints 109–110, 163, 197
   Exposed to different temperatures 197
   Flexible metal conduit (Type FMC). *See* Flexible metal
      conduit (Type FMC)
   Grounding 107, 122
   Induced currents 202–204
   Installed in shallow grooves 191
   Insulating bushings 191, 201
   Liquidtight flexible metal conduit (Type LFMC).
      See Liquidtight flexible metal conduit (Type LFMC)
   Liquidtight flexible nonmetallic conduit (Type LFNC).
      See Liquidtight flexible nonmetallic conduit (Type
      LFNC)

```
    Luminaires (fixtures) as .......................... 245
    Rigid metal conduit (Type RMC). See Rigid
      metal conduit (Type RMC)
    Rigid nonmetallic conduit (Type RNC). See Rigid
      nonmetallic conduit (Type RNC)
    Secured ..................................... 197–198
    Service. See Service raceways
    Support for nonelectrical equipment............ 198
    Supporting conductors, vertical ............ 201–202
    Surface metal. See Surface metal raceways
    Surface nonmetallic. See Surface nonmetallic
      raceways
    Underground................................... 194
    Wireways. See Wireways, metal; Wireways,
      nonmetallic
  Receptacles, cord connectors and attachment
      plugs (caps) ............................. 225–230
    Branch circuits ................................ 217
    Faceplates .................................... 228
    Grounding type ...................... 122–124, 217
    Insulated grounded terminals.................. 124
    Maximum cord- and plug-connected
      load to..................................... 219–222
    Minimum ratings .............................. 225
    Nongrounding-type, replacement.............. 122
    Outdoor. See Outdoor receptacles
    Ratings for various size circuits ............ 219, 220
    Show windows, in............................. 224
    Swimming pools ........................... 326–328
    Temporary installations.................... 19, 20–21
  Recessed luminaires (lighting fixtures) ........ 245–246
    Clearances, installation ....................... 246
    As raceways ............................. 238, 245
    Temperatures ............................ 245–246
  Rigid metal conduit (Type RMC) ............... 154–157
    Aluminum, in concrete and in earth ............ 196
    Bends ........................................ 155
    Bushings..................................... 157
    Cinder fill ................................... 154
    Couplings and connectors .................... 157
    Expansion joints ............................. 197
    Grounding ................................... 157
    Installation .............................. 154–157
```

Rigid metal conduit (Type RMC) *(continued)*
   Maximum number of conductors and fixture
      wires in 457–466
   Number of conductors....................360
   Reaming and threading...................155
   Supporting and securing..........135–136, 155–156
   Uses permitted 154
   Wet locations 154
Rigid nonmetallic conduit (Type RNC) 160–163
   Bends 162
   Bushings 163
   Construction specifications 160–163
   Expansion fittings 162–167, 197
   Joints 163
   Maximum number of conductors and
      fixture wires in 467–476
   PVC Schedule 80 194
   Supporting and securing 162
   Trimming ends 162
   Uses not permitted 161–162
   Uses permitted 161
Rod electrodes 99
Running threads 157

Separately derived systems 90, 92–98
Service conductors. *See* Conductors, service
Service drops
   Clearances 35–36
   Connections, service head 40
   Means of attachment 36
   Minimum size 34–35
   Point of attachment 36
   Supports over buildings 36
Service equipment
   Overcurrent protection 45–46
Service raceways
   Conductors, others permitted in 33
   Drainage, raintight 40
   Service head 40
   Underground 36–37
Service-entrance conductors 37–41
   Conductor sets, number of 37–38
   Considered outside of building 33

Index

    Disconnecting means........................ 41–44
    Insulation 38
    Overcurrent protection 45–46
    Physical damage 39
    Service head 40
    Size..................................... 38
    Splices 39
    Underground............................. 39
    Wiring methods 38–39

Service-entrance equipment
    Disconnecting means............... 41–44, 45–46
    Guarding................................. 41
    Overcurrent protection 41–46
    Panelboards as........................... 56–59

Services 31–46
    Emergency systems separate service 354
    Ground-fault protection..................... 45–46
    Insulation 34
    Number.................................. 32–33
    Overhead supply 34–36
    Supply to one building not through
        another 33
    Two or more buildings...................... 37
    Underground............................. 36–37

Snap switches
    Accessibility, grouping 60–61
    Motors 275–277
    Ratings 233–234

Space
    Cabinets and cutout boxes 142
    Lightning rods, conductor enclosures,
        equipment............................ 102, 113
    Working. *See* Working space

Spas and hot tubs 339–343
    Indoor installations 340–342
    Outdoor installations....................... 339–340
    Protection 340–343

Splices and taps
    Cabinets and cutout boxes 142
    Conduit bodies 133
    Construction sites......................... 19
    General provisions 8–10
    Surface raceways......................... 184, 185

Index

Splices and taps (continued)
    Underground................................194
    Wireways.............................181, 182
Support fittings fill, boxes.........................132
Supports. *See* articles on wiring and equipment
Surface metal raceways........................183–185
    Combination raceways.....................185
    Grounding................................185
    Installation..........................183–185
    Number of conductors or cables............184
    Size of conductors........................184
    Splices and taps..........................185
    Uses not permitted.......................184
    Uses permitted...........................184
Surface nonmetallic raceways...................185–186
    Combination raceways.....................186
    Grounding................................186
    Number of conductors or cables in.........186
    Size of conductors........................186
    Splices and taps..........................186
    Uses not permitted.......................186
    Uses permitted.......................185–186
Swimming pools, fountains, and similar
    installations...........................321–346
    Bonding...............................336–339
    Ceiling fans..........................327–328
    Fountains............................343–346
    Ground-fault circuit-interrupters.........327, 329, 333–343
    Grounding................322, 335–336, 345
    Junction boxes and enclosures.........333–336
    Lighting..............................326–333
    Overhead conductor clearances.............323
    Permanently installed................325–339
    Receptacles, location and protection....326–327
    Spas and hot tubs....................339–343
    Switching devices.........................328
    Transformers.........................328–329
    Underwater audio equipment..........328–333
    Underwater luminaires (lighting fixtures)...328–333
Switches. *See also* specific types
    of switches......................59–61, 231–234
    AC general use snap switch............233–234

Index

    Accessibility and grouping 60–61
    AC-DC general use snap switches. *See* AC-DC general use snap switches
    Air-conditioning and refrigerating equipment 293–296
    Circuit breakers used as 77
    Disconnecting means 41–44, 274–279
    Emergency systems 357
    Enclosures, installation in 317
    General use 231–234
    Identification 11
    Isolating, motors over 100 HP 277
    Manually operable 231–234
    Motor controllers 270–271
    Panelboards 58
    Service 41–44
    Snap. *See* Snap switches

Taps. *See also* Splices and taps
    Branch circuit 218
    Busways 177
    Feeders. *See* Feeders, taps
    Overcurrent protection 68–75
    Service-entrance conductors 39
Temperature limitations
    Conductors 197
    Nonmetallic raceways and tubing. *See* raceway or tubing type
    In outlet boxes for luminaires (fixtures) 238
Temporary installations 17–22
    All wiring installations 18
    Branch circuits 18–19
    Disconnecting means 19
    Feeders 18
    Ground-fault protection 20–21
    Guarding over 600 volts 22
    Lamp protection 19
    Protection from accidental damage 19–20
    Receptacles 19, 20–21
    Services 18
    Splices 19
    Terminations at devices 20
    Time constraints 18

Terminals
    Connections to 8–10, 90, 106
    Identification, wiring device 121
Transformer vaults 318–320
    Doorways 319
    Drainage 320
    Location 318
    Storage 320
    Ventilation openings 319–320
    Walls, roofs, and floors 318–319
    Water pipes and accessories 320
Transformers 309–320
    Autotransformers .50–51, 311–314
    Capacitors, installation 309–320
    Dry-type 318
    Electric discharge lighting systems, 1000 volts
      or less 246–248
    Grounding 317
    Guarding 316–317
    Installations, indoor and outdoor 318
    Instrument, grounding, connections
      at services 44
    Location, accessibility 317–318
    Overcurrent protection 310–311, 312
    Remote control circuits for 269
    Research and development 310
    Secondary ties 314–316
    Specific provisions 318
    Terminal wiring space 317
    Two-winding, underwater lighting 328–329
    Vaults 318–320
    Ventilation 317

Underground enclosures 8
Underground wiring
    Conductor types in raceways 207
    Ground movement and 195
    Liquidtight flexible metal conduit 159
    Minimum cover requirements 191
    Protection of 194, 195
    Rigid metal conduit 154
    Rigid nonmetallic conduit 160
    Services 36, 37

Splices and taps	194
Swimming pools	323, 325
Wet locations	207

Unused openings
- Boxes and fittings ... 7–8

Voltage and volts
- Electric discharge lighting ... 246–248
- Marking ... 77
- Swimming pool underwater lighting fixtures ... 329

Water pipe
- Bonding (metal) ... 111–112
- Connections ... 90
- As grounding electrode ... 98–100

Wet locations. *See also* Damp or wet locations
- Conductors, types ... 207
- Electrical metallic tubing ... 169–170
- Enameled equipment ... 196
- Luminaires (lighting fixtures) in ... 236
- Mounting of equipment ... 196–197
- Rigid metal conduit ... 154
- Rigid nonmetallic conduit ... 160
- Switches ... 60

Wires. *See also* Conductors
- In concrete footings, electrodes ... 99

Wireways, metal ... 179–181
- Dead ends ... 180
- Deflected insulated ... 180
- Extensions ... 181
- Insulated conductors ... 180
- Number of conductors ... 179
- Securing and supporting ... 180
- Splices and taps ... 180
- Uses not permitted ... 179
- Uses permitted ... 179

Wireways, nonmetallic ... 181–183
- Dead ends ... 183
- Expansion fittings ... 183
- Extensions ... 183
- Installation ... 181–183
- Insulated conductors ... 182
- Number of conductors ... 182

Securing and supporting	182
Splices and taps	183
Uses not permitted	181
Uses permitted	181
Wiring methods	188–205
Ducts	204–205
Planning	4
Temporary. *See* Temporary installations	
Within sight from. *See* In sight from	
Working space	
About electrical equipment	11–16
Switchboards	11–16, 55–56
Workmanlike installation	7–8